OPTIMIZATION METHODS FOR ENGINEERING PROBLEMS

Frontiers of Mechanical and Industrial Engineering

OPTIMIZATION METHODS FOR ENGINEERING PROBLEMS

Edited by

Dilbagh Panchal, PhD

Prasenjit Chatterjee, PhD

Mohit Tyagi, PhD

Ravi Pratap Singh, PhD

First edition published 2023

Apple Academic Press Inc.
1265 Goldenrod Circle, NE,
Palm Bay, FL 32905 USA
760 Laurentian Drive, Unit 19,
Burlington, ON L7N 0A4, CANADA

CRC Press
6000 Broken Sound Parkway NW,
Suite 300, Boca Raton, FL 33487-2742 USA
4 Park Square, Milton Park,
Abingdon, Oxon, OX14 4RN UK

Library and Archives Canada Cataloguing in Publication

Title: Optimization methods for engineering problems / edited by Dilbagh Panchal, PhD, Prasenjit Chatterjee, PhD, Mohit Tyagi, PhD, Ravi Pratap Singh, PhD.
Names: Panchal, Dilbagh, editor. | Chatterjee, Prasenjit, 1982- editor. | Tyagi, Mohit, editor. | Singh, Ravi Pratap, 1987- editor.
Description: First edition. | Series statement: Frontiers of mechanical and industrial engineering book series | Includes bibliographical references and index.
Identifiers: Canadiana (print) 20220446121 | Canadiana (ebook) 20220446172 | ISBN 9781774911303 (hardcover) | ISBN 9781774911310 (softcover) | ISBN 9781003300731 (ebook)
Subjects: LCSH: Mathematical optimization. | LCSH: Engineering—Mathematical models. | LCSH: Problem solving.
Classification: LCC QA402.5 .O68 2023 | DDC 519.6—dc23

Library of Congress Cataloging-in-Publication Data

CIP data on file with US Library of Congress

ISBN: 978-1-77491-130-3 (hbk)
ISBN: 978-1-77491-131-0 (pbk)
ISBN: 978-1-00330-073-1 (ebk)

Dedication

The editors would like to dedicate this book to our parents, life partners, children, students, scholars, friends, and colleagues.

ABOUT THE FRONTIERS OF MECHANICAL AND INDUSTRIAL ENGINEERING BOOK SERIES

Description of the Series

Mechanical engineering applies the principles of mechanics and materials science for analysis, design, manufacturing, and maintenance of mechanical systems, whereas industrial engineering involves figuring out how to make or do things better with minimum resource consumption. Mechanical and industrial engineering utilize a deep foundational understanding of mathematics, physics, and analysis to develop machines and systems. The former specializes in thermodynamics, combustion, and electricity to make complex machines. The latter focuses on designing workflow and making production efficient.

This series aims to report the cutting-edge developments to present novel innovations in the field of mechanical and industrial engineering. Volumes published in this series aim to embrace all aspects, subfields, and new challenges of mechanical and industrial engineering and to be of interest and use to all those concerned with research in various related domains of mechanical and industrial engineering disciplines. The series aims to provide comprehensive value-added resources to the body of the literature to serve as essential references for years ahead. Features of the volumes under this series will explore recent trends, model extensions, developments, solution to real-time problems, and case studies.

The coverage of the series is committed to provide a diverse yet rigorous research experience that builds on a solid knowledge of science and engineering fundamentals and application-oriented knowledge of mechanical and industrial engineering practices for efficient functioning in today's global multidisciplinary rapidly changing work environment. It includes all new theoretical and experimental findings in mechanical, industrial engineering, and allied engineering fields.

Topics and Themes:

The primary endeavor of this series is to introduce and explore contemporary research developments in a variety of rapidly growing research areas. The volumes will deal with the following topics but not limited to:

- Automotive engineering
- Aerospace technology
- Big data in mechanical / manufacturing / industrial engineering
- Computational methods for optimizing manufacturing technology, robotics, mechatronics
- Decision science applications in mechanical, manufacturing, and industrial engineering
- Design and optimization of mechanical components
- Dynamical systems, control
- Engineering design
- Engineering thermodynamics, heat, and mass transfer
- Evolutionary computation
- Finite element method (FEM) modeling/simulation
- Fluid mechanics
- Fuzzy logic and neuro-fuzzy systems for relevant engineering applications
- Green and sustainable engineering techniques for modern manufacturing
- Innovative approaches for modeling relevant applications and real-life problems
- Life cycle engineering
- Machinery and machine elements
- Manufacturing
- Mathematical concepts and applications in mechanical and industrial engineering
- Materials engineering
- Mechanical structures and stress analysis
- MEMS

- Modeling in engineering applications
- Nanotechnology and microengineering
- Non-conventional machining in modern manufacturing systems
- Precision engineering, instrumentation and measurement
- Numerical simulations
- Soft computing techniques
- Sustainability in mechanical, manufacturing, and industrial engineering
- Theoretical and applied mechanics
- Tribology and surface technology

The series aims to serve as a valuable resource for undergraduate, postgraduate, doctoral students, researchers, academicians, and industry professionals in mechanical, manufacturing, industrial engineering, product design, management, and more.

BOOKS IN THE SERIES

Optimization Methods for Engineering Problems
Editors: Dilbagh Panchal, PhD, Prasenjit Chatterjee, PhD, Mohit Tyagi, PhD, and Ravi Pratap Singh, PhD

Modern Manufacturing Systems: Trends and Developments
Editors: Rajiv Kumar Garg, PhD, Ravi Pratap Singh, PhD, Rajeev Trehan, PhD, and Ramesh Singh, PhD

Multi-Criteria Decision-Making Methods in Manufacturing Environments: Models and Applications
Editors: Shankar Chakraborty, PhD, Prasenjit Chatterjee, PhD, and Partha Protim Das, PhD

Supply Chain Performance Measurement in Textile Enterprises
Editors: Pranav G. Charkha, PhD, Santosh B. Jaju, PhD, Prasenjit Chatterjee

Optimization of Advanced Manufacturing Processes
Editors: Sandip Kunar, PhD, Prasenjit Chatterjee, PhD, and M. Sreenivasa Reddy, PhD

ABOUT THE EDITORS

Dilbagh Panchal, PhD

Assistant Professor, Department of Industrial and Production Engineering, Dr. B. R. Ambedkar National Institute of Technology Jalandhar, Punjab, India

Dilbagh Panchal, PhD, is currently working as an Assistant Professor in the Department of Industrial and Production Engineering, Dr. B. R. Ambedkar National Institute of Technology Jalandhar, Punjab, India, an Institute of National Importance. He works in the area of reliability and maintenance engineering, fuzzy decision making, supply chain management, and operation management. He obtained his bachelor degree (Hons.) in Mechanical Engineering from Kurukshetra University, Kurukshetra, India, in 2007 and a master's with a gold medal in Manufacturing Technology in 2011 from the Dr. B. R. Ambedkar National Institute of Technology Jalandhar, India. He earned his PhD from the Indian Institute of Technology Roorkee, India, in 2016. Presently, two PhD scholars are working under him. Five MTech dissertations have been guided by him, and two are in progress. He has published 20 research papers in SCI/Scopus indexed journals. Five book chapters and one book have also been published by him under a reputed publisher. He also edited a book on advanced multi-criteria decision-making for addressing complex sustainability issues. He is currently part of the Associate Editors team of the *International Journal of System Assurance and Engineering Management.* He is currently serving as an active reviewer of many reputed international journals published under Elsevier, Springer, and Inderscience publishers.

Prasenjit Chatterjee, PhD

Dean (Research and Consultancy), MCKV Institute of Engineering, West Bengal, India

Prasenjit Chatterjee, PhD, is currently the Dean (Research and Consultancy) at the MCKV Institute of Engineering, West Bengal, India. He has over

100 research papers published in international journals and peer-reviewed conferences. He has authored and edited more than 15 books on intelligent decision-making, supply chain management, optimization techniques, risk, and sustainability modelling. He has received numerous awards, including Best Track Paper Award, Outstanding Reviewer Award, Best Paper Award, Outstanding Researcher Award, and University Gold Medal. Dr. Chatterjee is the Editor-in-Chief of the *Journal of Decision Analytics and Intelligent Computing*. He has also been a guest editor of several special issues in different SCIE/Scopus/ESCI (Clarivate Analytics) indexed journals. He is also the Lead Series Editor of the book series Smart and Intelligent Computing in Engineering; Founder and Lead Series Editor of several book series: Concise Introductions to AI and Data Science; AAP Research Notes on Optimization and Decision-Making Theories; Frontiers of Mechanical and Industrial Engineering; and River Publishers Series in Industrial Manufacturing and Systems Engineering. Dr. Chatterjee is one of the developers of two multiple-criteria decision-making methods called Measurement of Alternatives and Ranking according to COmpromise Solution (MARCOS) and Ranking of Alternatives through Functional mapping of criterion sub-intervals into a Single Interval (RAFSI).

Mohit Tyagi, PhD

Assistant Professor, Industrial and Production Engineering Department, Dr. B. R. Ambedkar National Institute of Technology (NIT), Jalandhar, India.

Mohit Tyagi, PhD, is an Assistant Professor in the Industrial and Production Engineering Department at the Dr. B. R. Ambedkar National Institute of Technology (NIT), Jalandhar, India. His areas of research are industrial engineering, supply chain management, corporate social responsibility, performance measurement systems, data science, and fuzzy inference systems. He has around six years of teaching and research experience. He has guided 15 postgraduate dissertations and 13 undergraduate projects and is currently supervising several MTech and PhD scholars. Dr. Tyagi has over 55 publications in international and national journals and proceedings of international conferences to his credit. He is a reviewer for many international journals, including *International Journal of Industrial Engineering: Theory, Application and Practices, Supply*

Chain Management: An International Journal; International Journal of Logistics System Management; and *Journal of Manufacturing Technology Management.* He has organized two international conferences, one as a joint organizing secretary at Delhi Technological University, Delhi, and the other as a convener at NIT, Jalandhar. He has also organized three Technical Education Quality Improvement Programme-sponsored short-term courses and faculty development program in his area of expertise. Dr. Tyagi earned his BTech (Mechanical Engineering) (with honors) from Uttar Pradesh Technical University, Lucknow; his MTech (Product Design and Development) (with a gold medal) from Motilal Nehru National Institute of Technology, Allahabad; and his PhD from the Indian Institute of Technology, Roorkee, India.

Ravi Pratap Singh, PhD

Assistant Professor, Department of Industrial and Production Engineering, Dr. B. R. Ambedkar National Institute of Technology, Jalandhar, Punjab, India

Ravi Pratap Singh, PhD, is an Assistant Professor in the Department of Industrial and Production Engineering at the Dr. B. R. Ambedkar National Institute of Technology, Jalandhar, Punjab, India. His broad areas of research are advanced machining methods, ultrasonic machining, rotary ultrasonic machining, ceramics, composites, optimization methods, and micro-structure analysis. He has published about 35 research articles in several SCI/Scopus indexed journals and national and international conferences, as well as several book chapters. Dr. Singh is an editor and reviewer of several SCI/Scopus indexed journals. He has supervised two MTech scholars, and there are currently several master's and doctoral research work going on under his supervision. He is a life member of the Indian Institution of Industrial Engineering (IIIE), Mumbai, India, and Science and Engineering Institute (SCIEI), Los Angeles, USA. Dr. Singh has organized one international conference at NIT Jalandhar as convener. He received a Young Scientist in Mechanical Engineering VIRA-2019 Award and a Young Faculty in Engineering (major area: mechanical engineering) in VIFA-2019 awards from the Venus International Foundation. He earned his PhD from the National Institute of Technology, Kurukshetra, India.

CONTENTS

CONTRIBUTORS

Rajeev Agrawal
Malaviya National Institute of Technology, Jaipur, Rajasthan–302017, India

Saif Ahmad
Department of Mechanical Engineering, KIET Group of Institutions, Ghaziabad, Uttar Pradesh, India, E-mail: saifahmad1602@gmail.com

B. B. Ahuja
Director, Department of Manufacturing Engineering and Industrial Management, College of Engineering, Pune [COEP], Wellesley Road, Shivajinagar, Pune–411005, Maharashtra, India

Zafar Alam
Department of Mechanical Engineering, Indian Institute of Technology (Indian School of Mines), Dhanbad, Jharkhand–826004, India, E-mail: zafar@iitism.ac.in

Vijay Kumar Bajpai
Department of Mechanical Engineering, National Institute of Technology, Kurukshetra, Haryana, India

Harish Kumar Banga
Department of Mechanical Engineering, Guru Nanak Dev Engineering College, Ludhiana, Punjab, India, E-mail: drhkbanga@gmail.com

Tapan Kumar Barman
Department of Mechanical Engineering, Jadavpur University, Kolkata–700032, West Bengal, India

Soumya Bhattacharjya
Associate Professor, Department of Civil Engineering, Indian Institute of Engineering Science and Technology, Shibpur, West Bengal, India, E-mail: soumya@civil.iiests.ac.in

Bijoy Bhattacharyya
Department of Production Engineering, Jadavpur University, Kolkata, West Bengal, India, E-mail: bb13@rediffmail.com

Gian Bhushan
Department of Mechanical Engineering, National Institute of Technology, Kurukshetra–136119, Haryana, India

Pankaj Biswas
Department of Mechanical Engineering, IIT Guwahati–781039, Assam, India

Hullash Chauhan
KIIT Deemed to be University, Bhubaneswar, Odisha, India

Sandeep Chhabra
Department of Mechanical Engineering, KIET Group of Institutions, Ghaziabad, Uttar Pradesh, India

Debraj Das
Department of Mechanical Engineering, Tripura Institute of Technology, Narsingarh, Agartala–799009, Tripura, India, E-mail: er.debraj@gmail.com

Tushar Das
Assistant Professor, Department of Civil Engineering, Heritage Institute of Technology, Kolkata, West Bengal, India

Maneetkumar R. Dhanvijay
Associate Professor, Department of Manufacturing Engineering and Industrial Management, College of Engineering, Pune [COEP], Wellesley Road, Shivajinagar, Pune–411005, Maharashtra, India

Biswanath Doloi
Department of Production Engineering, Jadavpur University, Kolkata, West Bengal, India, E-mail: bdoloionline@rediffmail.com

Anshu Gupta
Department of Education, Kanohar Lal Girls Degree College, Meerut, Uttar Pradesh, India

Varun Gupta
Department of Electronics and Instrumentation Engineering, KIET Group of Institutions, Delhi-NCR, Ghaziabad, Uttar Pradesh, India, E-mail: vargup2@gmail.com

Faiz Iqbal
School of Engineering, University of Lincoln, Lincoln, LN6 7TS, United Kingdom

Anbesh Jamwal
Malaviya National Institute of Technology, Jaipur, Rajasthan–302017, India

Nikita Jatia
Instrumentation and Control Department, Dr. B. R. Ambedkar National Institute of Technology, Jalandhar, Punjab, India, E-mail: nikitajatia2604@gmail.com

J. V. Muruga Lal Jeyan
Professor, Department of Aerospace Engineering, School of Mechanical Engineering, Lovely Professional University, Punjab–144111, India, E-mail: jvmlal@ymail.com

J. S. Karajagikar
Department of Production Engineering and Industrial Management, College of Engineering (COEP), Pune, Maharashtra, India

Hareesh Kumar
Department of Mechanical Engineering, Guru Nanak Dev Engineering College, Ludhiana, Punjab, India

Puneet Kumar
Department of Mechanical Engineering, Guru Nanak Dev Engineering College, Ludhiana, Punjab, India

Pawandeep Singh Matharoo
Department of Mechanical Engineering, National Institute of Technology, Kurukshetra–136119, Haryana, India, E-mail: pawandeep_31806203@nitkkr.ac.in

Sameer Mishra
Department of Mechanical Engineering, KIET Group of Institutions, Ghaziabad, Uttar Pradesh, India

Jhuma Mitra
Department of Civil Engineering, Tripura Institute of Technology, Narsingarh, Agartala–799009, Tripura, India

Monika Mittal
Department of Electrical Engineering, National Institute of Technology, Kurukshetra, Haryana, India

Vikas Mittal
Department of Electronics and Communication Engineering, National Institute of Technology, Kurukshetra, Haryana, India

Ketan S. Mowade
Department of Mechanical Engineering, Michigan Technological University, Houghton, USA

Sanjay W. Mowade
Department of Mechanical Engineering, Smt. Radhikatai Pandav College of Engineering, Nagpur, Maharashtra, India

Arkadeb Mukhopadhyay
Department of Mechanical Engineering, Birla Institute of Technology, Mesra, Ranchi–835215, Jharkhand, India, E-mail: arkadebjume@gmail.com

Mayur Nikhar
PG Scholar, MTech VLSI, G. H. Raisoni College of Engineering, Nagpur, Maharashtra, India

Mohit Pandey
Department of Production Engineering, Jadavpur University, Kolkata, West Bengal, India, E-mail: mhtpnd93@gmail.com

Rohit Patki
MTech [Mechatronics] Scholar, Department of Manufacturing Engineering and Industrial Management, College of Engineering, Pune [COEP], Wellesley Road, Shivajinagar, Pune–411005, Maharashtra, India

Vyasmuni Prajapati
Department of Mechanical Engineering, KIET Group of Institutions, Ghaziabad, Uttar Pradesh, India

Ayush Purohit
Department of Mechanical Engineering, Guru Nanak Dev Engineering College, Ludhiana, Punjab, India

Mohammad Owais Qidwai
Department of Mechanical Engineering, Delhi Skill and Entrepreneurship University, Okhla-III campus, New Delhi–110020, India

Abhishek Raj
Mechanical Engineering Department, IIT (Banaras Hindu University), Varanasi, Uttar Pradesh, India, E-mail: abhishekraj.rs.mec17@itbhu.ac.in

Akhila Rupesh
PhD Scholar, Department of Aerospace Engineering, School of Mechanical Engineering, Lovely Professional University, Punjab–144111, India, E-mail: akhilarupesh56@gmail.com

Subash Chandra Saha
Department of Mechanical Engineering, NIT, Agartala–799046, Tripura, India

A. K. Sahoo
KIIT Deemed to be University, Bhubaneswar, Odisha, India

Prasanta Sahoo
Department of Mechanical Engineering, Jadavpur University, Kolkata–700032, West Bengal, India

Cherian Samuel
Mechanical Engineering Department, IIT (Banaras Hindu University), Varanasi, Uttar Pradesh, India, E-mail: csamuel.mec@itbhu.ac.in

Suchismita Satapathy
KIIT Deemed to be University, Bhubaneswar, Odisha, India, E-mail: suchismitasatapathy9@gmail.com

Abhishek Sen
Department of Mechanical Engineering, Calcutta Institute of Technology, Uluberia, West Bengal, India; Department of Production Engineering, Jadavpur University, Kolkata, West Bengal, India, E-mail: abhishek.sen1986@gmail.com

Aman Sharma
Department of Mechanical Engineering, National Institute of Technology, Kurukshetra, Haryana, India, E-mail: aman_6180081@nitkkr.ac.in

Sagar D. Shelare
Department of Mechanical Engineering, Priyadarshini College of Engineering, Nagpur, Maharashtra, India

Avtar Singh
Department of Computer Science and Engineering, Dr. B. R. Ambedkar National Institute of Technology, Jalandhar, Punjab, India

Sumit Singh
Department of Mechanical Engineering, KIET Group of Institutions, Ghaziabad, Uttar Pradesh, India

Himanshu Suyal
Department of Computer Science and Engineering, Dr. B. R. Ambedkar National Institute of Technology, Jalandhar, Punjab, India, E-mail: suyal.himanshu@gmail.com

Laxman Thakre
Department of Electronics Engineering, G. H. Raisoni College of Engineering, Nagpur, Maharashtra, India, E-mail: laxman.thakare@raisoni.net

Abhijeet Varma
Department of Production Engineering and Industrial Management, College of Engineering (COEP), Pune, Maharashtra, India, E-mail: varma21abhi@gmail.com

Karan Veer
Instrumentation and Control Department, Dr. B. R. Ambedkar National Institute of Technology, Jalandhar, Punjab, India, E-mail: veerk@nitj.ac.in

Subhash N. Waghmare
Department of Mechanical Engineering, Priyadarshini College of Engineering, Nagpur, Maharashtra, India

Alok Yadav
Malaviya National Institute of Technology, Jaipur, Rajasthan–302017, India

Ankit Yadav
Malaviya National Institute of Technology, Jaipur, Rajasthan–302017, India

ABBREVIATIONS

μm	micro-meter
3D	three-dimensional
AARM	adaptive autoregressive modeling
Acc	accuracy
AHCI	arts and humanities citation index
AI	artificial intelligence
ANN	artificial neural networks
ANOVA	analysis of variance
API	application programming interface
ARTFA	autoregressive time-frequency analysis
a-Si	amorphous silicon
BCI-S	book citation index–science
BCI-SSH	book citation index-social sciences and humanities
BEMRF	ball end magnetorheological finishing
BLW	baseline wander
BPM	beats per minute
BSE	backscattered electron
CCD	central composite design
CCD	charge-coupled device
CE	counter electrode
CFD	computational fluid dynamics
CIGS	copper indium gallium diselenide
CIS	copper-indium selenide
CLT	cross-laminated timber
CO	carbon monoxide
COF	coefficient of friction
CPC	center for pollution control
CPCI-S	conference proceedings citation index-science
CPCI-SSH	conference proceedings citation index-social sciences and humanities
CPV	concentrated photovoltaic
CWT	continuous wavelet transform
DCEN	direct current and electrode negative

DCSP	direct current electrode negative
DCV	direction control valve
DCV	door closing velocity
DMAB	dimethylamine borane
DN	nozzle diameter
DO	deterministic optimization
DOE	design of experiment
Dr	detection rate
DSP	digital signal processing
DSSC	dye-sensitized solar cell
DWT	discrete wavelet transform
EC	entropy criterion
ECG	electrocardiogram
EDS	energy dispersive spectroscopy
EM	Expectation-Maximization
EMD	empirical mode decomposition
FDM	fused deposition modeling
FEA	finite element analysis
FESEM	field emission scanning electron microscope
FFD	full factorial design
FFT	fast Fourier transform
FHS	fetal heart sound
FIR Filter	finite impulse response filter
Fpcg	fetal phonocardiogram
FRL	filter, regulator, and lubricator unit
GBMO	gases Brownian motion optimization
GRA	grey relational analysis
GRC	grey relational coefficient
GRG	grey relational grade
GTAW	gas tungsten arc welding
HAZ	heat affected Zone
HB	higher-the-better
HCW	healthcare waste
HMI	human-machine interface
HRV	heart rate variability
IMFs	intrinsic mode function
IoT	internet of things
IT	information theory

LHS	Latin hypercube sampling
LPTN	lumped parameter thermal network
LSM	least squares method
MCS	Monte Carlo simulation
MLSM	moving least squares method
m-Si	mono crystalline Silicon
MV	majority voting
OA	orthogonal array
ODDD	optimally designed digital differentiator
OLE	object linking and embedding
Pa	ambient pressure
PCA	principal component analysis
PCG	phonocardiogram
PHC	primary health care centers
Pi	pressure of inlet
PLA	polylactic acid
PLI	power line interference
Pp	positive predictivity
PPM	probabilistic propagative method
p-Si	polycrystalline Silicon
QEC	quick exposure check
RANS	Reynolds averaged Navier Stokes equations
RBO	reliability-based optimization
RD	redundant design
RDO	robust design optimization
RSM	response surface methodology
RULA	rapid upper limb assessment
SCIE	science citation index expanded
SD	saturated design
Se	sensitivity
SJR	SCImago journal rank
SMCBWD	sound metric and have developed the critical band wavelet decomposition
SNIP	source normalized impact per paper
SNR	signal-to-noise ratios
SSCI	social sciences citation index
SSWT	synchro squeezed wavelet transform
STL	stereolithographic

SVM	support vector machine
Tc	temperature of cooling
TF	time-frequency
Ti	temperature of the inlet
TIG	tungsten inert gas
TiO_2	titanium dioxide
UD	uniform design
VBA	visual basic for applications
WoS	web of science
WT	wavelet transform
XRD	X-ray diffraction

ACKNOWLEDGMENTS

The editors wish to express their warm thanks and deep appreciation to those who provided input, support, constructive suggestions, and comments and assisted in the editing and proofreading of this book.

This book would not have been possible without the valuable scholarly contributions of authors across the globe. The editors avow the endless support and motivation from their family members and friends.

The editors are very grateful to all the members who served in the editorial and review process of the book. Mere words cannot express the editors' deep gratitude to the entire Apple Academic Press team for keeping faith and showing the right path to accomplish this high-level research book.

Finally, the editors use this opportunity to thank all the readers and expect that this book will continue to inspire and guide them in their future endeavors.

—**Editors**

PREFACE

In recent years, optimization methods have gained supreme importance due to their ability to augment the design and performance of various engineering systems. Generally, any optimization method includes the application of mathematical models for optimizing engineering problems under various constraints. An objective function is formed, and more than one mathematical solution can be derived in order to identify global optimum values. Alternatively, it can be said that optimization in engineering designs uses mathematical formulation for a particular design problem and thus support the selection of optimal design among many alternatives. In the current era of industrial development, optimization of engineering problems has become intricate, which can be attributed to the presence of several social, economic, technical, and reliability-based dimensions.

The current volume offers a platform for discussion among investigators, industry professionals, stakeholders, and economic strategists and then argues the possibility of determining new ways of solving optimization problems related to different industrial sectors. Optimization methods deal with both operative conditions in the process or in service industries, and emerging research areas are explored by researchers towards the implementation of optimization algorithms for the enhancement of system performance as well as system effectiveness. In the field of engineering, it is the need of the current era to reduce the time-dependent factors as well as different types of cost-based clusters. In view of this, it is required to optimize the dependent as well as independent factors of different models, which are implemented towards profit maximization. The role of the optimization method is not only for engineering applications. However, it is going to use in the medical, food, oil, textile, energy, and agriculture sectors, etc. Thus, the present book has collected high-quality papers that present different optimization model-based engineering problems that address valuable inputs.

The main advantage of this volume is that it delivers a wide range of topics pushing forwarded to cater to the requirements of researchers in different engineering domains using novel mathematical modeling-based optimization methods for solving real-life problems. Discussions the implementation of optimization tools may lead to affordable product prices with high quality, which is essential for a company to be in competition with other competitors in the market. In this book, different constructive demonstration of optimization modeling has been proposed in order to support a strong framework to finally get the need of the engineering sectors.

ORGANIZATION OF THE BOOK

This book consists of 21 chapters on a diverse range of optimization themes and applications. Concise discussions about these chapters are previewed below:

Chapter 1 presents an efficient analysis of the electrocardiogram (ECG) signal is still a challenge due to large variations in its morphology. Therefore, it requires the proper utilization of digital signal processing (DSP) techniques to analyze raw ECG signals. Presently, cardiologists need the active involvement of computers equipped with efficient DSP techniques. In this chapter, pre-processing is performed using wavelet transform (WT) due to its better time-frequency resolution. Adaptive autoregressive modeling (AARM) is used for extracting features for time-dependent variable parameters. Support vector machine (SVM) is also considered for classification due to its high modeling stability, even for non-linear data. The proposed model is evaluated in accordance with different parameters, and model accuracy of 99.93%, the sensitivity of 99.95%, positive predictivity of 99.95%, and detection rate of 99.95% have been observed.

Chapter 2 presents an application related to the pneumatic door slam platform for design development using a PLC-based controller to check the structural and functional integrity of vehicle door components. An overview of different components whose functionality can be checked using this platform is explored. This work also explains the pneumatic circuit for developing a door slam platform. Pneumatic door slam platform, measurement of the door opening and closing velocity, and carryout required number of continuous cycles are considered in this chapter.

Chapter 3 investigates ergonomic hazard levels for farmers working in hot-climatic conditions in India by utilizing the RULA tool.

Chapter 4 discusses that health care waste disposal is one of the essential concerns for any developing country. People are traveling around the whole world, and the increasing population of the world has led to the healthcare sector's rising prominence. Increment in the healthcare sector also increases healthcare waste. Due to the increment in waste, their disposal is a crucial matter. The health care waste generated from different hospitals is not entirely hazardous, so proper segregation must avoid transportation of all the waste to the incineration center. Varanasi is a highly dense city with lots of hospitals. A case study was performed in Varanasi to find the collection centers and optimize routes in Varanasi. A total of 20 collection centers were found, which will cover almost 200 hospitals in the city. This study will help the hospital management and the government take necessary action for effective management.

Chapter 5 aims to summarize the research trends, advancements, and future research direction in life cycle assessment. The literature reported in this study is based on a research database ranging from 2000 to 2019 and is further categorized into different sections. Different types of analysis have been done in this chapter, such as author analysis, trend analysis, country analysis, discipline-wise analysis, and source analysis. It has been found that in the last five years, the adoption of LCA has become popular in the manufacturing industries, which helps to promote sustainability. Also, the Govt. policies among the global and customer pressure forced many industries to adopt sustainable production. Also, some business has been seen in both of the databases because of common authors with highly cited publications in both databases. A list of top 10 highly cited papers is also considered.

Chapter 6 focuses on how we can save time by reducing the number of mouse clicks and giving assurance of design quality by automating the task and tracking its results which will give good CAD quality with minimum design time and CAD completion percentage. So, using CAD tool in combination with VBA, a macro has been built which does all these repetitive activities performed in a single click which saves 50-60% of the time for repetitive tasks. By using all the results and tracked data, the project CAD Completion percentage can be calculated. This results in design time optimization and cost optimization as well.

Chapter 7 aims to develop 3D printable body-fueled prosthetic Finger for the grown-up in our nation that permits a parametric plan. The upside of the parametric structure is that it tends to be customized for each client, and each grown-up can be fitted with a prosthesis that almost looks like the size of his/her sound hand. A factual examination has been directed to comprehend which parameters are more appropriate for parametric structure. The plan was done in solid work, and it was associated with an outer record that permits us to change and adjust the structure without the requirements to open the CAD document. Acknowledgment of these gadgets is dependent upon the solace of the client, which relies intensely upon the size, weight, and in general, tastefulness of the gadget. As found in various applications, parametric displays can be used to create clinical gadgets that are explicit to the patient's needs.

Chapter 8 compares some of the existing truth inference algorithms to make a comparative study of the algorithms in real-time datasets. In the world, the popularity of artificial intelligence enlarged the researcher tries to incorporate human behavior into the machine. Due to this, it led to too many problems that machines cannot solve alone. Researchers believe that machines and humans can act together, which led to the new field called crowdsourcing. Crowdsourcing is used to address harder problems that require human intelligence. Increased requirements of the crowd (called workers) led to create low-quality data and redundant data due to the availability of the low-quality workers. To solve this problem, many redundancy-based algorithms can be used by assigning the tasks to workers to find the correctness of answers.

Chapter 9 aims to examine the designed ventilation system of the car parks using CFD Simulations. Automobile parking either be completely enclosed or it can be partially open. Completely closed parking are generally underground and needs a ventilation system. Partially open car park garages are often above decks that have an open side. Natural and Mechanical ventilation, along with their consolidated use, can provide ajar car park. The specific aim is to check the ventilation of car parking garages for predefined positions of inlet and outlets. It involves 3D modeling of the car park area, meshing into finite volumes and carrying out the simulation using CFD tools to have quick analysis of the CO concentration and velocity profile of the flow in the domain. The results emphasize that the designed ventilation system was able to extract CO to maintain its safe level in the car park.

Chapter 10 deals with a comparative study between the single-, double-, and four-inlet conditions of a rocket nozzle.

Chapter 11 focuses on the investigation and efficiency of vortex tube, which works on the principle of the hot gas stream through one end and a cold gas flow from the further end without any external source of energy. The nozzle design is the prime concern of the research as it will give a greater cooling effect as compared to the inlet and outlet orifice of the vortex tube. The geometrical parameters have been analyzed to get the better and more efficient design of the improved vortex tube.

Chapter 12 mainly relates with fetal phonocardiogram (fPCG) processing of signal. Signal denoising is always a major task after recording the signal, so in this chapter basically, two types of denoising methods that are finite impulse response filter (FIR filter) and empirical mode decomposition (EMD) methods are used and compared. For testing, or comparing these two, recordings of real dataset were used, and the estimate depends on cognitive observation and signal advancement after performing both methods. The results in this chapter proved that both the methods assisted for improving fetal PCG signal by de-noising the signal. On the ground of the results, in the end we concluded that EMD as a suitable method for denoising and processing the fetal PCG signal.

Chapter 13 deals with the prediction of aerodynamics in a flow as it moves over a streamlined body. This is executed with the use of wind tunnels of appropriate Mach numbers, with the influence of computational analysis. But it is mandatory to verify the theoretical results to the experimental results carried out in a wind tunnel. Before testing, it is necessary that the flow parameters, namely the velocity distribution, angularity of the flow, turbulence, are evaluated and are within the margins. This is important when we are evaluating the flow in three dimensions. Thus, it ensures reliability of the experimental results. In order to evaluate the truthiness of the flow, a two-hole spherical flow analyzer is being used. The truthiness of two wind tunnels across India is being evaluated by a two-hole spherical flow analyzer. From the experiment, the data map will be plotted for the flow parameters using the yaw head constant and the respective orientations of the two-hole spherical flow analyzer. Their nature is evaluated, and the findings are discussed.

Chapter 14 attempts to realize hot spot cooling, *heat* transfer *path* (thermal path), and the effectiveness of a considered cooling system. A comparison case without gaps is simulated to check the claim, that even, a negligible flow rate may improve the overall working condition of the electromagnet. Based on the analysis of the results, some improvements are suggested for the design of the cooling system.

Chapter 15 presents a comprehensive review of current state of the art of PV technology, material, manufacturing techniques, and their efficiency and also highlights different studies done globally in this area of research. Solar PV technology is demonstrating its application all over the globe. With the research efforts, the power generation efficiency has increased a lot from a mere 6% in the beginning. The technology has enormous scope for increasing its efficiency with integration and applications of various methods. The introduction of new generation solar cells has widened this scope. The second, third and fourth generation PV cells have properties that make them viable for commercial use.

Chapter 16 focuses on strategies and steps taken to improve farming by focusing on technical knowledge and development to make the agricultural sector more reliable and easier for the farmers by predicting the suitable crop by using Machine learning algorithms.

Chapter 17 investigates the high-temperature tribological performance of Ni-B-W coating. Taguchi's L_{27} orthogonal array was adopted to perform experiments. Grey relational analysis (GRA) has been implemented to optimize the tribo-test parameters.

Chapter 18 aims to focus on utilization of a nanosecond pulsed fiber laser system to generate micro-holes on quartz. A combination of pulse frequency of 65 kHz, a duty cycle of 50%, laser power of 37 W, and air pressure of 2.50 kgf/cm^2 leads to the maximum circularity of 0.88 at the entry side. Effect of laser power, pulse frequency, duty cycle, and air pressure on the entry hole circularity of micro-holes on quartz are considered for further analyses using response surface methodology (RSM).

Chapter 19 employs a multi-response optimization technique using on grey-based Taguchi method to optimize weld bead geometry and process parameters for Tungsten Inert Gas (TIG) bead-on-plate welding of IS 2062B mild steel.

Chapter 20 presents a new model of the moving least squares method (MLSM) for Robust Design Optimization (RDO) for different case studies.

Chapter 21 attempts to laser mark a geometrical figure on stainless steel 304 using a multi-diode pumped fiber laser. Desirability function analysis has also been adopted for optimization of process parameters for the optimum value of responses of mark intensity and circularity.

—Editors

CHAPTER 1

ADAPTIVE AUTOREGRESSIVE MODELING BASED ECG SIGNAL ANALYSIS FOR HEALTH MONITORING

VARUN GUPTA,[1] MONIKA MITTAL,[2] VIKAS MITTAL,[3] and ANSHU GUPTA[4]

[1]*Department of Electronics and Instrumentation Engineering, KIET Group of Institutions, Delhi-NCR, Ghaziabad, Uttar Pradesh, India, E-mail: vargup2@gmail.com*

[2]*Department of Electrical Engineering, National Institute of Technology, Kurukshetra, Haryana, India*

[3]*Department of Electronics and Communication Engineering, National Institute of Technology, Kurukshetra, Haryana, India*

[4]*Department of Education, Kanohar Lal Girls Degree College, Meerut, Uttar Pradesh, India*

ABSTRACT

In the present era, efficient analysis of electrocardiogram (ECG) signal is still a challenge due to large variations in its morphology. Therefore, it requires the proper utilization of digital signal processing (DSP) techniques to analyze raw ECG signals. Presently, cardiologists need the active involvement of computers equipped with efficient DSP techniques

Optimization Methods for Engineering Problems. Dilbagh Panchal, Prasenjit Chatterjee, Mohit Tyagi, Ravi Pratap Singh (Eds.)

like pre-processing, feature extraction, and classification. In this chapter, pre-processing is performed using wavelet transform (WT) due to its better time-frequency resolution. Adaptive autoregressive modeling (AARM) is used for extracting features as its parameters are allowed to vary in time. And support vector machine (SVM) technique is considered for classification due to its high modeling stability even for non-linear data. The performance of the proposed technique is evaluated on the basis of parameters such as sensitivity (Se), positive predictivity (Pp), accuracy (Acc), and detection rate (Dr). The proposed technique has secured Se of 99.95%, Pp of 99.95%, Acc of 99.93%, and Dr of 99.95%. The authors expect that the proposed technique may be successful in classifying all major types of arrhythmias that were not classified by the existing methodologies single-handedly.

1.1 INTRODUCTION

A timely and accurate diagnosis of the heart condition has great importance in controlling the death rate due to cardiac diseases [1, 2]. Cardiac diseases are sometimes called "Arrhythmia" [3–6]. In this era, Electrocardiogram (ECG) emerges as a primitive and less expensive essential tool in biomedical signal processing (BSP) [7–9] for capturing the electrical activity of the heart [10–12]. Using the specific lead arrangement, ECG signal is plotted on the chart paper using ECG machine [13–15], which is not only random but even quasi-periodic in nature [16–20]. It is presented in the form of three main waves – P-wave, QRS-wave (or QRS-complex), and T-wave [21–23]. Figure 1.1 shows the schematic of a subject with ECG signal recording and related treatment. This figure shows that ECG signal acquisition during subject health treatment with complete setup. In QRS-wave, R-wave has maximum amplitude (1–2 mVolt) [24–27]. Among these three waves, QRS-wave contains most essential clinical information, which is extracted by estimating heart rate (H.R) [28, 29]. The root cause of arrhythmia is the discrepancies in H.R, which results further into chaotic electrical activity [30–35]. The heart specialists usually base their decision in a particular situation by investigating amplitude, frequency, and polarity of R-wave [22, 36–38]. In this chapter, different techniques [39–44], such as wavelet transform (WT) for pre-processing, adaptive autoregressive modeling (AARM) for extracting features, and Support vector machine (SVM) technique is considered for classification purpose.

This chapter is structured as follows; Section 1.2 presents the research background; Section 1.3 presents research methods; Section 1.4 shows a case study of considered databases; Section 1.5 showcases the results and discussion over those; and finally, Section 1.6 concludes the chapter.

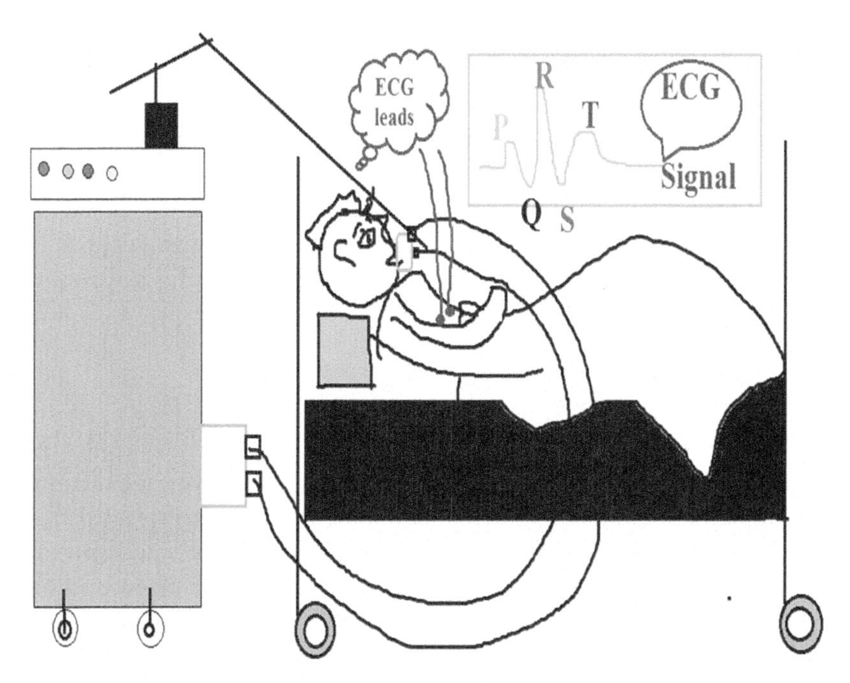

FIGURE 1.1 Subject during ECG signal recording and related treatment.

1.2 RESEARCH BACKGROUND

In Ref. [28], Sharma and Sharma proposed a synchrosqueezed wavelet transform (SSWT) for QRS detection. A synchrosqueezed wavelet transform (SSWT) is carried out by synchrosqueezing to the continuous wavelet transform (CWT). They obtained Se of 99.92%, Pp of 99.93% and error rate (E.R) of 0.15%, respectively. In Ref. [45], Cleetus et al. proposed Granger causality to interpret respiratory and ECG signals. They have used frequency domain Granger causality for healthy subjects by recording cardiac and respiratory signal during postural change from supine to standing. In Ref. [30], Alickovic and Subasi proposed autoregressive (AR) modeling for extracting features of the recorded ECG signal. They have obtained %Acc of 99.93. In Ref. [29] Rekik and Ellouze proposed Entropy Criterion (EC) of the Wavelet Transform (WT) for finding R-peaks. The WT has been used at the analysis window with a first derivative Gaussian wavelet. In Ref. [9], Nayak et al. proposed an optimally designed digital differentiator (ODDD) for precise detection of QRS complex. For optimization purpose, gases Brownian motion optimization (GBMO) was considered. In the detection stage, Hilbert transform was used. In Ref. [46], Gupta and Mittal used principal component analysis (PCA), fast Fourier transform (FFT), and autoregressive time-frequency analysis (ARTFA) for analyzing the considered respiratory signal by estimating principal components for differentiating normal and sinus subjects. In Refs. [47–53], authors have used Fourier-based analysis for a variety of applications such as BLW removal, PLI removal, etc.

1.3 RESEARCH METHODS

An efficient diagnostic research always requires an effective computational algorithm [54, 55]. In this chapter, different methods are selected from various domains for efficient analysis in healthcare. For extracting useful information in cardiology, ECG is the primary authentic diagnostic tool worldwide, which is obtained in the form of three basic waves namely; P, QRS, and T [56, 57]. It is taken from the body surface of the subject using electrodes. QRS is the main reference wave for diagnosis of subject. Unfortunately, QRS detection is not an easy task due to its time-varying nature.

1.3.1 SIGNAL ACQUISITION

Signal acquisition [58] is performed in 3-lead arrangement after taking consent at KIET Group of Institutions, Delhi-NCR, Ghaziabad in Virtual NI/Biomedical Lab. The clinical (pathological) databases are obtained from the cardiology lab of Anand Hospital, Meerut, as Figure 1.1 shows the schematic with a complete setup.

1.3.2 WAVELET TRANSFORM

In ECG signal analysis, QRS-complex is the most essential wave. Unfortunately, various obstacles are there in its analysis, such as electrode artifacts, respiration artifacts, bad quality of gel, muscular noise, baseline drift (or baseline wander (BLW)), power line interference (PLI) or Hum, low signal-to-noise ratios [46, 47, 59]. BLW is generated due to respiration and PLI comes due to single operating frequency of the country introducing the narrow-band noise [60–62]. Therefore, pre-processing is required using some specialized tools [60, 63]. In this chapter, discrete wavelet transform (DWT) has been considered for pre-processing of considered clinical databases. Mathematically, it is represented as [63]:

$$Y_{k,l} = \langle u(t), B_{k,l}(t) \rangle = \int_{-\infty}^{\infty} u(t) B_{k,l}(t)\,dt \quad (1)$$

where; *k* and *l* are the scale factor index and location of wavelet coefficient, respectively in which *u*(*t*) denotes ECG signal; and $B_{k,l}(t)$ denotes wavelet basis function.

1.3.3 ADAPTIVE AUTOREGRESSIVE MODELING (AARM)

Adaptive autoregressive modeling (AARM) is used for extracting features as its parameters are allowed to vary in time. Mathematically, AARM process for order m is defined as [64]:

$$c[k] = \sum_{i=1}^{m} \alpha_i c[k-i] + \in[k] \quad (2)$$

where; $\in[k]$ is the time sequence, related to independent variables with E[*] = 0. Spectral density can be expressed using Eqn. (2) as:

$$S(f) = \frac{\sigma^2}{\left|1 - \sum_{i=1}^{m} \alpha_i e^{-j2\pi fi}\right|^2} \quad (3)$$

where; $w = 2\pi f$ for $-0.5 < f < 0.5$ or $-\pi < w < \pi$.

1.3.4 SUPPORT VECTOR MACHINE (SVM)

Support vector machine (SVM) presents an optimal solution comprising support vectors that lie near the hyperplane that is also known as decision surface [21, 65]. Figure 1.2 shows SVM classification using hyperplane. In this figure, two types of classes are shown – one is shown by diamond shaped and the other is shown by oval shaped. Left figure shows unclassified data (i.e., X-zigzag in fashion) and the right figure shows classified data (F-final classified data by hyperplane). Its main advantage over existing classification techniques is the higher flexibility to modify to fit the training data.

- **Support Vectors:** These show the upper and lower points, which are expressed as:

$$H_1 : wx_i + b = +1 \tag{4}$$

$$H_2 : wx_i + b = -1 \tag{5}$$

$$\text{In general, for } w^T x + b \geq 0 \text{ for } d_i = +1 \tag{6}$$

$$w^T x + b < 0 \text{ for } d_i = -1 \tag{7}$$

where; $d_i = +1$ is the shortest distance towards positive support vector (H_1); $d_i = -1$ is the shortest distance towards negative support vector (H_2); *w* denotes weight vector; *x* denotes input vector; *b* denotes bias.

The middle plane is defined as:

$$w^T x + b = 0 \tag{8}$$

Optimal hyperplane is defined as higher margin of separation between H_1 and H_2. This margin is known as "gutter."

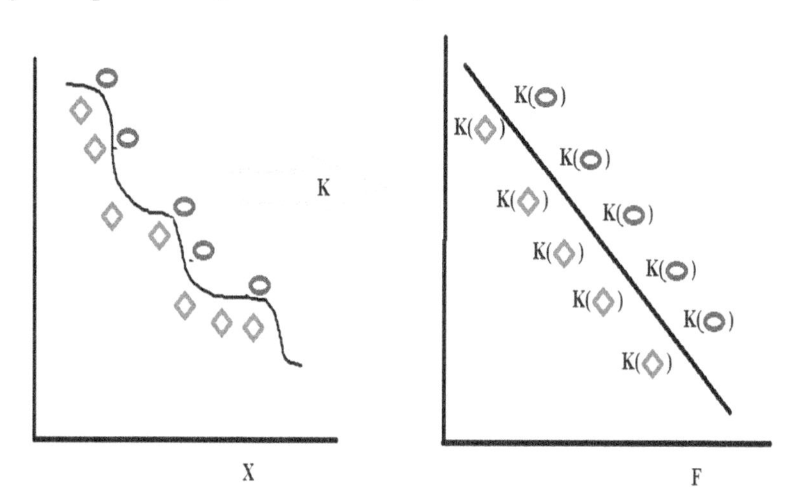

FIGURE 1.2 SVM classification using hyperplane.

1.3.5 PERFORMANCE EVALUATING PARAMETERS

In this chapter, different parameters such as sensitivity (Se), positive predictivity (Pp), accuracy (Acc), and detection rate (Dr) [28, 29, 36, 37, 54, 55] have been selected for estimating performance of the proposed technique.

1.4 CASE STUDY

In this chapter, different clinical (pathological) databases (subject recordings) have been used for effectively identifying the underlying arrhythmia such as infantile *apnea (*bradycardia), and atrial fibrillation. For analyzing such datasets, first-of-all pre-processing is performed by selecting Daubechies wavelet of order 4 (db4), in the second step, feature extraction is done using AARM, and finally, peaks are classified using SVM.

Results from Figures 1.2 and 1.3 demonstrate that AARM is powerful for effectively identifying the underlying arrhythmia in clinical (pathological) databases. For infantile *apnea (*bradycardia) subject recording, heart rate was less than 60 beats per minute, and for clinical atrial fibrillation subject recording, sinus rate was 377 beats per minute.

Table 1.1 shows that the reported Se were 0.9946 (or 99.46%), 0.9970 (or 99.70%), 0.9993 (or 99.93%), 0.9979 (or 99.79%), 0.9750 (or 97.50%), 0.9990 (or 99.90%) in Kaya and Pehlivan [1]; Mehta and Lingayat [67]; Gupta and Mittal [41]; Gupta et al. [43]; Elgendi et al. [20]; Gupta et al. [12]; respectively, Pp were 0.9991 (or 99.91%), 0.9775 (or 97.75%), 0.9997 (or 99.97%), 0.9990 (or 99.90%) in Kaya and Pehlivan [1]; Mehta and Lingayat [67]; Gupta and Mittal [41]; Elgendi et al. [20]; respectively, Dr were 0.9866 (or 98.66%), 0.9991 (or 99.91%), 0.9990 (or 99.90%), 0.9981 (or 99.81%) in Mehta et al. [66]; Gupta and Mittal [41]; Gupta et al. [43]; Gupta et al. [12]; respectively, Acc were 0.9969 (or 99.69%), 0.9984 (or 99.84%), 0.9993 (or 99.93%), 0.9991 (or 99.91%), 0.9879 (or 98.79%) in Kaya and Pehlivan [1]; Nayak et al. [9]; Alickovic and Subasi [30]; Mehta and Lingayat [67]; respectively, and the proposed work reported Se of 0.9995 (or 99.95%), Pp of 0.9995 (or 99.95%), Dr of 0.9995 (or 99.95%) and Acc of 0.9993 (or 99.93%) in total beats of 23,761 (True Positives =23,749).

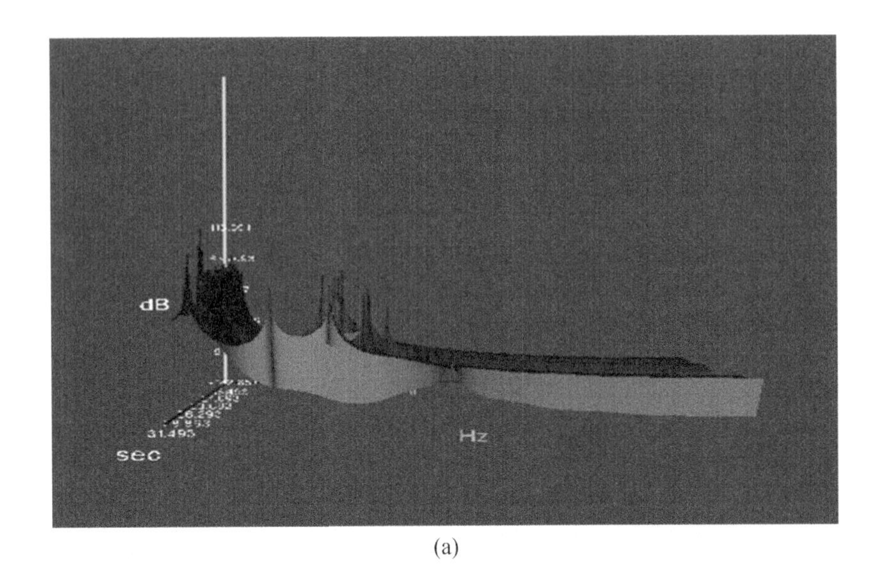

(a)

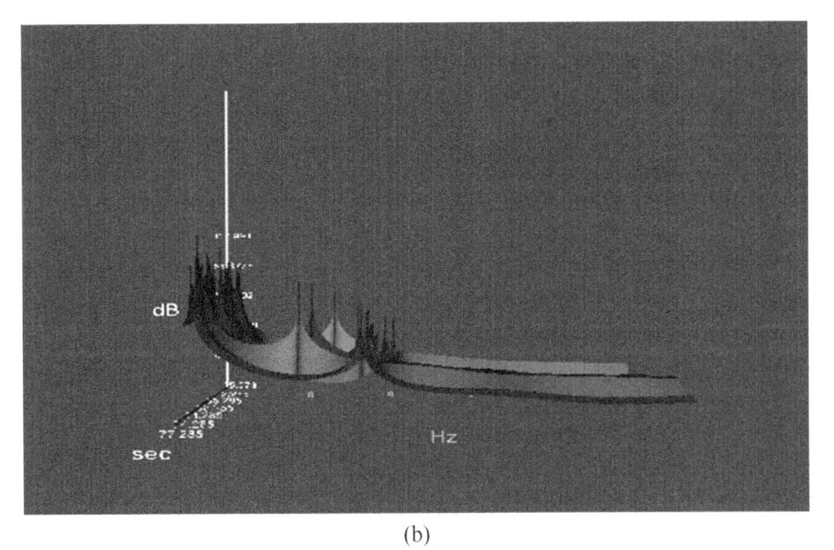

(b)

FIGURE 1.3 AARM of clinical–(a) infantile *apnea* (bradycardia); and (b) atrial fibrillation subject (patient).

1.4.1 APPLICATION OF RESEARCH METHODS

The proposed algorithm is shown to successfully detect R-peaks in various clinical and real-time conditions (varying QRS morphologies).

TABLE 1.1 Comparison between Proposed and Existing State-of-the-Art Methods based on Acc, Se, Pp, and Dr

Method	Acc (%)	Se (%)	Pp (%)	Dr (%)	References
Adaptive autoregressive modeling (AARM)	99.93	99.95	99.95	99.95	[63]
Genetic algorithm with KNN	99.69	99.46	99.91	–	[1]
Optimally designed digital differentiator (ODDD)	99.84	–	–	–	[9]
Autoregressive modeling (ARM)	99.93	–	–	–	[30]
K-means algorithm	–	–	–	98.66	[66]
Signal entropy (single lead ECG signal)	–	99.70	97.75	–	[67]
Short time Fourier transform	99.91	99.93	99.97	99.91	[41]
ARTFA with KNN	–	99.79	–	99.90	[43]
Maximal overlap wavelet packet transform (MOWPT)	98.79	–	–	–	[26]
Dynamic thresholds	–	97.50	99.90	–	[20]
CWT, spectrogram, and ARTFA	–	99.90	–	99.81	[12]

1.5 RESULT DISCUSSION

Figure 1.4 shows that based on Se and Pp parameters, the proposed technique outperforms the existing techniques. For instance, Kaya and Pehlivan [1] reported Se of 0.9946 (or 99.46%) and Pp of 0.9991 (or 99.91%) as compared to Se of 0.9995 (or 99.95%) and Pp of 0.9995 (or 99.95%) obtained by the proposed technique.

Figure 1.5 shows that based on Acc and Dr parameters, the proposed technique outperforms the existing techniques. For instance, Gupta and Mittal [41] reported Acc of 0.9991 (or 99.91%) and Dr of 0.9991 (or 99.91%) as compared to Acc of 0.9993 (or 99.93%) and Dr of 0.9995 (or 99.95%) obtained by the proposed technique.

In the existing techniques, both Alickovic and Subasi [30] and the proposed technique achieved Acc of 0.9993 (or 99.93%). But AARM has the capability of adaptively tune the AR coefficients, unlike ARM. Also, the proposed technique is able to yield consistently high values of all the performance parameters, i.e., Se of 0.9995 (or 99.95%), Pp of 0.9995 (or 99.95%), Dr of 0.9995 (or 99.95%), and Acc of 0.9993 (or 99.93%).

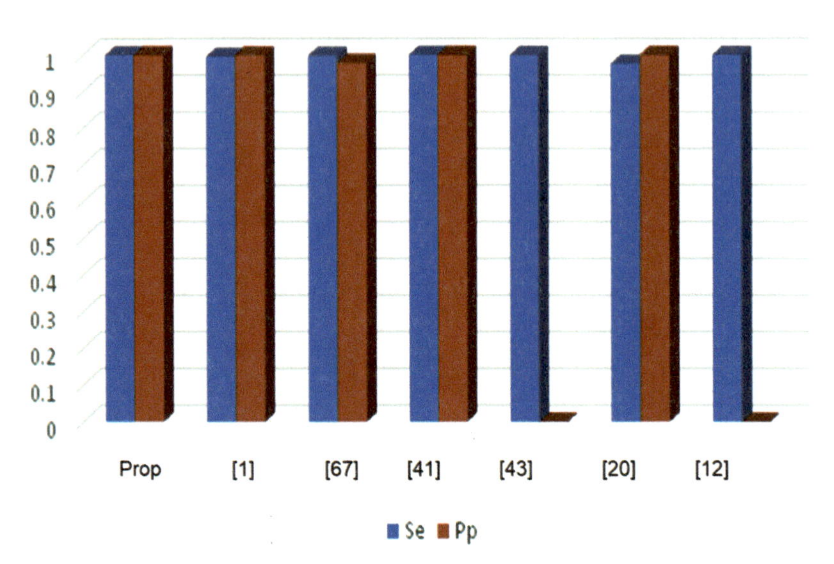

FIGURE 1.4 Comparison between proposed and existing state-of-the-art methods (based on Se and Pp).

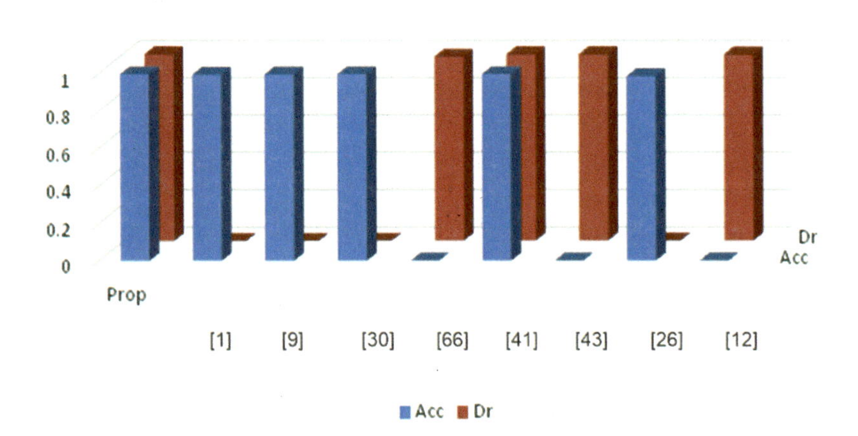

FIGURE 1.5 Comparison between proposed and existing state-of-the-art methods (based on Acc and Dr).

1.6 CONCLUSION

Quick, efficient, and accurate analysis are still the main challenges in health informatics. The proposed AARM reduces the burden over SVM classifier and still yield effective results for almost all of the considered pathological datasets. The proposed work reported consistently high values of all the considered performance parameters, i.e., Se of 99.95%, Pp of 99.95%, Dr of 99.95%, and Acc of 99.93% unlike other existing techniques where not all the values are high. Also, it has been demonstrated that the proposed technique can be applied even for signals with low signal-to-noise ratios (SNR). In future, AARM can be extended to other standard databases to make it universally acceptable as a robust, accurate, and economical technique.

KEYWORDS

- **adaptive autoregressive modeling**
- **arrhythmia**
- **digital signal processing**
- **electrocardiogram**
- **positive predictivity**
- **support vector machine**

REFERENCES

1. Kaya, Y., & Pehlivan, H., (2015). Feature selection using genetic algorithms for premature ventricular contraction classification. In: *9th International Conference on Electrical and Electronics Engineering (ELECO)*. Bursa, Turkey.
2. Gupta, V., Mittal, M., & Mittal, V. (2020). Chaos theory: An emerging tool for arrhythmia detection. *Sensing and Imaging, 21*(10), 1–22.
3. Agneeswaran, V. S., Mukherjee, J., Gupta, A., Tonpay, P., Tiwari, J., & Agarwal, N., (2013). Real-time analytics for the healthcare industry: Arrhythmia detection. *Big Data, 1*(3), 176–182. doi: 10.1089/big.2013.0018.
4. Gupta, V., & Mittal, M., (2020). Arrhythmia detection in ECG signal using fractional wavelet transform with principal component analysis. *J. The Inst. Eng. (India): Series B*. Springer. doi: 10.1007/s40031-020-00488-z.

5. Gupta, V., & Mittal, M., (2020). Efficient R-peak detection in electrocardiogram signal based on features extracted using Hilbert transform and burg method. *J Inst Eng. India Ser. B*. https://doi.org/10.1007/s40031-020-00423-2.
6. Kaur, H., & Rajni, R., (2017). On the detection of cardiac arrhythmia with principal component analysis. *J. Wireless Pers. Commun., 97*, 5495–5509.
7. Sharma, M., Tan, R. S., & Acharya, U. R., (2018). A novel automated diagnostic system for classification of myocardial infarction ECG signals using an optimal biorthogonal filter bank. *Computers in Biology and Medicine, 102*, 341–356.
8. Anurudhya, K., & Mohan, N. M., (2019). Analysis of a contactless ECG monitoring system. *IETE Journal of Research*. doi: 10.1080/03772063.2018.1562386.
9. Nayak, C., Saha, S. K., Kar, R., & Mandal, D., (2018). An efficient QRS complex detection using optimally designed digital differentiator. *Circ. Sys. Sig. Processing, 38*(5), 716–749.
10. Helen, M. C., et al., (2020). Changes in scale-invariance property of electrocardiogram as a predictor of hypertension. *International Journal of Medical Engineering and Informatics (IJMEI), 12*(3), 228–236.
11. Elgendi, M., Eskofier, B., Dokos, S., & Abbott, D., (2014). Revisiting QRS detection methodologies for portable, wearable, battery-operated, and wireless ECG systems. *PLoS One, 9*(1), e84018. doi: 10.1371/ journal.pone.0084018.
12. Gupta, V., Mittal, M., & Mittal, V., (2021). ECG signal analysis using CWT, spectrogram and autoregressive technique. *Iran J. Comput. Sci*. https://doi.org/10.1007/s42044-021-00080-8.
13. Sharma, M., Tan, R. S., & Acharya, U. R., (2019). A new method to identify coronary artery disease with ECG signals and time-Frequency concentrated antisymmetric biorthogonal wavelet filter bank. *Pattern Recognition Letters, 125*, 235–240.
14. Rajankar, S. O., & Talbar, S. N., (2019). An electrocardiogram signal compression techniques: A comprehensive review. *Analog Integrated Circuits and Signal Processing, 98*(1), 59–74.
15. Gupta, V., Monika, M., & Vikas, M., (2019). R-peak detection using chaos analysis In standard and real time ECG databases. *IRBM, 40*(6), 341–354.
16. Lin, C., Yeh, C. H., Wang, C. Y., Shi, W., Serafico, B. M. F., Wang, C. H., Juan, C. H., et al., (2018). Robust fetal heart beat detection via r-peak intervals distribution. *Trans. Biomed Eng., 66*(12), 3310–3319.
17. Gupta, V., et al., (2020). Attractor plot as an emerging tool in ECG signal processing for improved health informatics. *International Conference on Future Technologies 2020 (ICOFT 2020) in Manufacturing, Automation, Design and Energy (MADE@NITPY)*. National Institute of Technology Puducherry Karaikal, India.
18. Elgendi, M., (2013). Fast QRS detection with an optimized knowledge-based method: Evaluation on 11 standard ECG databases. *PLoS One, 8*(9), e73557. doi: 10.1371/journal.pone.0073557.
19. Xingyuan, W., & Juan, M., (2009). Wavelet-based hybrid ECG compression technique. *Analog. Integr. Circ. Sig. Process, 59*(3), 301–308.
20. Elgendi, M., Jonkman, M., & Boer, F. D., (2009). Improved QRS detection algorithm using dynamic thresholds. *International Journal of Hybrid Information Technology, 2*(1), 65–80.

21. Mehta, S. S., & Lingayat, N. S., (2008). Development of SVM based ECG pattern recognition technique. *IETE J. Res., 54*(1), 5–11.
22. Gupta, V., & Mittal, M., (2019). R-peak detection in ECG signal using yule–walker and principal component analysis. *IETE J. Res.* https://doi.org/ 10.1080/03772063.2019.1575292.
23. Gupta, V., & Mittal, M., (2019). R-peak based arrhythmia detection using Hilbert transform and principal component analysis. In: *2018 3rd International Innovative Applications of Computational Intelligence on Power, Energy and Controls with their Impact on Humanity (CIPECH)*. Ghaziabad, India.
24. Gupta, V., & Mittal, M., (2018). ECG (electrocardiogram) signals interpretation using chaos theory. *J. Adv. Res. Dyn. Cont. Sys (JARDCS), 10*(2), 2392–2397.
25. Gupta, V., & Mittal, M., (2015). Principal component analysis & factor analysis as an enhanced tool of pattern recognition. *Int. J. Elec & Electr. Eng. &Telecoms, 1*(2), 73–78.
26. Huang, J. S., Chen, B. Q., Zeng, N. Y., et al., (2020). Accurate classification of ECG arrhythmia using MOWPT enhanced fast compression deep learning networks. *J. Ambient Intell. Human Comput.* https://doi.org/10.1007/s12652-020-02110-y.
27. Kim, J. S., Kim, S. H., & Pan, S. B., (2020). Personal recognition using convolutional neural network with ECG coupling image. *J. Ambient Intell. Human Comput., 11*, 1923–1932 https://doi.org/10.1007/s12652-019-01401-3.
28. Sharma, T., & Sharma, K. K., (2016). QRS complex detection in ECG signals using the synchro squeezed wavelet transform. *IETE J. Res., 62*(6), 885–892.
29. Wang, Z., Zhu, J., Yan, T., & Yang, L., (2019). A new modified wavelet-based ECG denoising. *Computer Assisted Surgery*, 174–183.
30. Alickovic, E., & Subasi, A., (2015). Effect of multiscale PCA de-noising in ECG beat classification for diagnosis of cardiovascular diseases. *J. Cir. Sys. and Sig. Proc., 34*, 513–533.
31. Gupta, V., Monika, M., & Vikas, M., (2020). R-peak detection based chaos analysis of ECG signal. *Analog Integrated circuits and Signal Processing, 102*, 479–490.
32. Kora, P., & Krishna, K. S. R., (2016). ECG based heart arrhythmia detection using wavelet coherence and bat algorithm. *Sens Imaging, 17*(12), 1–16.
33. Gupta, V., & Mittal, M., (2018). Blood pressure and ECG signal interpretation using neural network. *International Journal of Applied Engineering Research, 13*(6), 127–132.
34. Singh, J., Sharma, M., & Acharya, U. R., (2019). Hypertension diagnosis index for discrimination of high-risk hypertension ECG signals using optimal orthogonal wavelet filter bank. *Int. J. Environ. Res. Public Health, 16*, 40–68.
35. Gupta, V., & Mittal, M., (2019). Investigation of normal and abnormal blood pressure signal using Hilbert transform, Z-transform, and modified Z-transform. *Inter. J Computational Medicine and Healthcare (IJCMH)*. (in press).
36. Halder, B., Mitra, S., & Mitra, M., (2019). Classification of complete myocardial infarction using rule-based rough set method and rough set explorer system. *IETE J. Res*. doi: 10.1080/03772063.2019.1588175, In press.
37. Jung, W. H., & Lee, S. G., (2017). An arrhythmia classification method in utilizing the weighted KNN and the fitness rule. *IRBM*. http://dx.doi.org/10.1016/j.irbm.2017.04.002.IRBM, *38*(3), 138–148.

38. Jog, N. K., (2013). *Electronics in Medicine and Biomedical Instrumentation* (2nd edn., pp. 85–109). PHI.
39. Gupta, V., & Mittal, M., (2018). Dimension reduction and classification in ECG signal interpretation using FA and PCA: A comparison. In: *Proceedings of the Jangjeon Mathematical Society*. South Korea. 1542050872-pjms21-4-18.pdf (jangjeonopen.or.kr).
40. Gupta, V., & Mittal, M., (2018). ECG signal analysis: Past, present and future. In: *Proceedings of the 8th IEEE Power India International Conference*. NIT Kurukshetra. https://doi:10.1109/POWERI.2018.8704365.
41. Gupta, V., & Mittal, M., (2019). QRS complex detection using STFT, chaos analysis, and PCA in standard and real-time ECG databases. *J. The Inst. Eng. (India): Series B, 100*, 489–497.
42. Gupta, V., & Mittal, M., (2018). KNN and PCA classifier with autoregressive modeling during different ECG signal interpretation. *Procedia Computer Science, 125*, 18–24.
43. Gupta, V., et al., *(*2019*)*. Auto-regressive time-frequency analysis (ARTFA) of electrocardiogram (ECG) signal. *International Journal of Applied Engineering Research, 13*(6), 133–138.
44. Gupta, V., et al., *(*2011*)*. Principal component and independent component calculation of ECG signal in different posture. *AIP Conference Proceedings, 1414,* 102–108.
45. Cleetus, H. M. M., Singh, D., & Deepak, K. K., (2019). Assessment of interaction between cardio-respiratory signals using directed coherence on healthy subjects during postural change. *IRBM, 40*(4), 167–173.
46. Gupta, V., & Mittal, M., (2016). Respiratory signal analysis using PCA, FFT and ARTFA. In: *Proceedings of the 2016 International Conference on Electrical Power and Energy Systems (ICEPES)*. Bhopal India.
47. Singhal, A., Singh, P., Fatimah, B., & Pachori, R. B., (2020). An efficient removal of power-line interference and baseline wander from ECG signals by employing Fourier decomposition technique. *Biomedical Signal Processing and Control, 57*, 101741. ISSN: 1746-8094. https://doi.org/10.1016/j.bspc.2019.101741.
48. Fatimah, B., Singh, P., Singhal, A., & Pachori, R. B., (2020). Detection of apnea events from ECG segments using Fourier decomposition method. *Biomedical Signal Processing and Control, 61*, 102005. https://doi.org/10.1016/j.bspc.2020.102005.
49. Singh, P., Joshi, S. D., Patney, R. K., & Saha, K., (2017). The Fourier decomposition method for nonlinear and non-stationary time series analysis. *Proc. R. Soc. A, 473*, 20160871. http://dx.doi.org/10.1098/rspa.2016.0871.
50. Singh, P., & Pachori, R. B., (2017). Classification of focal and nonfocal EEG signals using features derived from Fourier-based rhythms. *J. Mech. Med. Biol., 17*(7), 1–16. https://doi.org/10.1142/S0219519417400024.
51. Mehla, V. K., Singhal, A., & Singh, P., (2020). A novel approach for automated alcoholism detection using Fourier decomposition method. *Journal of Neuroscience Methods, 346*, 108945. https://doi.org/10.1016/j.jneumeth.2020.108945.
52. Singh, P., & Joshi, S. D., (2019). Some studies on multidimensional Fourier theory for Hilbert transform, analytic signal and AM–FM representation. *Circuits, Systems, and Signal Processing, 38*, 5623–5650. Https://Doi.Org/10.1007/S00034-019-01133-X.

53. Singh, P., (2020). Novel generalized Fourier representations and phase transforms. *Digital Signal Processing, 106*, 102830. ISSN: 1051-2004. https://doi.org/10.1016/j.dsp.2020.102830.
54. Rekik, S., & Ellouze, N., (2017). Enhanced and optimal algorithm for QRS detection. *IRBM, 38*(1), 56–61.
55. Padmavathi, K., & Ramakrishna, K. S., (2015). Detection of atrial fibrillation using autoregressive modeling. *International Journal of Electrical and Computer Engineering (IJECE), 5*(1), 64–70.
56. Kaya, Y., Pehlivan, H., & Tenekeci, M. E., (2017). Effective ECG beat classification using higher order statistic features and genetic feature selection. *J. Biomedical Research, 28*, 7594–7603.
57. Daamouche, A., Hamami, L., Alajlan, N., & Melgani, F., (2012). A wavelet optimization approach for ECG signal classification. *Biom. Sig. Proc. Contr., 7*(4), 342–349.
58. Alemi, F., Avramovic, S.,; Schwartz, M. D., (2018). Electronic health record-based screening for substance abuse. *Big Data, 6*(3), 214–224. doi: 10.1089/big.2018.0002.
59. Gupta, V., & Mittal, M., (2020). A novel method of cardiac arrhythmia detection in electrocardiogram signal. *International Journal of Medical Engineering and Informatics, 12*(5), 489–499.
60. Das, M., & Ari, S., (2013). Analysis of ECG signal denoising method based on S-transform. *IRBM, 34*(6), 362–370.
61. Gupta, V., & Mittal, M., (2021). R-peak detection for improved analysis in health informatics. *International Journal of Medical Engineering and Informatics, 13*(3), 213–223.
62. Gupta, V., & Mittal, M., (2019). A comparison of ECG signal pre-processing using FrFT, FrWT and IPCA for improved analysis. *IRBM, 40*(3), 145–156.
63. Gupta, V., Mittal, M., & Mittal, V., (in press). Performance evaluation of various pre-processing techniques for R-peak detection in ECG signal. *IETE Journal of Research*. https://doi.org/10.1080/03772063.2020.1756473.
64. Arnold, M., Miltner, W. H. R., Witte, H., Bauer, R., & Braun, C., (1998). Adaptive AR modeling of nonstationary time series by means of Kalman filtering. *IEEE Transactions on Biomedical Engineering, 45*(5), 553–562.
65. Mehta, S. S., & Lingaya, N. S., (2008). SVM-based algorithm for recognition of QRS complexes in electrocardiogram. *IRBM, 29*, 310–317.
66. Mehta, S. S., Shete, D. A., Lingayat, N. S., & Chouhan, V. S., (2010). K-means algorithm for the detection and delineation of QRS-complexes in electrocardiogram. *IRBM, 31*, 48–54.
67. Mehta, S. S., & Lingayat, N. S., (2008). SVM based QRS detection in electrocardiogram using signal entropy. *IETE Journal of Research, 54*(3), 231–240.

CHAPTER 2

DEVELOPMENT OF PLC-BASED CONTROLLER FOR DOOR SLAM PLATFORM

MANEETKUMAR R. DHANVIJAY,[1] ROHIT PATKI,[2] and B. B. AHUJA[3]

[1]*Associate Professor, Department of Manufacturing Engineering and Industrial Management, College of Engineering, Pune [COEP], Wellesley Road, Shivajinagar, Pune–411005, Maharashtra, India*

[2]*MTech [Mechatronics] Scholar, Department of Manufacturing Engineering and Industrial Management, College of Engineering, Pune [COEP], Wellesley Road, Shivajinagar, Pune–411005, Maharashtra, India*

[3]*Director, Department of Manufacturing Engineering and Industrial Management, College of Engineering, Pune [COEP], Wellesley Road, Shivajinagar, Pune–411005, Maharashtra, India*

ABSTRACT

This chapter illustrates the design and development aspects of a pneumatically operated door slam platform using a PLC-based controller for analyzing and checking the structural and functional integrity of vehicle door components. An overview of different components whose functionality can be checked using this platform is explored. This work also explains the pneumatic circuit for developing door slam platform.

Optimization Methods for Engineering Problems. Dilbagh Panchal, Prasenjit Chatterjee, Mohit Tyagi, Ravi Pratap Singh (Eds.)

The proposed PLC automation system involves the sequential control of a number of processes carried out for pneumatic door slam platform, measurement of door opening and closing velocity, and to carry out the required number of continuous cycles. The control system based on Micrologix 1,400 Series-A from Allen-Bradley and Ladder logic program is coded in RS-Logix 500 to analyze the proposed control system. Imitation of the PLC program was undertaken using RS-Logix emulate-500.

2.1 INTRODUCTION

The door system of a vehicle is one of the few systems that is used frequently but has a high construction complexity. Door shutting or slamming produces a sound which is a critical parameter for vehicle purchase. As a result, slam tests are carried out on vehicle doors so as to ascertain and confirm their structural and functional integrity in different environmental conditions when the vehicle is in the stage of prototyping. Cost and lead time are of prime importance in today's competitive environment; hence, any design alterations during this phase are bound to increase these two parameters. Moreover, development cycle time can be disrupted by the use of up-front analytical prediction tools and customer-focused quality issues such as warranty costs as well as the online inspection of door fitting and finishing.

Yang et al. [1] have done extensive studies on the vehicle door slam sound considering the traditional and subjective sound quality metrics. They have adopted critical band wavelet decomposition, which is based on sound metric, and have developed the critical band wavelet decomposition (SMCBWD) method for analyzing the quality of door slamming sound. Su et al. [2] have used a CAE virtual environment, called key life test to evaluate the durability of plastic door trim components. Such an environment helped to reduce the product development time and cost at an early stage even before prototyping and testing.

The process of slam of a door assembly may cause the outer panel to experience buckling resulting in flutter or oil-canning. It could also induce localized strains in the inner panel, which may cause fatigue damage [3]. Vehicle door is subjected to repetitive loading and transient dynamic impact for door slam analysis. Appreciable shocks/vibrations are caused as a result of the impact loading thereby inducing acceleration as well as inertial accelerations because of rigid body motion. The door slam analysis is a transient dynamic impact as well as a repeated type of loading in nature. The impact loading produces appreciable shock or vibration and induces accelerations in addition to inertial accelerations due to rigid body motion [4]. Plourde et al. [5] have studied the effect of light door design utilizing multiple materials and observed that light weight front and rear door does not affect the structural performance.

Pneumatic power is cheap and easy to control and hence utilized in a number of processes. Sajaysurya and Saravana [6] have synthesized logic equations for the control of pneumatic cylinders. They have devised a method to convert a fixed sequence operation to a flexible one by using a Genetic Algorithm and genetic programming. Singh and Verma [7] have designed a MicroLogix 1,000 PLC-based controller for bearing press work which reduces the delay occurring at every stage while reducing the maintenance time and improving operator safety. Unadkat et al. [8] have adopted a dynamic analysis (LSDyna) to predict the behaviors of automobile door slam test, which is a highly complex in terms of simulation results. Door slam velocities were varied in the range of 1.1 m/s to 1.6 m/s with varied glass window openings to study its effect on the door strain. Mahadule and Chavan [9] have built a mathematical model to predict the door closing velocity (DCV), which is an important parameter in the overall durability of the vehicle. All major components of the door and their properties, such as seal stiffness, door latch, seal air cavity, etc., are included in the MS-Excel application, which gives the energy requirements for a typical front door of a sedan vehicle. These studies will be helpful to co-relate with the CAE simulation results. An FEM evaluation of door slam is conducted by Patil and Dhuri [10] considering door acceleration, stress, strain, and buckling energy. It is observed that the outer door panel experiences complex buckling pattern during slam operation. Also, maximum acceleration and maximum principal strain is observed at location No. 2. Song and Tan [11] have done a comparative evaluation of linear and non-linear stress methodology using LSDyna. They conclude that the nonlinear methodology is moiré accurate in terms of stress and fatigue life for low cycle fatigue problems. Wahab and Adarsha [12] designed and developed a door slam test rig to eliminate human effort and fatigue for carrying out the slam cycles repeatedly. Their results show that damage was 0.1 considering the ratio of achieved to the designed cycles.

The objective of this work is to develop a PLC-based pneumatic door slam platform which is capable of carrying out a required number of slam cycles while measuring door opening and closing velocity for every cycle. This platform can carry out door slam test on front and rear door in one cycle. HMI interfacing with PLC makes the system user-friendly and easy to operate.

2.2 DEVELOPMENT OF DOOR SLAM FIXTURE

The work begins with the design and development of a fixture for the door slam test. Standard door slam cycle consists of opening and closing of one front door and one rear door which is in diagonally opposite direction to front door. Door slam cycle is as follows:

- Opening the front door latch by pulling the door handle;
- Opening the front door fully with required velocity specified by manufacturer's standard;
- Opening the latch on the rear door;
- Opening the rear door fully with required velocity specified by manufacturer's standard;
- Closing front door with its specified velocity;
- Closing the rear door with its specified velocity.

So, a mechanical structure was developed which is capable of all these motions and strong enough to carry out designated cycles without structural failure. Fixture is also capable of Opening and closing the vehicle door completely, also it can be customized for all types of passenger cars such as hatchback, sedan, SUV, etc. The working span of fixture ranges from 1.5 m–3 m for opening and closing of doors, which can be customized for specific vehicle according to requirement. The fixture can be divided in two different parts, such as:

1. **Door Latch Opening Fixture:** Basically, it is the same fixture on which door latch opening and full door opening actuators are mounted, but to understand the fixture easily, it has been divided into two different parts.

   As shown in Figure 2.1, this setup unlocks the door from its latch provided that central lock system of the vehicle is not initiated. Motion and force required for opening of the latch of the door is provided by a pneumatic cylinder of stroke 50 mm and diameter 20 mm. This cylinder is mounted on to fixture and it is connected to vehicle door handle through belt.

2. **Full Door Opening and Closing Fixture:** The fixture for door performance test (Figure 2.2) is developed so that it should be able to carry outdoor performance test on any passenger vehicle. The fixture comprises of a heavy mounting which is further connected to swiveling arm which rotates around the hinge. A screw is mounted on heavy mount to limit the rotation angle to maximum opening angle of vehicle door to avoid extra stress on doors during the test.

FIGURE 2.1 Door latch opening mechanism.

FIGURE 2.2 Full door opening and closing mechanism.

2.3 PNEUMATIC CIRCUIT FOR DOOR SLAM PLATFORM

All the actuations required to carry out door slam test are done by using a pneumatic circuit. This pneumatic circuit consists of four different actuators which carryout four different actuations.

2.3.1 *PNEUMATIC COMPONENTS*

All the pneumatic components required for building operational door slam circuit are as follows:

- Double acting pneumatic cylinder of stroke 1,000 mm: 2 (Qty);
- Double acting pneumatic cylinder of stroke 50 mm: 2;
- Double solenoid operated direction control valve: 4;
- Flow control valve: 8;
- Filter, regulator, and lubricator unit (FRL): 1;
- Compressor: 1;
- Pneumatic pipe: 20 m.

2.3.2 *PNEUMATIC CIRCUIT*

Pneumatic connections to carry out slam cycle are shown in Figure 2.3.

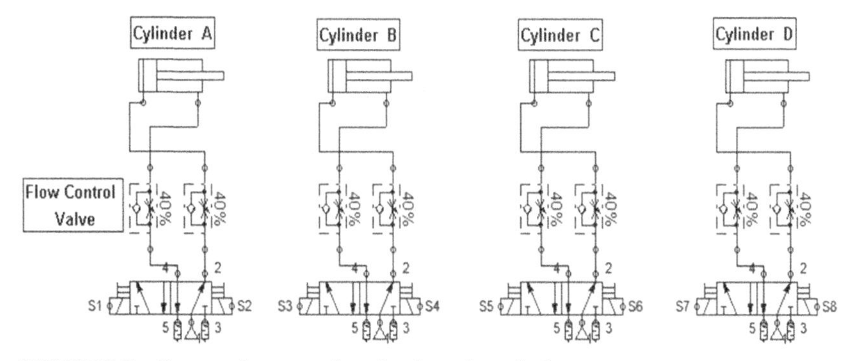

FIGURE 2.3 Pneumatic connections for door slam platform.

Designations for actuators are as follows:

➢ **Cylinder A:** Actuator for opening front door latch.

- **Cylinder B:** Actuator for full opening and closing movement of front door.
- **Cylinder C:** Actuator for opening rear door latch.
- **Cylinder D:** Actuator for full opening and closing movement of rear door.

This circuit gives a brief idea about all the pneumatic connections required to complete door slam platform test. As it can be observed from Figure 2.1, velocity of extension and retraction of each circuit can be controlled manually with a flow control valve. Cylinder A and B are used for opening the door lock, so these cylinders have 50 mm extension, which is enough to open the door lock. Cylinder B and D are responsible for complete opening and closing of door hence these cylinders have stroke of 1,000 mm. All the operation of direction control valves can be controlled using PLC signals.

2.4 AUTOMATION OF DOOR SLAM PLATFORM

2.4.1 NEED OF AUTOMATION

The following points explain in brief the need of automation for slam test:

- Door is a frequently used component of vehicle so to check its durability, a slam test is conducted on the door and is to be performed for around 75,000–80,000 cycles which is not possible to be carried out manually.
- For every cycle, door opening and door closing velocity needed to be measured to ensure the required force is applied on the door while opening and closing which can be easily done using PLC.
- A consistent force needed to be applied on doors to open and close them with constant velocity through all the cycles.
- Keeping the record of test is important to analyze the final results and to know the test is carried out correctly.
- Automation reduces workforce requirement needed to carry out whole slam test.

Door slam platform is automated using Allen-Bradley Micrologix 1,400 series A PLC. It carries out door slam cycle required number of times while giving feedback of door velocity for every cycle so slamming condition can be controlled.

2.4.2 SEQUENCE OF OPERATION

Sequence of operation of all pneumatic cylinders to carry out door slam cycle is explained in the following points:

- As start pushbutton is pressed PLC sends signal to DCV (Direction Control Valve) of cylinder a when operated results in retraction of cylinder A which opens front door latch. After a delay of 0.5 seconds Cylinder A is extended back to its original position.
- After ensuring successful opening of door latch, signal sent by PLC to DCV of Cylinder B which triggers extension of Cylinder B which in turn opens front door fully with predetermined velocity.
- Once front door is fully open then Cylinder C is retracted which results in opening latch of rear door. After latch opening with a delay of 0.5 seconds, Cylinder C is extended back to its default position.
- As soon as the rear door latch is opened, Cylinder D is extended, which results in full opening of the rear door with the required velocity.
- As rear door fully opens, the front door starts closing with predetermined velocity. After closing of front door Cylinder D starts retracting to its original position which results in closing of rear door.

2.5 PLC-BASED CONTROL SYSTEM FOR DOOR SLAM PLATFORM

All the controls of door slam tests are handled by AB Micrologix 1,400 PLC. Different operations done by PLC are explained in subsections.

2.5.1 DOOR VELOCITY MEASUREMENT

Door durability performance is analyzed by Door closing velocity (DCV) which is one of the important design parameters. The physical properties as well as the door design parameters dictate the closing velocity of the door. When the overall performance of vehicle durability is a concern, in such a case the door closing effort variation may lead to an increase/decrease in the door durability and the durability of the door components.

Inductive type proximity sensors with sensing range of 10 mm are utilized as input sensors for detecting the location of every component of the system during the cycle. These same proximity sensors are also used for measuring the velocity of door opening and closing.

As it can be observed from Figure 2.4, two proximity sensors are mounted at a fixed distance and as door passes through them, time is recorded for door to travel from first sensor to next and since distance between the sensors is constant velocity of door can be calculated with recorded time. All the calculations required to find velocity are done by PLC itself and stored in a memory location which can be read with HMI.

FIGURE 2.4 Velocity measurements with proximity sensors.

2.5.2 DIGITAL INPUTS AND OUTPUTS

Inductive type proximity sensors feed required information to PLC to carry out slam cycle and required output signals are given by PLC to different solenoids of DCV of actuators to get desired motion. All the inputs to PLC and outputs from PLC are shown in Figure 2.5.

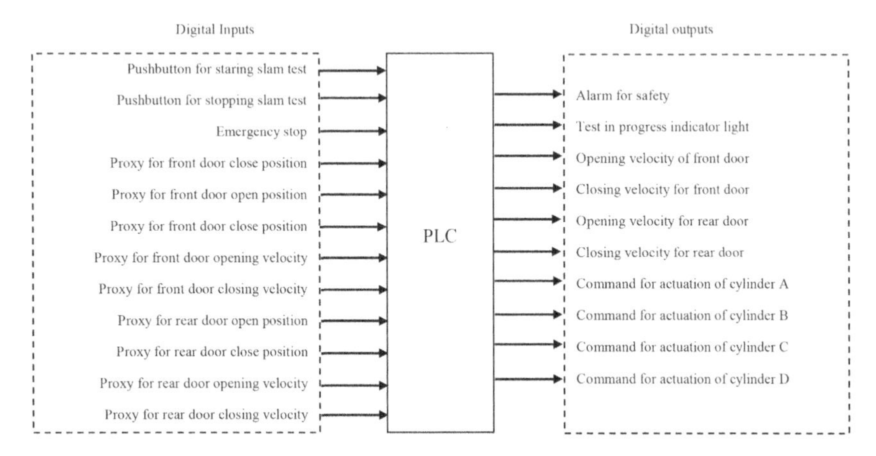

FIGURE 2.5 Input/output flowchart of PLC.

2.5.3 PLC PROGRAM FLOWCHART

Designations used in flowchart (Figure 2.6):

- **Proxy 1:** Front door closed position.
- **Proxy 2:** Front door fully open position.
- **Proxies 3 and 4:** Proximity sensor to measure front door opening velocity.
- **Proxies 5 and 6:** Proximity sensor to measure front door closing velocity.
- **Proxy 7:** Rear door closed position.
- **Proxy 8:** Rear door fully open position.
- **Proxies 9 and 10:** Proximity sensor to measure rear door opening velocity.
- **Proxies 11 and 12:** Proximity sensor to measure rear door closing velocity.

2.5.4 INTERFACING HMI WITH PLC

HMI (human-machine interface) is a tool that enables semi-skilled work-force to interact with PLC. The HMI talks to the PLC and reads its data to populate the highly graphical screens. The screen (Figure 2.7) provides animation and colors that imitates the process and showed the real-time data

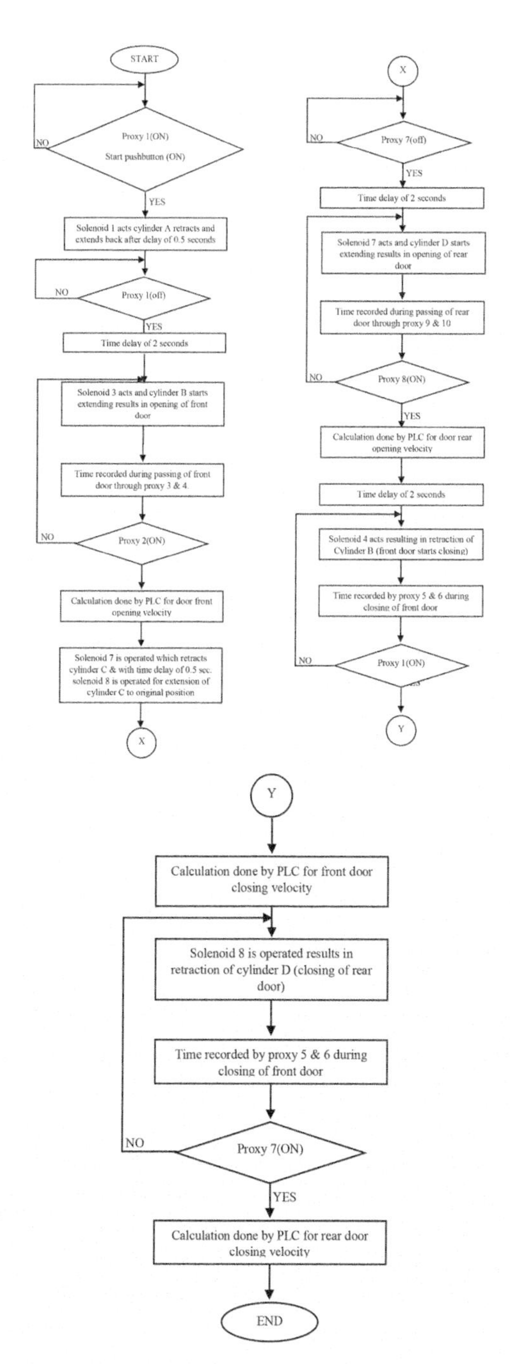

FIGURE 2.6 Flowchart of PLC.

on-screen to provide better feedback to operators. HMI used communicate with AB-PLC is Delta DOP B-05 HMI. Screen for door slam platform is shown in the following figure.

This screen is utilized for monitoring door slam platform. This screen is also used to monitor different parameters of door slam platform which are as follows:

- Front door opening velocity;
- Front door closing velocity;
- Rear door opening velocity;
- Rear door closing velocity;
- Selecting number of cycles to be completed by platform before stopping;
- Monitor cycles completed by platform;
- Starting and stopping the tests as per requirement;
- Returning back to the main menu.

Measurement of door velocities is done by PLC itself and those stored in its memory. HMI just reads respective memory bits to show respective door velocities. Remaining functions of HMI for door slam platform are the same as other platforms.

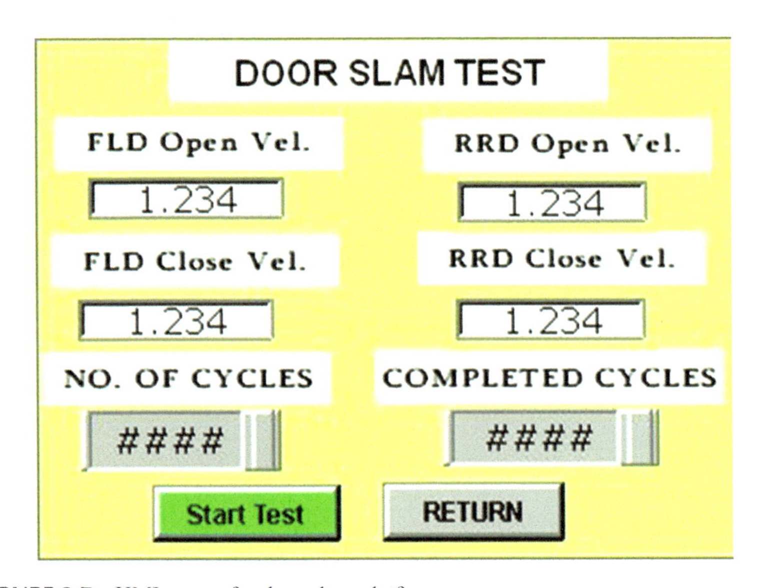

FIGURE 2.7 HMI screen for door slam platform.

2.6 CONCLUSIONS

i. PLC-based door slam platform is successfully developed and tested to check durability of any Light duty passenger vehicle.
ii. Slam test on one front and one rear door can be carried out simultaneously with this developed platform.
iii. Designed pneumatic circuit is able to change velocities of extension and retraction of all cylinders using flow control valves.
iv. HMI screen enables operator to monitor slam cycle continuously for door opening and closing velocities.

KEYWORDS

- **automation**
- **door closing velocity**
- **door slam platform**
- **human-machine interface**
- **pneumatic circuit**
- **programmable logic controller**

REFERENCES

1. Yang, C., Yu, D., & Xia, B., (2014). Research on the sound metric of door-slamming sound based on leaky integration and wavelet decomposition. *International Journal of Automotive Technology, 15*(5), 853–860.
2. Hong, S., Chuck, D., & Alex, K., (2003). *CAE Virtual Door Slam Test for Plastic Trim Components, 2003 SAE World Congress Detroit*. Michigan.
3. Mohan, I. R., & Chang, T., (2018). "*A Comprehensive Study of Door Slam*" *at Kings Meadow Campus*. SAE Technical Paper series 2004-01-0161.
4. Baskar, S., GM-Midsize and Luxury Car Group, (2018). *Door Structural Slam Durability Inertia Relief Approach*. Birmingham City Univ, SAE Technical paper series 982309.
5. Plourde, L., Azzouz, M., Wallace, J., & Chellman, M., (2015). *MMLV: Door Design and Component Testing*. SAE Technical paper 2015-01-0409. doi: 10.4271/2015-01-0409.

6. Sajaysurya, G., & Saravana, K. G., (2017). Evolutionary algorithms for programming pneumatic sequential circuit controllers. In: *27th International Conference on Flexible Automation and Intelligent Manufacturing, FAIM 2017*. Modena, Italy.
7. Unadkat, S., Kangde, S., Burkul, M., & Badireddy, M., (2016). *Closure Slam CAE Method Investigation for Automobiles.* SAE Technical paper 2016-01-1349. doi: 10.4271/2016-01-1349.
8. Ranjeetha, S., & Verma, H. K., (2018). Development of PLC based controller for pneumatic pressing machine in engine bearing manufacturing plant. In: *6th International Conference on Smart Computing and Communication, ICSCC 2019 Procedia Computer Science* (Vol. 125, pp. 449–458).
9. Mahadule, R., & Chavan, J., (2016). *Evaluation of Minimum Door Closing Velocity Using Analytical Approach.* SAE technical paper 2016-01-0434. doi: 10.4271/2016-01-0434.
10. Sunil, S. P., & Gautami, U. D., (2020). Assessment of automobile door slam using finite element method. *International Journal of Innovative Technology and Exploring Engineering (IJITEE)* (Vol. 9, No. 4). ISSN: 2278-3075.
11. Song, G., & Tan, C., (2015). *Door Slam CAE Method Investigation.* SAE technical paper 2015-01-1324. doi: 10.4271/2015-01-1324.
12. Abdul, W., & Adarsha, H., (2014). Design and development of automated door slam test rig for cars. *International Journal of Engineering Research and Informatics, 1*(7), 1–42.

CHAPTER 3

ERGONOMIC RISKS ASSESSMENT OF FARMERS IN ADVERSE CLIMATIC CONDITIONS

HULLASH CHAUHAN, SUCHISMITA SATAPATHY, and A. K. SAHOO

KIIT Deemed to be University, Bhubaneswar, Odisha, India, E-mail: suchismitasatapathy9@gmail.com (S. Satapathy)

ABSTRACT

Individuals have constrained abilities, including muscle ability to work with varying loads. Because of the rising temperatures just as expanding as atmosphere variability, including dynamic visits and genuine remarkable climatic changes, the overall impacts of ecological change on provincial and nourishment frameworks are altogether looking at various risks. In this way, an exertion is made in this study to investigate the ergonomic hazard levels for farmers working in hot-climatic conditions in India by utilizing the RULA tool.

3.1 INTRODUCTION

Creating rules for safe human presentation to tropical hot situations is basic. The universal agencies have created rules and criteria documents in diminishing dangers of warmth incited diseases and enhancing breaking point of exposures in hot conditions. The viable warmth load on people has

Optimization Methods for Engineering Problems. Dilbagh Panchal, Prasenjit Chatterjee, Mohit Tyagi, Ravi Pratap Singh (Eds.)

the two aspects of outside or natural, and interior or metabolic conditions; nonetheless, the data on the impacts of warmth on well-being, execution, and security for the laborers occupied with tropical cultivating practices. Human have constrained capacities, including muscle capacity to convey or work with loads. Prompting rising temperatures and expanding atmosphere changeability, including progressively visit and serious extraordinary climate occasions, the worldwide effects of environmental change on rural and food systems are significantly facing of numerous dangers. Therefore, an effort is made in this study to explore the ergonomic risk levels for farmers working in hot-climatic conditions in India.

According to Oka [1] a psychogenic fever refers to stress-related and psychosomatic-disease particularly found in young women, developing extreme body temperature up to 41°C when exposed to emotional events. Whereas, a persistent low-grade temperature up to 37–38°C is found in others during situations of chronic-stresses. Heidari et al. [2] have carried out a cross-sectional study of 79 farmers working in diverse areas including agriculture, husbandries, and horticulture in north of Iran. The evaluation of the heat stress was conducted by the use of the "Wet Bulb Glob Temperature" index according to ISO-7243 and heat-strain was evaluated by the use of the "individual physiological responses" including "oral, aural, and mean-skin temperatures" according ISO-9886. Al-Saleh [3] has made an attempt to update the RULA and Quick Exposure Check (QEC) software to facilitate and to speed up the process in getting the scores for every task. Bhandare et al. [4] have used RULA and REBA method to report fatigue analysis that included direct observations and assessments through video recordings without disrupting work in confined workspaces. Ansari and Sheikh [5] have made an evaluation of postures using RULA and REBA worksheets for the workers engaged in diverse activities in a small-scale industry at MIDC Wardha of Maharashtra, India. It was found using RULA that the majorities of the workers were under high levels of risk and also required immediate changes. Using REBA method, it was seen that some workers were under lower-level as well as majority at higher risk levels. Thus, it was concluded that a lack of ergonomic awareness and understanding existed in small-scale industries leading to musculoskeletal disorders. MdYusop et al. [6] have used RULA analysis to analyze the postures of learners during the welding process by using "CATIA V5R19" software. The proposed design provided a better result reducing the RULA score to 2 from 6 when the color turned to green from orange (an acceptable position).

Vijay and Kumar [7] have made an effort to study farmers' postures in growing maizes in the traditional methods of farming and found the newly proposed sowing equipment to be improving the working efficiency by reducing work-related injuries. Susihono et al. [8] have reported of the musculoskeletal complaints of a filling section before and after the work activity as 39.3 ± 10.6 and 76.5 ± 14.9, respectively. Weichelt et al. [9] have developed a web-based platform for farmers and physicians for timely and safe return to work. Budhathoki and Zander [10] have tried to estimate the health impacts, cold/heat stress levels and losses in labor productivity. They suggested that community engagement or community-based education/ programs could be developed in order to assist in adaptations. This chapter concerns with study of ergonomics and fatigue during the manual process of working in hot climates in India. The manual process was recorded with a high-definition video camera. This video is analyzed and discussions are made with workers regarding health issues during the process. The Rapid Upper Limb Assessment (RULA) tool was then used to evaluate postural loading on the whole body.

3.2 METHODOLOGY

The farmers working in their fields during hot climates were recorded, observed, analyzed, and documented. One of the established tools viz Rapid Upper Limb Assessment (RULA) was used for ergonomic assessment of these tasks. Task times and repetitions of task were recorded from videos. Neck, leg, trunk, arm-angle were examined. Scores were calculated from RULA survey using "Ergo Fellow 3.0 Software." Depending upon final scores, the recommendations were made for the respective tasks and postures.

3.3 RESULTS AND DISCUSSION

It was observed from Figure 3.1, that the female farmers' upper arms was 20°, her wrist was twisted 15° and wrist is band away from mid-line her neck was bent more than 20° and side bend her legs and feet well supported and in an evenly balanced poster, her lower arm working across the middle line of the body or out to the side, her wrist twisted away from handshake position her trunk was bent more than 60° twist and side bend also, 5th upper arm, lower arm and wrist muscles where found to be held for longer than 1 minute and the load carried was less than 2 kg similarly her neck trunk and legs muscles where found to be more than 4 times per minutes with a load of less than 2 kg on the basis of these input, a RULA score of 6 was found. Hence it was concluded that "further investigation and changes are required" (Figures 3.1–3.11).

FIGURE 3.1 Farmer removing weeds.

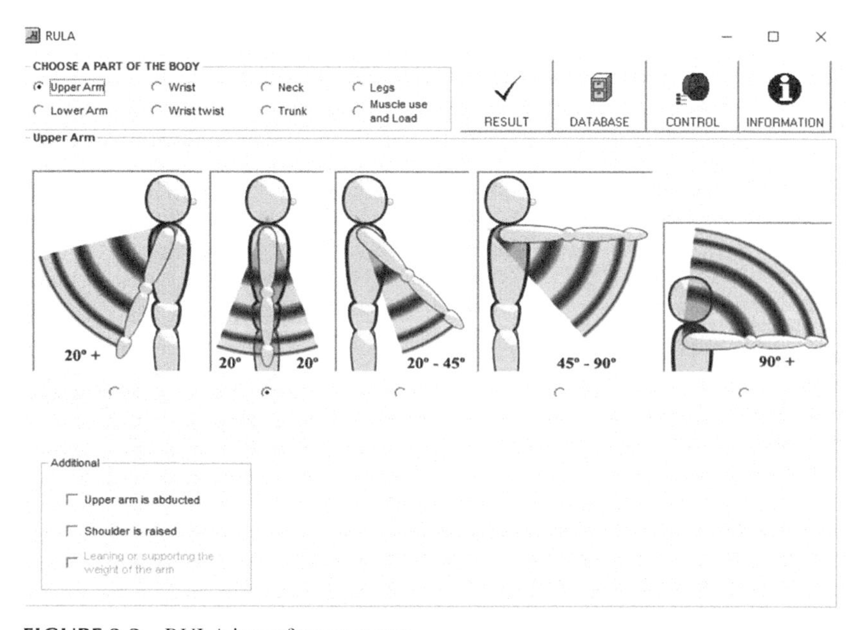

FIGURE 3.2 RULA input for upper arm.

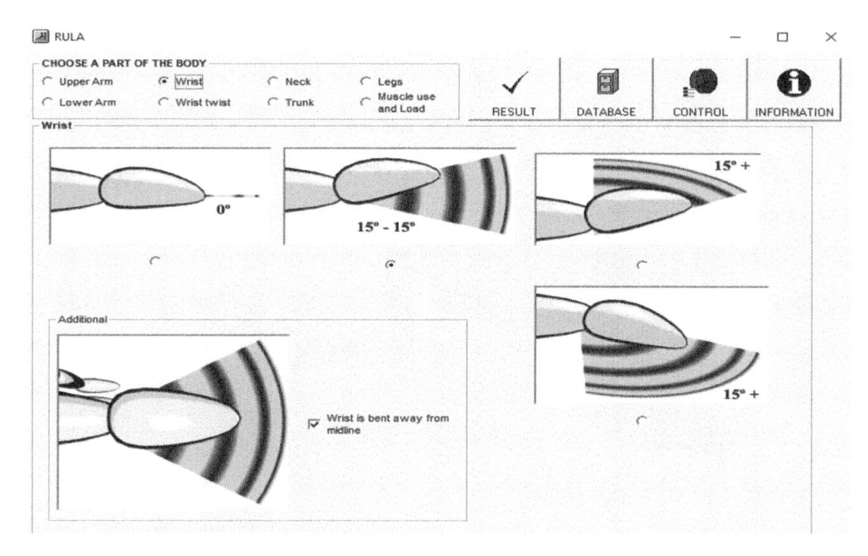

FIGURE 3.3 RULA input for wrist.

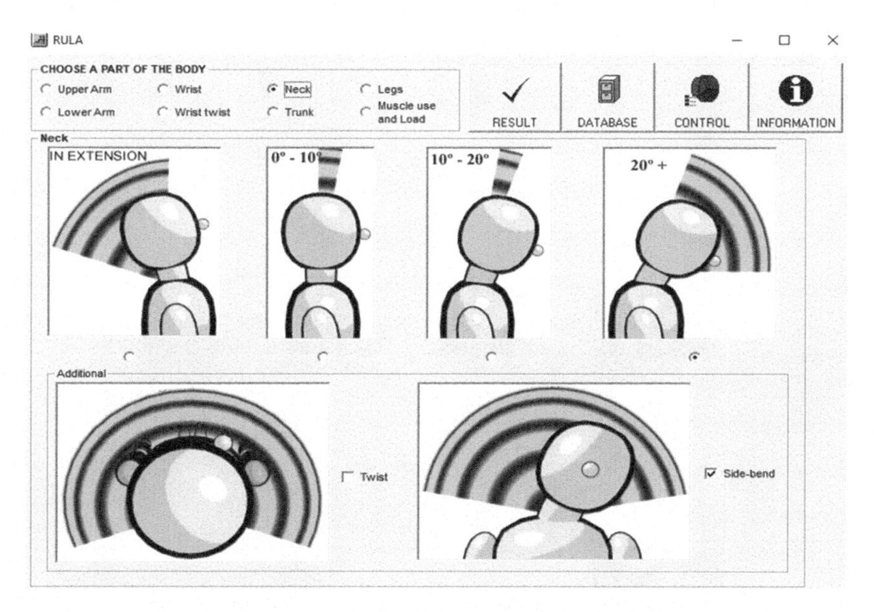

FIGURE 3.4 RULA input for neck.

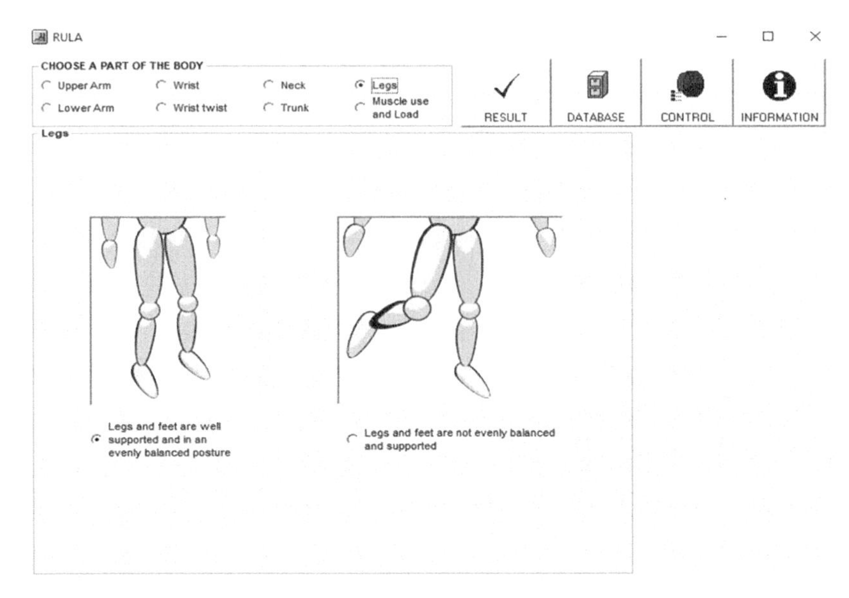

FIGURE 3.5 RULA input for legs.

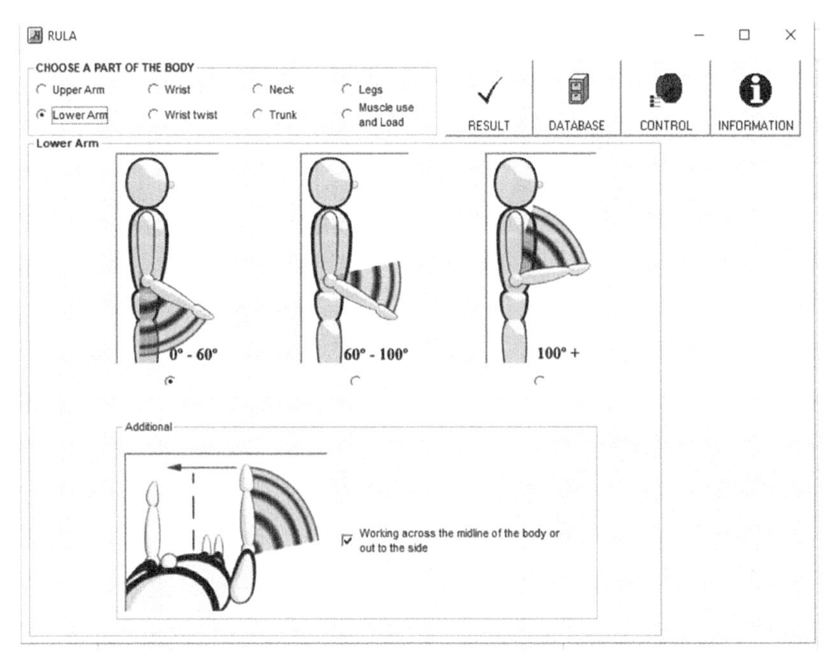

FIGURE 3.6 RULA input for lower arm.

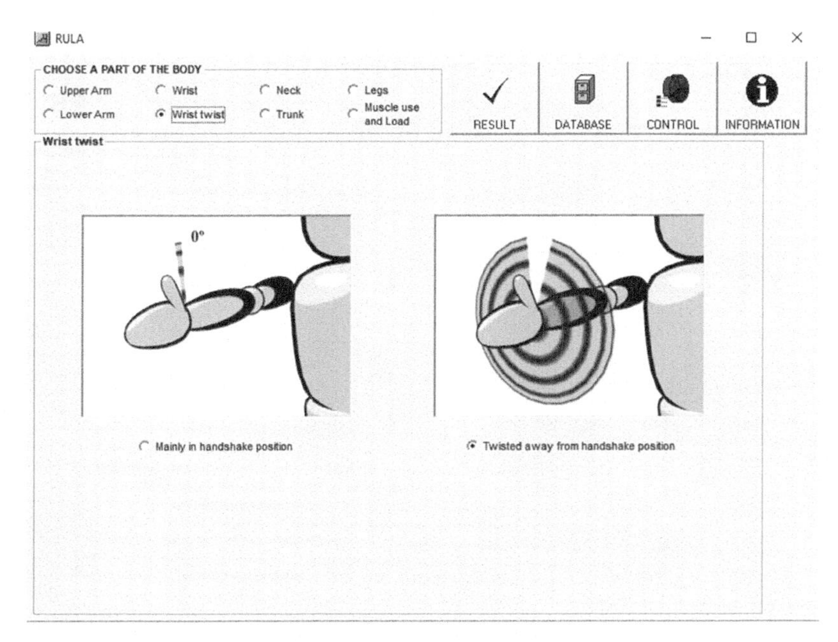

FIGURE 3.7 RULA input for wrist.

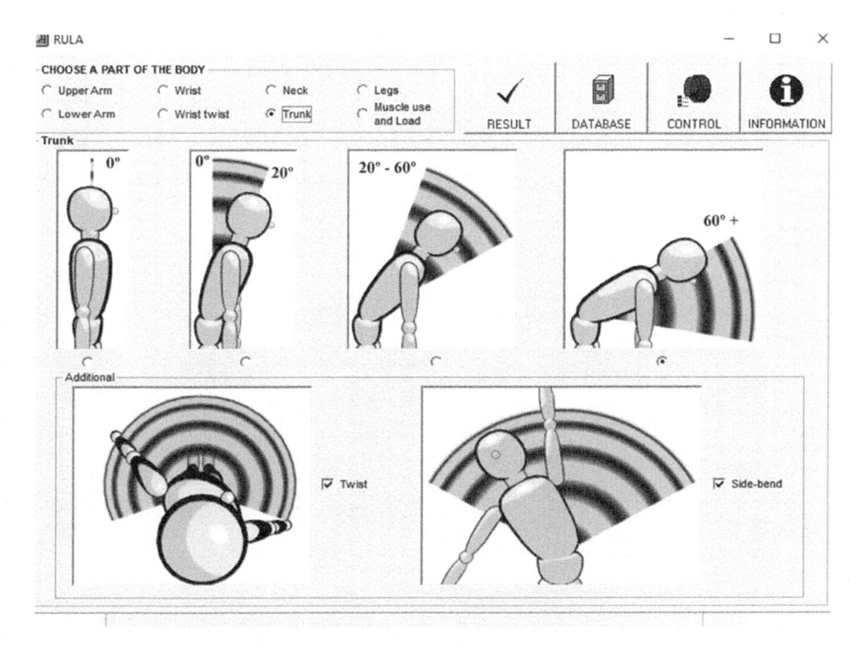

FIGURE 3.8 RULA input for trunk.

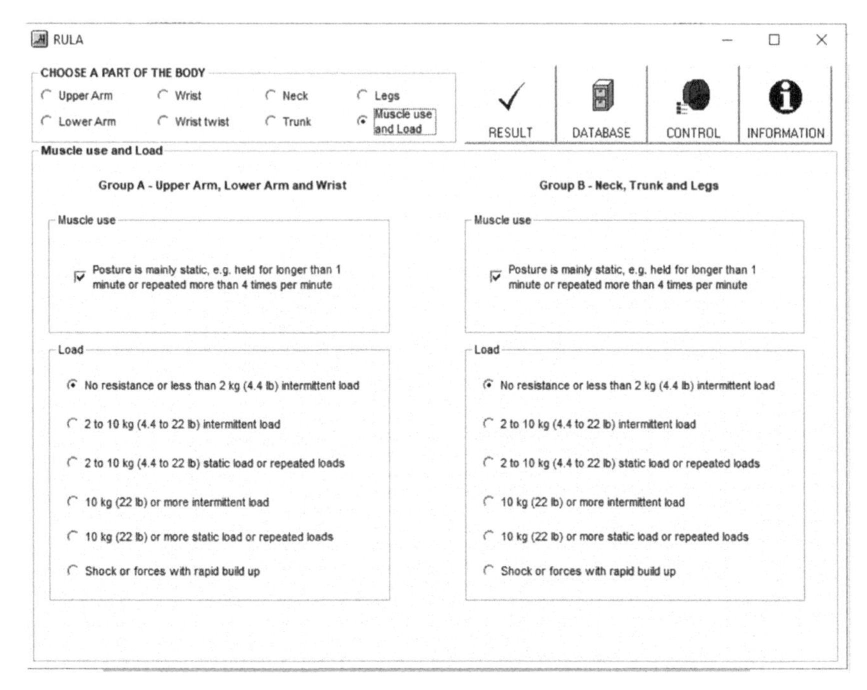

FIGURE 3.9 RULA input for muscles use and load.

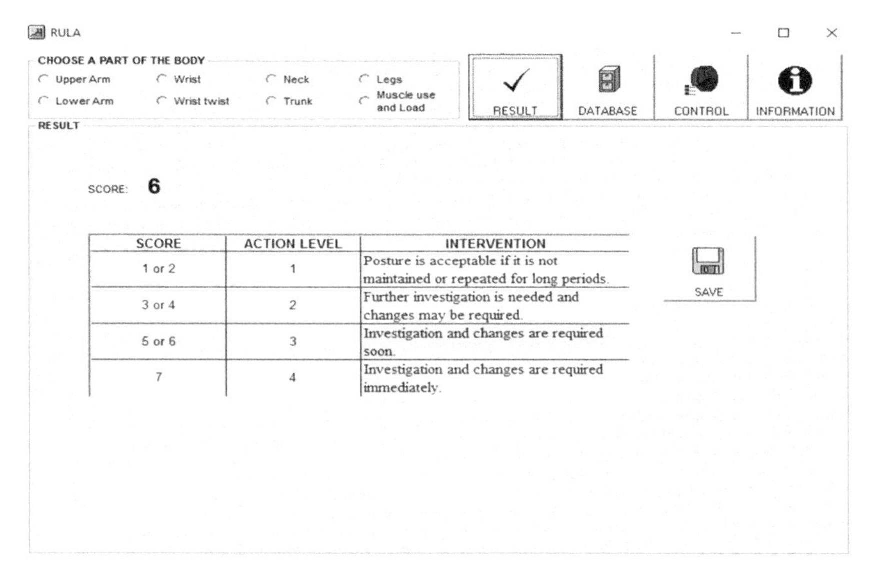

FIGURE 3.10 RULA final result.

FIGURE 3.11 Farmer carrying seeds.

As shown in Figure 3.11, a women worker was found of carrying seeds to the farm site during hot climate, it was seen that RULA score of totals of 7 was indicated representing "investigations and changes are instantly required (Figures 3.12 and 3.13).

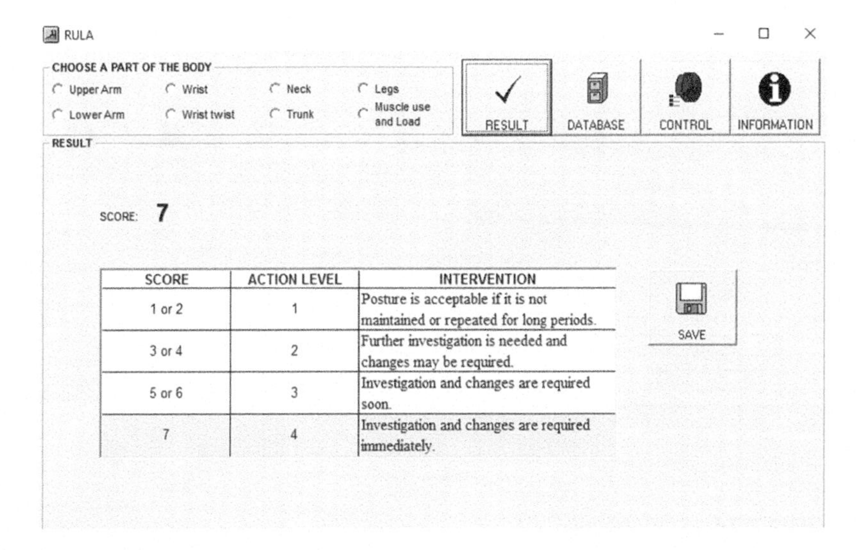

SCORE	ACTION LEVEL	INTERVENTION
1 or 2	1	Posture is acceptable if it is not maintained or repeated for long periods.
3 or 4	2	Further investigation is needed and changes may be required.
5 or 6	3	Investigation and changes are required soon.
7	4	Investigation and changes are required immediately.

FIGURE 3.12 RULA result.

FIGURE 3.13 Male worker removing weeds.

As shown in Figure 3.13, a male worker was found of removing weeds at during hot climate it was found that RULA total of 7 is indicates "investigations and changes are instantly required (Figures 3.14 and 3.15).

RULA

CHOOSE A PART OF THE BODY

Upper Arm | Wrist | Neck | Legs
Lower Arm | Wrist twist | Trunk | Muscle use and Load

RESULT | DATABASE | CONTROL | INFORMATION

RESULT

SCORE: **7**

SCORE	ACTION LEVEL	INTERVENTION
1 or 2	1	Posture is acceptable if it is not maintained or repeated for long periods.
3 or 4	2	Further investigation is needed and changes may be required.
5 or 6	3	Investigation and changes are required soon.
7	4	Investigation and changes are required immediately.

SAVE

FIGURE 3.14 RULA result.

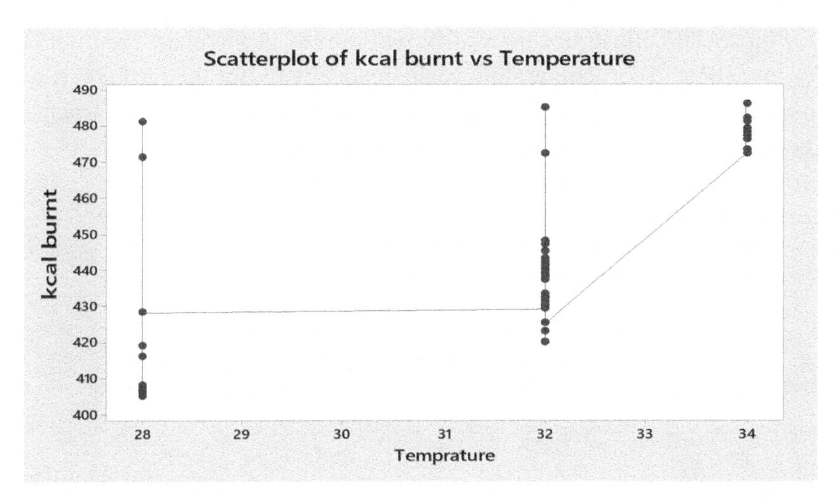

FIGURE 3.15 Kcal burnt vs. temperature.

Figure 3.15 shows the scatter plot for the Kcal burnt with respect to the temperature scale as measured in the farming field for the selected 53 farmers. It can be seen that there exists a high correlation between temperatures as well as Kcal burnt such that as the temperature increases, the Kcal burnt of farmer's also increases.

Similarly, Figure 3.16 shows the scatter plot for the pulse rate with respect to the temperature scale for 53 farmers. It can be seen that there also exists a high correlation between temperature as well as pulse rate such that as the temperature increases, the pulse rate of farmer's also increases.

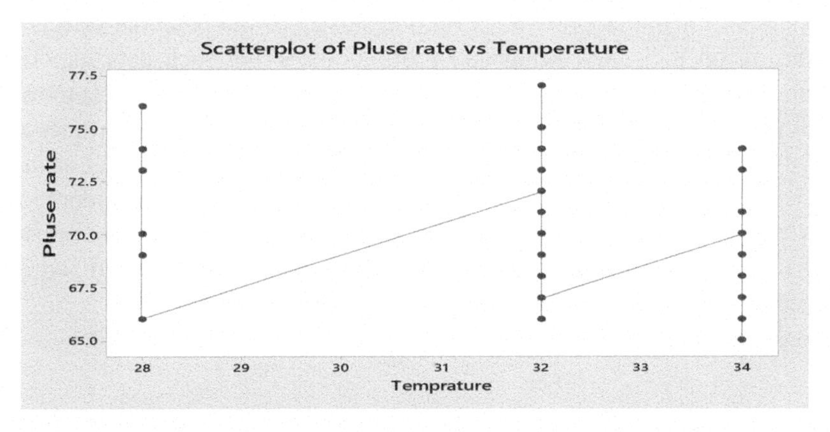

FIGURE 3.16 Pulse rate vs. temperature.

Figure 3.17 illustrates three different curves in the chart for high BP, For low BP, and for temperature scale, respectively for the farmers. It was observed that as the temperature increases, the BP level also rises irrespective of high or low.

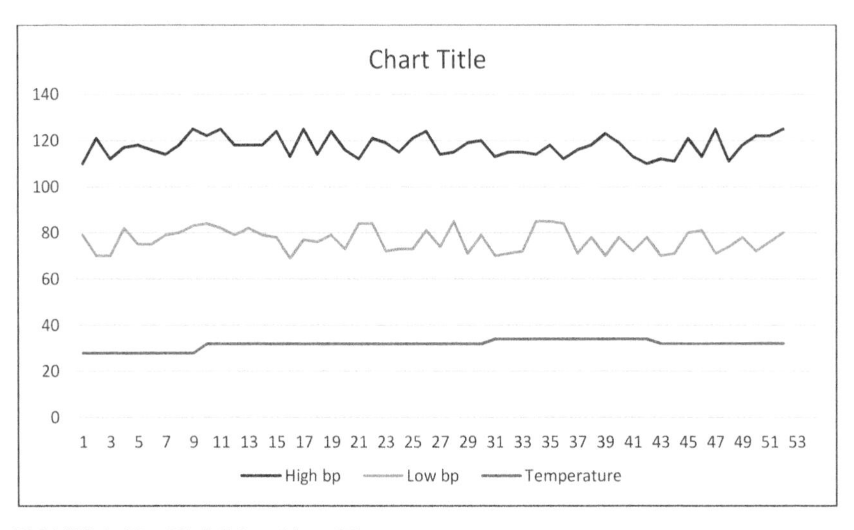

FIGURE 3.17 High BP and low BP vs. temperature.

3.4 CONCLUSION

To limit over-the-top warmth conditions in the work environment, it is desirable for the laborers for a routine survey of the potential effects of warmth on their well-being and profitability. From such data, laborers can receive the best warmth counteractive action technique and empower savvy and safe work practices. Warmth-related work limit misfortunes are a significant avocation for progressively dynamic environmental change alleviation strategies and projects all around the globe. Due consideration, investigation, and orders should be taken because of this environmental change and well-being challenges. Any program endeavoring to address medical problems related with atmospheric conditions needs to be considered.

KEYWORDS

- **agriculture**
- **ergonomics**
- **farmers**
- **hot climate**
- **psychosomatic-disease**
- **quick exposure check**
- **rapid upper limb assessment**
- **RULA**

REFERENCES

1. Oka, T., (2015). Psychogenic fever: How psychological stress affects body temperature in the clinical population. *Temperature, 2*(3), 368–378. doi: 10.1080/23328940.2015.1056907.
2. Heidari, H., Golbabaei, F., Shamsipour, A., Forushani, A. R., & GAEINI, A., (2015). Evaluation of heat stress among farmers using environmental and biological monitoring: A study in North of Iran. *International Journal of Occupational Hygiene, 7*, 1–9.
3. Al-Saleh, K., (2004). Computerizing RULA and QEC. *Proceedings of the 2nd IIEC-2004*. Riyadh, Kingdom of Saudi Arabia.
4. Bhandare, A., Bahirat, P., Nagarkar, V., & Bewoor, A., (2013). Postural analysis and quantification of fatigue by using RULA and REBA techniques. *International Journal of Mechanical and Production Engineering, 1*(3).
5. Ansari, N. A., & Sheikh, M. J., (2014). Evaluation of work posture by RULA and REBA: A case study. *IOSR Journal of Mechanical and Civil Engineering, 11*(4), 18–23.
6. MdYusop, M. S., Mat, S., Ramli, F. R., Dullah, A. R., Khalil, S. N., & Case, K., (2018). *Design of Welding Armrest Based on Ergonomics, 13*(1).
7. Vijay, K., & Kumar, S. M., (2018). Ergonomic evaluation of maize seed sowing method in small holdings and its impact on musculoskeletal disorders by using digital human model and virtual ergonomics techniques. *International Journal of Agriculture Sciences, 10*(2), 5056–5060.
8. Susihono, W., Ariesca, Suryanawati, Mirajiani, & Gunawan, G., (2018). Design of standard operating procedure (SOP) based at ergonomic working attitude through

musculoskeletal disorders (Msd's) complaints. *MATEC Web of Conferences, 218,* 04019. https://doi.org/10.1051/matecconf/201821804019.

9. Weichelt, B., Ray, W., & Keifer, M., (2019). Development of an occupational health safe return to work prototype application and ergonomics data set for agricultural tasks. *Safety, 5,* 40. doi: 10.3390/safety5020040.
10. Budhathoki, N. K., & Zander, K. K., (2019). Socio-economic impact of and adaptation to extreme heat and cold of farmers in the food bowl of Nepal. *Int. J. Environ. Res. Public Health, 16,* 1578. doi: 10.3390/ijerph16091578.

CHAPTER 4

ORGANIZED HEALTH CARE WASTE COLLECTION AND ROUTE OPTIMIZATION: A CASE STUDY

ABHISHEK RAJ and CHERIAN SAMUEL

Mechanical Engineering Department,
IIT (Banaras Hindu University), Varanasi, Uttar Pradesh, India,
E-mails: abhishekraj.rs.mec17@itbhu.ac.in (A. Raj),
csamuel.mec@itbhu.ac.in (C. Samuel)

ABSTRACT

Health care waste disposal is one of the essential concerns for any developing country. People are traveling in the whole world, and the increasing population of the world leads to the healthcare sector's rising prominence. Increment in the healthcare sector also increases healthcare waste. Due to the increment in waste, their disposal is a crucial matter. The health care waste generated from different hospitals is not entirely hazardous, so proper segregation must avoid transportation of all the waste to the incineration center. Varanasi is a highly dense city with lots of hospitals. A case study was performed in Varanasi to find the collection centers and optimize routes in Varanasi. A total of 20 collection centers were found, which will cover almost 200 hospitals in the city. This study will help the hospital management and the government take necessary action for effective management.

Optimization Methods for Engineering Problems. Dilbagh Panchal, Prasenjit Chatterjee, Mohit Tyagi, Ravi Pratap Singh (Eds.)

4.1 INTRODUCTION

Health care is one of the utmost critical areas of society. Therefore, healthcare waste management also deserves much attention. Healthcare waste (HCW) is a by-product of healthcare, comprised of sharps, non-sharps, body parts, chemicals, and radioactive materials. Poor management of HCW uncovers healthcare workers, waste handlers, and the community to contagions, toxic effects, and injuries. As the world population is multiplying, there is an increment in the requirement of hospitals. Increment in hospitals leads to an increase in the amounts of waste. In the standard case, only 30% of the generated waste is hazardous. The rest of the 70% of the waste is non-hazardous, but due to improper knowledge, many money wastes at their disposal. So proper management of health care waste is mandatory to avoid any adverse impacts. It is not easy for healthcare organizations to handle healthcare waste because these organizations have varieties of barriers. As a result of this, several healthcare sectors worldwide are facing problems related to healthcare waste management. Various research on HCWM problems in different countries has already been done [1–3]. Similarly, if there is no good coordination of hospitals with the pollution control board and municipal corporation, then collection, handling, and transportation will not be accessible [4, 5]. Hospitals must have to give special attention to health care waste and disinfectant techniques, and they have to be aware of the type and quantity of the waste, which is generating in their hospitals. There must be appropriate disinfectant techniques to reduce the waste's harmful effects [6]. For proper handling of the waste, there must be the availability of the handling tools, and the handler in the Organization must be adequately trained to operate handling tools [5, 7]. The waste that is easy to recycle must be recycled in the hospitals' recycling centers, which reduce the amount of waste that is to be incinerated. The recycled product can be easily usable [8, 9]. The mostly handling and disposal methods implemented for HCW management in developing countries are open dumps or uncontrolled landfills with incineration without adequate measures to deal with toxic emissions to air, soil, water, with serious health hazards to humans.

However, specific research related to HCWM barriers is limited for developing countries like India. Thus, this chapter's main contribution lies in satisfying this research gap by intensifying the narrow body of knowledge. This is done by considering the effects of organizational, waste handling, and human resource barriers on healthcare waste management, particularly for healthcare sectors located in India, one of the world's highly populated countries.

4.2 RESEARCH BACKGROUND

In the present scenario, developed and developing countries are trying to solve their healthcare waste handling issues. Several countries have already imposed strict regulations and policies for the appropriate disposal of all infectious waste. Adequate placement of contagious waste is an essential concern for developing countries. The waste management act is satisfactory for the country, but proper training and awareness are also mandatory, so the strict implementation must occur. Different countries are facing different types of issues regarding health care waste management. A conceptual model was established to optimize the design of a collection, transfer, treatment, and disposal scheme for infectious medical waste in the Greek Region. In this model, a collection center concept was included where storage is temporary when the incineration plant is busy [10]. Healthcare waste management is a matter of deep concern among all developed countries. Despite the severe nature of this issue, there is diminutive attention given to it in Asian developing countries, all of which are in dearth of waste management strategies, firm policies, proper knowledge, awareness, strict regulation, sufficient funds, and their implementation [11].

4.3 RESEARCH METHODOLOGY

The problem in this study is based on proposing an optimal waste-collecting strategy for Varanasi, India. As the location's geographical conditions are not apt for the transportation of waste using huge vehicles because of narrow roads, small vehicles are preferred to large vehicles. Waste cannot be stored for a more extended period as it is contagious and has terrible impacts on human beings and society. A model of the p-median approach is to be proposed rather than a simple waste collection approach.

The p median problem is one of the best NP-hard discrete location problems, which works to locate p facilities among a set of given locations and assign the other sites or nodes to these p facilities. The p median problem gives us the location of p-facilities. For each collection center, some other nodes are allocated to them, which provides service to these collection centers. The p median issue's objective is to minimize the maximum weighted distance for all demand points from the individual nodes to the corresponding collection center. The p-median problem has many real-life applications, such as locating fire stations in an urban area, locating ambulance stations, hospitals in a city.

To locate collection and separation centers, each hospital is convenient to give their health care waste to appoint collection centers where the separation of waste is done because 70 to 80% of waste is non-hazardous. Collection centers reduce the waste collection cost and separation center reduces the treatment cost. All the other hospitals are allotted to these collection centers. This is done so that all the hospitals must be assigned to at least one collection center, and it is also to be ensured that no hospital gets allotted to more than one collection center.

The vehicles' optimal route to smoothly go to the collection centers to collect waste directly without going to each hospital door; this reduces total transportation cost.

The following actions were performed in this study:

- To collect data from the Center for Pollution Control (CPC) private limited about the hospitals they cover daily;
- To find appropriate places which are suitable for the collection center.

➢ **Assumptions:**

- Each vehicle will start and end its journey from the depot;
- The system is assumed to be measured on daily operation;
- Each vehicle is of equal capacity;
- Only one vehicle visits a waste bin every time;
- The total amassing capacity of a vehicle must not exceed its maximum limit.

To find out the appropriate place for the collection center, the problem is considered a p-median facility location problem.

➢ **Indices:**

i, j: Nodes.

➢ **Parameters:**

d_{ij} = Distance between two nodes.

P = Numbers of collection center to locate.
C = Cost per unit distance.
EC = Fixed establishment cost of collection center.

- **Decision Variables:**

$$X_j = \begin{cases} 1 & \text{if we locate a collection center at} \\ 0 & \text{if not} \end{cases}$$

$$Y_j = \begin{cases} 1 & \text{if demand at node } i \in I \text{ is served by the facility at } j \in J \\ 0 & \text{if not} \end{cases}$$

- **Objective Function:**

$$Z = \Sigma_{j \in I} EC * X_j + C * \Sigma\Sigma d_{ij} Y_{ij}$$

- **Constraints:**

  - $Yij \leq Xj$ $\quad i \in I; j \in J$
  - $\sum Yij = 1$ $\quad i \in I; j \in J$
  - $\sum Xj = P$
  - $Xj \in 0, 1$ $\quad j \in J$
  - $Yij \in 0, 1$ $\quad i \in I; j \in J$

4.4 CASE STUDY

Two private organizations are working on the management of health care waste in Varanasi. Effective waste management is done by the Center for Pollution Control (CPC), which has four vehicles of 100 kg capacity each, and all these vehicles cover almost 200 healthcare units of all types of hospitals, Nursing homes, and pathology labs.

In the present scenario, all the hospitals planning to dispose of their wastes are covered by the CPC organization's vehicles, and their incineration plant is located at Mohansarai, Varanasi. Every four vehicles have to go to every hospital to collect the healthcare waste. In this way, vehicles have to travel a considerable distance, and traveling cost is also high due to this type of collection system. The concept of a collection center is introduced in this study to reduce overall waste handling cost in a significant amount.

4.4.1 APPLICATION OF RESEARCH METHODOLOGY

At present, four vehicles are covering 45, 36, 50, and 60 hospitals, respectively. To find collection centers, p-median approach is applied, and the results are written below:

- For Vehicle 1: (Hospital 2, Hospital 5, Hospital 20, Hospital 28, Hospital 41).
- For Vehicle 2: (Hospital 4, Hospital 15, Hospital 22, Hospital 28, Hospital 34).
- For Vehicle 3: (Hospital 1, Hospital 15, Hospital 33, Hospital 35, Hospital 47).
- For Vehicle 4: (Hospital 1, Hospital 7, Hospital 14, Hospital 30, Hospital 46).

Following hospitals are working as the collection centers for other hospitals, and vehicles have to go to these collection centers to collect waste accordingly.

- **For Vehicle 1:**
  - Jeevan Jyoti Children hospital;
  - Anand Seva Sadan;
  - Shiv welfare hospital;
  - Mani hospital;
  - Amanya skin and hair clinic.
- **For Vehicle 2:**
  - R.K. Netralay eye hospital (Mahmoorganj);
  - Shubham Seva Sadan;
  - OM Chikitsalay;
  - Shivangam Hospital;
  - Jhumawati Hospital.
- **For Vehicle 3:**
  - A S G hospital;
  - Deva Mental healthcare;
  - Sanjivani Hospital;
  - Jivan Anmol Hospital;
  - Maa Nursing Home.
- **For Vehicle 4:**
  - Maxwell Hospital;

- Surbhi Hospital;
- Jamea Hospital;
- Vivek Netralay;
- Shaurya Neurology Center.

In this way, each vehicle has to cover five collection centers in their routes to collect complete waste of the hospitals.

From Figure 4.1, it is clear that vehicles have to start and end their journey from the incineration plant, which is working as a depot. After that, vehicles have to go to each assigned collection center (C1 and C2), where waste from other nearby hospitals (H1, H2, H3, H4, H5, H6, …, Hn) has already been collected. After covering all collection centers, vehicles have to end the journey in the incineration plant. Waste after completing the incineration process disposes to the landfill as ordinary waste.

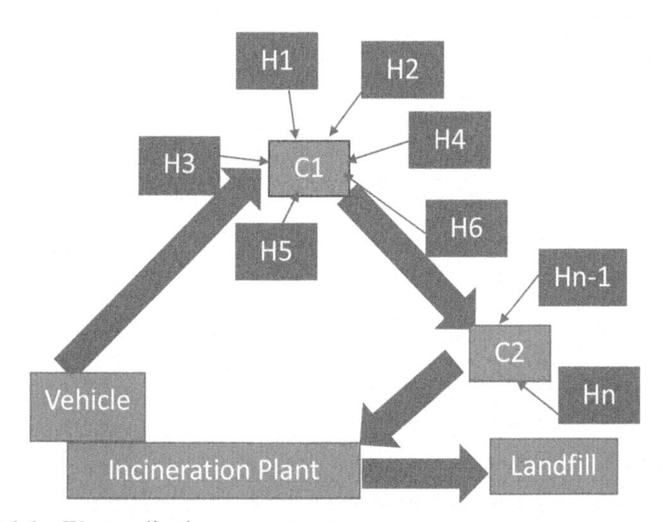

FIGURE 4.1 Waste collection.

Note: C1 and C2 are the collection centers; H1, H2, H3, H4, H5, H6, Hn–1, and Hn denote hospitals.

4.5 RESULTS AND DISCUSSION

For the collection of waste from different hospitals, now vehicles have to travel only to collection centers. After implementing the collection center and optimal route criteria, there are improvements in reducing cost, times, and resources. Each waste collection vehicles have to travel only to the assigned five collection centers in place of several hospitals. The distance has to cover fewer kilometers than earlier cases, reducing travel kilometers, fuel consumption, and travel time in a meaningful manner.

4.6 CONCLUSION

As each vehicle covers different numbers of hospitals on various routes, they have to go to each hospital to take the waste; thus, vehicles have to cover more distance. This increases transportation costs. Now on finding collection centers, it is easy to handle every hospital's waste by just going to their collection centers; this reduces transportation costs because it covers less distance than earlier cases.

According to the model proposed, the Organization has to appoint a certain number of nodes as collection centers that are given the duty to collect the waste from those assigned to it and store the waste till the vehicle collects the waste from the p collection center. By this, some people will get a job in the collection center located at the p points, which provides relief to at least some people as unemployment are a significant problem nowadays.

KEYWORDS

- **center for pollution control**
- **collection centers**
- **healthcare organizations**
- **healthcare waste management**
- **methodology**
- **vehicle routing**

REFERENCES

1. Mbongwe, B., Mmereki, B. T., & Magashula, A., (2008). Healthcare waste management: Current practices in selected healthcare facilities, Botswana. *Waste Management, 28*(1), 226–233.

2. Abdallah, A., (2014). Implementing quality initiatives in healthcare organizations: Drivers and challenges. *International Journal of Health Care Quality Assurance, 27*(3), 166–181.
3. Chaerul, M., Tanaka, M., & Shekdar, A. V., (2008). A system dynamics approach for hospital waste management. *Waste Management, 28*(2), 442–449.
4. Delmonico, D. V. D. G., Santos, H. H. D., Pinheiro, M. A., De Castro, R., & De Souza, R. M., (2018). Waste management barriers in developing country hospitals: Case study and AHP analysis. *Waste Management & Research, 36*(1), 48–58.
5. Caniato, M., Tudor, T. L., & Vaccari, M., (2016). Assessment of healthcare waste management in a humanitarian crisis: A case study of the Gaza strip. *Waste Management, 58*, 386–396.
6. Bdour, A., Altrabsheh, B., Hadadin, N., & Al-Shareif, M., (2007). Assessment of medical wastes management practice: A case study of the northern part of Jordan. *Waste Management, 27*(6), 746–759.
7. Aung, T. S., Luan, S., & Xu, Q., (2019). Application of multi-criteria-decision approach for the analysis of medical waste management systems in Myanmar. *Journal of Cleaner Production, 222*, 733–745.
8. Tudor, T. L., Noonan, C. L., & Jenkin, L. E. T., (2005). Healthcare waste management: A case study from the National Health Service in Cornwall, United Kingdom. *Waste Management, 25*(6), 606–615.
9. Chauhan, A., Singh, A., & Jharkharia, S., (2018). An interpretive structural modeling (ISM) and decision-making trial and evaluation laboratory (DEMATEL) method approach for the analysis of barriers of waste recycling in India. *Journal of the Air & Waste Management Association, 68*(2), 100–110.
10. Mantzaras, G., & Voudrias, E. A., (2017). An optimization model for collection, haul, transfer, treatment and disposal of infectious medical waste: Application to a Greek region. *Waste Management, 69*, 518–534.
11. Khan, B. A., Cheng, L., Khan, A. A., & Ahmed, H., (2019). Healthcare waste management in Asian developing countries: A mini review. *Waste Management & Research, 37*(9), 863–875.
12. Al-Khatib, I. A., Khalaf, A. S., Al-Sari, M. I., & Anayah, F., (2020). Medical waste management at three hospitals in Jenin district, Palestine. *Environmental Monitoring and Assessment, 192*(1), 10.
13. Chau, P., (1997). Reexamining a model for evaluating information center success using a structural equation modeling approach. *Decision Sciences, 28*(2), 309–334.
14. Cheng, Y. W., Sung, F. C., Yang, Y., Lo, Y. H., Chung, Y. T., & Li, K. C., (2009). Medical waste production at hospitals and associated factors. *Waste Management, 29*(1), 440–444.
15. Doylo, T., Alemayehu, T., & Baraki, N., (2019). Knowledge and practice of health workers about healthcare waste management in public health facilities in Eastern Ethiopia. *Journal of Community Health, 44*(2), 284–291.
16. Guerrero, L. A., Maas, G., & Hogland, W., (2013). Solid waste management challenges for cities in developing countries. *Waste Management, 33*(1), 220–232.
17. Jain, V., & Ajmera, P., (2019). Modeling of the factors affecting lean implementation in healthcare using structural equation modeling. *International Journal of System Assurance Engineering and Management, 10*(4), 563–575.

18. Korkut, E. N., (2018). Estimations and analysis of medical waste amounts in the city of Istanbul and proposing a new approach for the estimation of future medical waste amounts. *Waste Management, 81*, 168–176.
19. Minoglou, M., & Komilis, D., (2018). Describing health care waste generation rates using regression modeling and principal component analysis. *Waste Management, 78*, 811–818.
20. Elliott Steen Windfeld & Marianne Su-Ling Brooks (2015). Medical waste management: A review. *Journal of Environmental Management, 163*, 98–108.

CHAPTER 5

LIFE CYCLE ASSESSMENT RESEARCH: A REVIEW BY BIBLIOMETRIC ANALYSIS

RAJEEV AGRAWAL, ANKIT YADAV, ALOK YADAV, and ANBESH JAMWAL

Malaviya National Institute of Technology, Jaipur, Rajasthan–302017, India

ABSTRACT

In the last few decades, life cycle assessment has been a topic of interest growing widely. Life cycle assessment plays an important role in assessing the carbon emissions as well as environmental impacts generated from the manufacturing or production activities during the life cycle of a product. This research work aims to summarize the research trends, advancements, and future research direction in life cycle assessment. The literature reported in this study is based on a database of Scopus and Web of Science (WoS) from 2000 to 2019 and is further categorized into different sections. Different types of analysis have been done in this chapter, such as author analysis, trend analysis, country analysis, discipline-wise analysis, and source analysis. It has been found that in the last five years adoption of LCA has become popular in the manufacturing industries which help to promote sustainability. Also, the Government policies among the global and customer pressure forced many industries to adopt sustainable production. Also, some business has been seen in

Optimization Methods for Engineering Problems. Dilbagh Panchal, Prasenjit Chatterjee, Mohit Tyagi, Ravi Pratap Singh (Eds.)

both of the databases because of common authors with highly cited publications in both databases. A list of the top 10 highly cited papers is also assessed from both the Scopus and web of sciences databases. The present results study is beneficial for the researchers to decide their future research direction in the research area of life cycle assessment and to trace its evolution globally.

5.1 INTRODUCTION

Worldwide there is the problem of global warming due to changes in climate conditions and increasing levels of carbon dioxide. According to earth's CO_2 data, the amount of CO_2 in November 2019 is 410.27 parts per million which is more than safe level and also increasing day by day, also October 2019 is the 2nd warmest October after 1880 [1]. It requires urgent solution so that environment will be safe for future generations. Many researchers are working on this and LCA can be considered as one of the tools used to evaluate total environmental impacts in quantitative data [2]. This may be done either for a single step of a product life cycle or for whole steps. LCA is a popular method used globally in every field; with the help of this researcher, get to know the top contributor in GHG emission and specific action are taken to reduce the emissions [3]. Many software is also used for LCA, such as GABI, Semipro, open LCA, etc. LCA is one of the hottest topics in the area of manufacturing science from the emergence of sustainability. However, life cycle assessment is important for industries and includes four main steps as stated in the ISO 14040 (2006). The first is to define the goal and scope, the second is inventory analysis means collection of data, the third one impact assessment, and the final one to analyze the results [4]. In the past few years, many research or case studies have been reported in the LCA by a number of researchers, organizations, and journals. Some authors summarized the articles of LCA by review papers in different sectors. The objective of this chapter is to collect and summarize the papers data related to LCA. In order to understand the research field of life cycle assessment means to find the trend, to know what are the top journals of LCA? Top countries contributing in research field of LCA, to know hot topics. A quantitative analysis of all articles related to LCA is performed, result of this analysis help the researcher to know top journals, top institutes working in the field of LCA, top authors in this field and make them able to trace the growth of LCA studies in past two decades from 2000 to 2019.

LCA in terms of sustainability or triple bottom line can be considered as the product's factual analysis [5]. Generally, LCA evaluates all about the extraction stage when the materials are extracted from the natural resources or any other modes, the manufacturing stage, where the manufacturing operations are performed on the material to convert it into the final shape. In the end, material which is converted into the product is delivered to the consumers, where the during and post-use impacts are evaluated. This is the whole process for the life cycle analysis evaluation of the product from the first stage to disposal stage [6, 7].

LCA results can help you improve your product development, marketing, strategic planning and even policymaking. Consumers can learn how sustainable a product is [8]. A purchasing department of a company can learn which suppliers have the most sustainable products and methods. In the organizations product designers are important as they can explore the opportunities in which sustainability can be affected by the designs.

In the initial stages of the life cycle evaluation, the material is recovered from the disposal stage, or it is extracted from the natural resources. In the second stage manufacturing operations are done on the material in which material is converted into the final product form. In the manufacturing stage many emissions are released into the environment in the form of GHG or water emissions. These emissions can be minimized by the effective process planning approaches. However, in some cases scheduling approaches are also useful to minimize carbon emissions. In the next stage, the material is transported or delivered to the customer, where the emissions during the transportation cycle are evaluated and minimized. In the last stage, the material is disposed of after using a specific time period [9–13].

In all stages, materials and products or operations are done on the products interact with the environment in different ways and emit various types of emissions. A different type of strain is represented by environment in every stage. For the LCA to be perfect each and every aspect must be considered and not only a few. All the minor and major issues and energy flows need to be considered during the evaluation of environmental assessment.

5.1.1 ISO STAGES OF LCA

Life cycle assessment added in ISO-14000 series in 1997. According to ISO:14040 Life cycle assessment classified into four phases which is given as follows:

- **Stage 1: Goal and Scope:** This stage is used to identify what product or services or what chunk of the life cycle of product, process, and services will be analyzed. In this, we target the functional basis for comparison and specify the required details. The goal should refer to the reasons to perform assessment, its various possible applications, target audience and whether it will use or not for comparative studies in public.
- **Stage 2: Inventory Analysis:** This stage helps us to understand energy and material flows within the system and its interaction with the surrounding environment. It covers all the data required for LCA. This stage allows us to find all inputs and outputs related to the concerned product, process, and services.
- **Stage 3: Impact Assessment:** This stage allows us to use the data from inventory analysis to broadly classify the resource used and emissions generated in order of their potential impacts. After that, classify them in various impact categories and figure out their relative importance with reference to goal.
- **Stage 4: Interpretation Assessment:** In this stage, we approach results in terms of data quality, data sensitivity, and its contribution and relevance to the field. We also aspire and look for opportunities to systematically reduce negative impacts on the environment by the targeted product, process, and services [14–18].

5.2 RESEARCH METHODOLOGY

Bibliometric investigation is the quantitative analysis of data [19]. Scopus database is a trademark of Elsevier BV (Netherlands Private Limited Company) and web of science is a publication of Thomson Reuters company. Both web of science (WOS) and Scopus database are important for Bibliometric investigation in the science field, both databases have consistent and standardized records, also both databases have broad scope of LCA related publications and top journals related to LCA that is why these databases are considered for analysis [20]. It is easy to retrieve the publications by author name and title of paper plus advance search option in both Scopus and web of science.

5.2.1 PAPER RETRIEVAL METHOD

Publication's data are collected from the web of science and Scopus. Scopus is a multiple database platform that provide four types of quality measure for each title; those are h-Index, Cite Score, SCImago Journal Rank (SJR), and Source Normalized Impact per Paper (SNIP). Secondly, web of science which includes Science Citation Index Expanded (SCIE), Social Sciences Citation Index (SSCI), Conference Proceedings Citation Index-Science (CPCI-S), Arts and Humanities Citation Index (AHCI), Conference Proceedings Citation Index-Social Sciences and Humanities (CPCI-SSH), Book Citation Index-Social Sciences and Humanities (BCI-SSH), Book Citation Index–Science (BCI-S) [21].

Search strategy is as follows: Title search is "Life cycle assessment within the timespan 2000 to 2019. This process is done on 2019.10.30. Language is limited to English (Table 5.1).

TABLE 5.1 Search Strategy for Data Collection

Options	Input
Title	Life cycle assessment
Type	Article, review papers
Language	English
Time span	2000–2019
Citation index	SCIE, SSCI, CPCI-S, AHCI, CPCI-SSH

5.2.2 DATA COLLECTION

A total of 5,355 publications are found in the web of science database, and 8,769 for Scopus database. These are the total publications in the last two decades, and this data is used for further Bibliometric investigation by dividing the data according to Countries, Years, Journals, Authors, and

Universities. Also, contribution of India is analyzed separately, a total of 87 publications are found in last two decades, LCA is a topic of interest in India also which is growing extensively, Top Indian institute working on LCA are also listed in this chapter. The search results that only 37 papers are published in 2005 worldwide, this number increase to 654 in 2019 for WOS database, similarly number increased from 88 to 979 in Scopus database. The journal with a maximum number of publications is "Journal of cleaner production" (Figure 5.1).

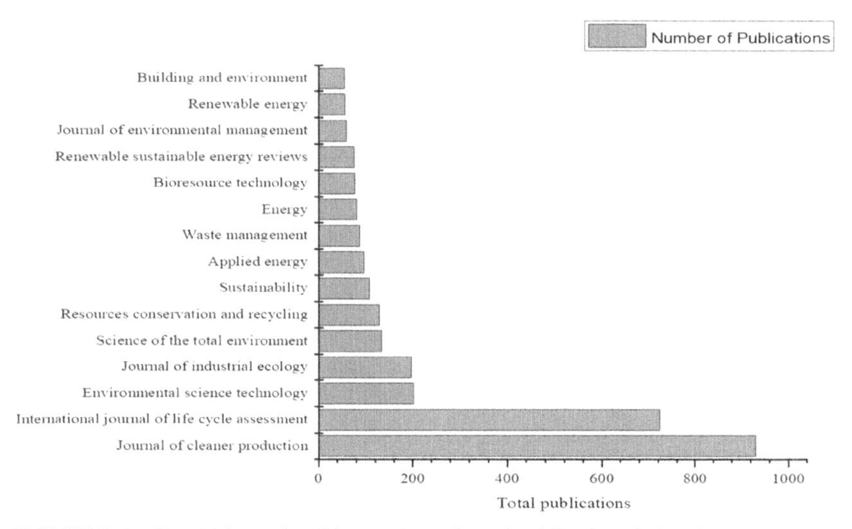

FIGURE 5.1 Top 15 journals with a total number of publications in last 20 years.

5.3 RESULTS AND DISCUSSIONS

5.3.1 YEAR-WISE ANALYSIS FROM 2000 TO 2019

In last 20 years studies on LCA are growing extensively, there is an exponential growth can be seen in Figure 5.2 from 2000 to 2019 for web of science database. 37 publications in 2000 reached to 674 in 2019.

From the graph, it can be clear that studies are increasing yearly, similar results are obtained from the Scopus database, only a change in data otherwise same trend is found. Overall, research interest is increasing globally.

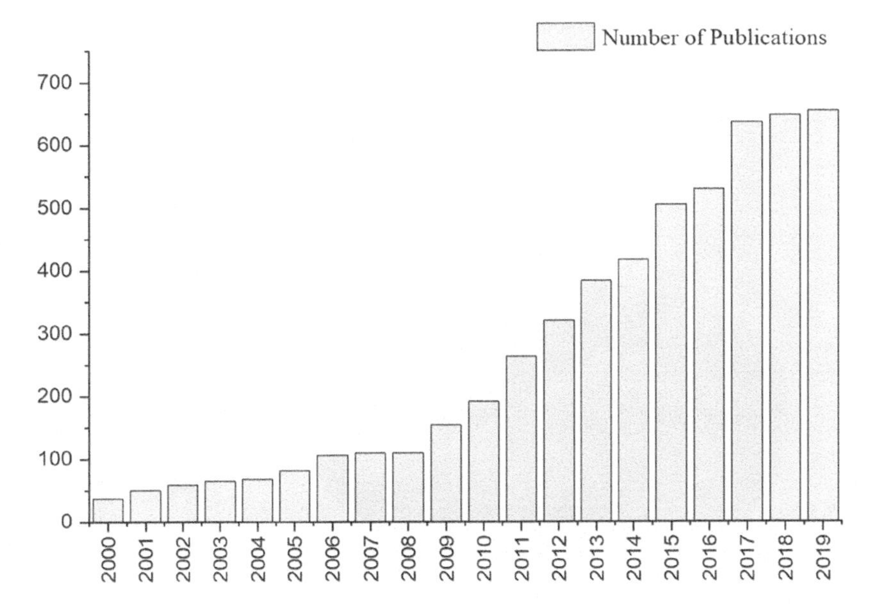

FIGURE 5.2 Year wise number of publications related to LCA according to WOS database.

5.3.2 COUNTRY WISE DISTRIBUTION FROM 2000 TO 2019

It is important to know that which country is providing maximum contribution. In this research field United States hold the first position with maximum number of publications, i.e., 1,125. Here top 15 countries are considered, if it is assumed that top 15 countries are contributing 100% then USA contribution is percent, the second one after this is China with 536 number of publications according to WOS database All other countries contribution is shown by a pie chart. The same results are founded by the Scopus database US and China are the top 2 countries with 1,743 and 987 numbers of publications (Figure 5.3).

5.3.3 WORLDWIDE MOST PRODUCTIVE TOP 15 INSTITUTES IN THE FIELD OF LCA WITH MORE THAN 65 PAPERS

From Table 5.2, it can be seen that technical university of Denmark holds top position globally in field of LCA with 189 publications according to web of science. Also, out of 15 total 4 of them belong to the United States alone, as it is also discussed that country wise United States contribution is at top position.

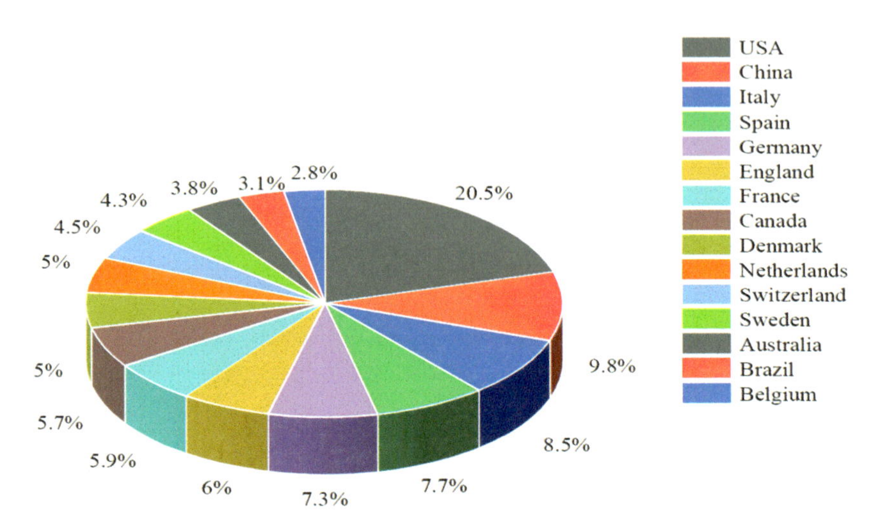

FIGURE 5.3 Top 15 Countries with highest number of publications according to WOS database.

TABLE 5.2 Top 15 Organizations Focused on LCA Studies According to WOS Database

SL. No.	Organizations	Country	Publications
1.	Technical University of Denmark	Denmark	189
2.	University of California System	United States	142
3.	Norwegian University of Science Technology	Norway	106
4.	Eth Zurich	Switzerland	91
5.	National Institute for Agricultural Research	French	91
6.	Chalmers University of Technology	Sweden	82
7.	University of Montreal	Canada	82
8.	Polytechnique Montreal	Canada	79
9.	National Center for scientific research	French	77
10.	Radboud University Nijmegen	Netherlands	75
11.	United States Department of Energy	United States	75
12.	University of Michigan	United States	72
13.	University of Michigan System	United States	72
14.	University of Santiago de Compostela	Spain	68
15.	Leiden University	Netherlands	65

5.3.4 TOP RESEARCHER'S WITH MAXIMUM NUMBER OF PUBLICATIONS ON LCA

A number of researchers are working on life cycle assessment to reach toward sustainability goals, here top 15 authors in the field of LCA are listed, this data will be helpful for endeavor working on this topic in future because better communication with experts help future researchers to reach their goal easily and efficiently plus gain in knowledge. List is based on a web of science database. Data of citation are as on 30 October 2019 (Table 5.3).

TABLE 5.3 List of Top 15 Authors Having Contribution in Researches Related to LCA

SL. No.	Author Name	Publications	Citation	H-Index
1.	Huijbregts Maj	58	23,436	64
2.	Jolliet O	52	16,410	60
3.	Moreira MT	48	11,511	63
4.	Feijoo G	43	12,458	65
5.	Margni M	40	11,500	47
6.	Hellweg S	39	15,796	58
7.	Hauschild MZ	37	21,006	66
8.	Hong JL	37	1,988	26
9.	Gonzalez-Garcia S	34	3,687	38
10.	Finkbeiner M	33	5,512	40
11.	Heijungs R	33	24,206	62
12.	Iribarren D	31	2,614	30
13.	Benetto E	30	2,924	31
14.	Stromman AH	30	7,555	41
15.	Dewulf J	29	9,279	52

5.3.5 TOP 15 CATEGORIES AND DOCUMENT TYPE ACCORDING TO WEB OF SCIENCE

Around 31% publications are from environmental sciences category, then engineering environmental category and third one is green sustainable science technology, Means LCA is tool used for sustainability. LCA is used to compare different products, different raw materials of a product for eco-efficient planning [6]. Majorly used in engineering fields such as chemical, civil, agricultural, and material science. In civil engineering, LCA is mostly used to find emission from construction material also in buildings. To find most sustainable raw materials and processes for a building so that goal of sustainable building can be achieved, a number of advancements have been done in recent days on building materials some examples are – use of cross-laminated timber (CLT) as a framing material, its practical example can be seen in the United States. CLT buildings is a new concept so for now 8 to 10 story buildings are found, but in future it will be constructed on large scale worldwide with more than 10 story. Other sustainable construction material is green concrete, use of fly ash bricks in place of red bricks. Fly ash bricks consist 50% of recyclable material, it contains quicklime, gypsum, cement, fly ash, water, and aluminum powder. Manufacturing method of fly ash brick reduce energy requirement and less mercury pollution Also these bricks are lighter and stronger as compare to clay bricks. Researcher are working hard to produce more sustainable materials plus sustainable methods, that will contribute for better environment, also help to achieve the goal of only 2° change in global temperature till 2,100. It is only possible with sustainability approach in all the sectors (Tables 5.4 and 5.5).

TABLE 5.4 Top Categories and Document based on WOS Database

SL. No.	Categories	Publication	Document Type	Publication
1.	Environmental sciences	3,176	Article	4768
2.	Engineering environmental	2,580	Review	310
3.	Green sustainable science technology	1,558	Proceedings paper	161
4.	Energy fuels	888	Editorial material	134
5.	Engineering chemical	333	Meeting abstract	60
6.	Engineering civil	266	Correction	48
7.	Environmental studies	227	Letter	31
8.	Construction building technology	215	Early assess	17
9.	Biotechnology applied microbiology	212	Book chapter	2
10.	Chemistry multidisciplinary	181	News item	2
11.	Agricultural engineering	145	Book review	1
12.	Materials science multidisciplinary	143	Software review	1
13.	Thermodynamics	143	–	–
14.	Water resources	108	–	–
15.	Food science technology	76	–	–

TABLE 5.5 Most Cited Article of Each Year According to Web of Science Database

SL. No.	Title	Citation	Year
1.	Application of Life Cycle Assessment in Municipal Solid Waste Management: A Worldwide Critical Review	22	2019
2.	Sustainable Conversion of Carbon Dioxide: An Integrated Review of Catalysis and Life Cycle Assessment	235	2018
3.	ReCiPe2016: A Harmonized Life Cycle Impact Assessment Method at Midpoint and Endpoint Level	107	2017
4.	Techno-Economic and Life Cycle Assessment on Lignocellulosic Biomass Thermochemical Conversion Technologies: A Review	107	2016
5.	Perovskite Photovoltaics: Life-Cycle Assessment of Energy and Environmental Impacts	219	2015
6.	Life Cycle Assessment (LCA) and Life Cycle Energy Analysis (LCEA) of Buildings and the Building Sector: A Review	387	2014
7.	Comparative Environmental Life Cycle Assessment of Conventional and Electric Vehicles	438	2013
8.	Optimal Design of Sustainable Cellulosic Biofuel Supply Chains: Multiobjective Optimization Coupled with Life Cycle Assessment and Input-Output Analysis	366	2012
9.	Life Cycle Assessment: Past, Present, and Futures	471	2011
10.	Comparing Environmental Impacts for Livestock Products: A Review of Life Cycle Assessments	461	2010
11.	Recent Developments in Life Cycle Assessment	1,308	2009

On the basis of maximum citations this table contain year wise most cited article of the last 11 years. It can be seen from the table that LCA is a topic of interest in sectors like automobile, buildings, and waste management. Other some article is review papers that summarize the studies that are previously completed in this field and what are the research gaps for future study that will help the society for better environmental conditions. Building itself is a wide topic it requires attention because globally construction sectors produce 40–50% greenhouse gas emissions, also worldwide building construction sectors consumes 40% of material entering the economy. Reduction in GHG emission by the construction sector will have a great impact on total GHG emissions. Another paper is on electrical vehicles versus conventional vehicles, which provide the data of environmental impact created by both vehicles [7].

5.3.6 CONTRIBUTION OF INDIAN RESEARCHER IN LCA ACCORDING TO WOS DATABASE

From the bibliometric analysis, it is found that the United States is the top country with the greatest number of Publications in LCA. Indian researchers are also working in this field, a total of 87 papers are published in top journals by Indian researchers. This number is increasing every year. Only one paper in 2007 and now it is increased to 16 in 2019. India holds 19th position. Indian Institute of Technology Bombay is the top institute working on LCA-related studies (Figure 5.4).

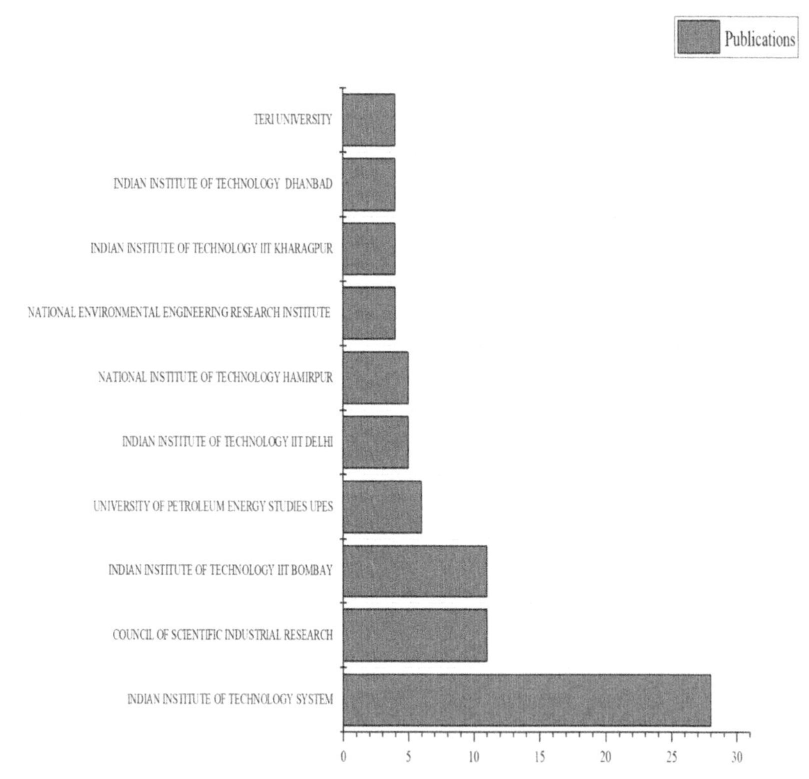

FIGURE 5.4 Top contributing institutes of India in LCA research.

Most cited paper based on LCA from India is" Simplified method of sizing and life cycle cost assessment of building integrated photovoltaic system" (2009). Topics covered in most cited papers are Life cycle assessment of solid waste management, LCA of fuel sources used in India, etc. Overall interest in this topic is growing.

5.3.7 TOP KEYWORDS USED IN LAST 20 YEAR ACCORDING TO SCOPUS AND WOS DATABASE

There are some keywords which frequently used in most of the paper. These keywords also give an idea about what is the trending or important topics. Here data is collected using Scopus and WOS database and top 15 keywords are listed in Table 5.6. Mostly used keyword is "Life cycle" that represents fundamental concept and method of thinking regarding LCA. Other keywords such as greenhouse gases, global warming, Carbon dioxide, Eutrophication represent the environmental related issues during the life cycle. Another group of keywords such as environmental impact assessment, environmental management relate to method of reducing total emissions. The degree of occurrence of keyword sustainable development is 1,269, that means sustainability is the main goal and LCA provides a direction toward sustainability by providing data of total environmental emissions. Keyword co-occurrence is analyzed by VOSviewer software for WOS database and it is found that result from both databases is almost same (Figure 5.5).

TABLE 5.6 Top 15 Keywords from Scopus Database

SL. No.	Keyword	Occurrence
1.	Life cycle	5,494
2.	Life cycle assessment	4,021
3.	Life cycle assessment (LCA)	3,990
4.	Environmental impact	3,678
5.	Life cycle analysis	2,168
6.	Article	1,471
7.	Sustainable development	1,269
8.	Global warming	1,237
9.	Priority journal	1,113
10.	Greenhouse gases	1,066
11.	Environmental impact assessment	999
12.	LCA	958
13.	Environmental management	955
14.	Carbon dioxide	945
15.	Eutrophication	865

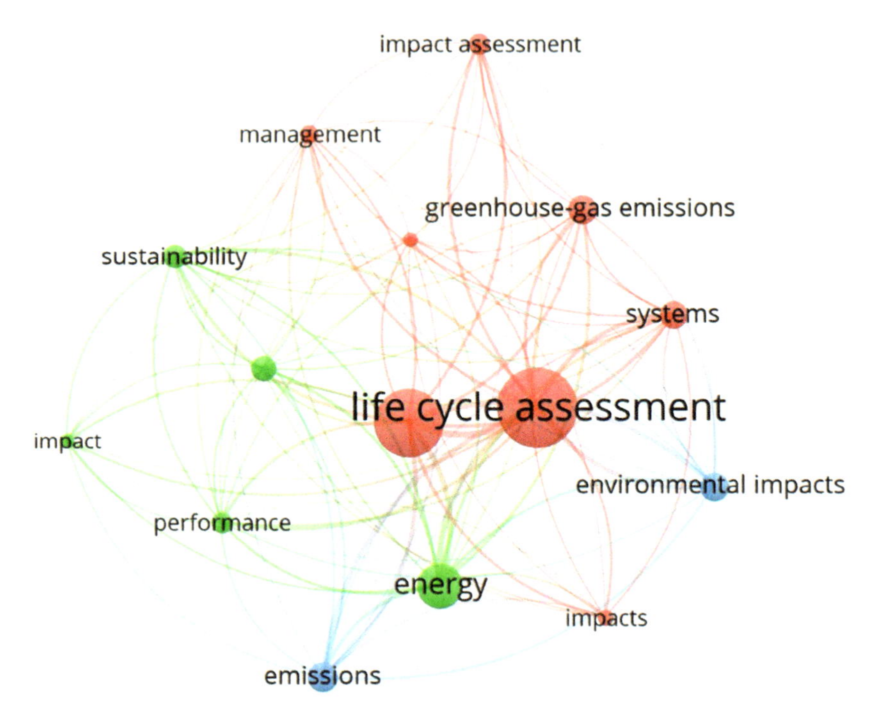

FIGURE 5.5 Top 15 co-occurring keywords in last 20 years according to WOS database.

5.4 CONCLUSIONS

LCA is a systematic tool used worldwide to access total emissions from product or process life cycle. However, in the past, a very few systematic and chronological studies are performed on LCA. This Bibliometric paper shed some light in this matter to detect the status and trend of LCA. On the basis of 5,355 publication retrieved from WOS database plus 8,769 publication of Scopus database, this Bibliometric study provides an overview of research in LCA by summarizing the institute contribution, top countries in the LCA research domain, top author working in this field, how the growth of LCA research vary year wise, top keywords mostly used in publications, top journal for LCA, top Indian institutes working in LCA field, also most cited paper of each year from 2009 to 2019. This study traces the current state and current evolution of LCA research studies by considering data from both Scopus and web of science.

In terms of subject categories of LCA research, environmental science, engineering, and energy contain the most bibliographic record. However, a research focus upon sustainability is emerging in the past five years. Several reputed journals have published significant LCA research findings, including the journal of cleaner production, international journal of life cycle assessment and environmental science technology. Most of the highly cited papers belong to these journals. After analysis a total of 5 hot topic are found which are sustainability, LCA of residential and non-residential buildings, how to reduce emissions during the building life cycle, etc., overall, after analyzing the statistical data the result is:

- This bibliometric investigation is helpful for both researcher and practitioner to understand the status and future agenda of LCA research. This chapter provides insight of top authors and institutes working in this area.
- Graph of year-wise contribution shows that LCA-related research got a huge increment in past 20 years. Most of the publications are belong to top journals such as Journal of cleaner production and international journal of life cycle assessment. Top country working in this field is United States and top institute with highest publication in last 20 year is technical university of Denmark.
- Progress report of India in this field is also positive increasing yearly. Chief core institute with top contribution from India is IIT Bombay, among all NITs, NIT Hamirpur holds peak position. A researcher interested in this topic can collaborate with these institutes and their authors to reach their goal.

The aim of this study is to provide an understanding of present scenario and trend of life cycle assessment studies. Future researchers get insight of complete development of LCA research in past 20 years from 2000 to 2019. By following this researcher would extend the body of life cycle assessment research and move toward sustainability for better environmental conditions in the future.

ACKNOWLEDGMENT

Author(s) received financial support from TEQIP-3 sponsored project "Exploration of sustainable manufacturing in Indian manufacturing industries" for this research work.

KEYWORDS

- **arts and humanities citation index**
- **bibliometric investigation**
- **conference proceedings citation index-science**
- **ISO 14040**
- **life cycle assessment (LCA)**
- **science citation index expanded**
- **SCImago journal rank**

REFERENCES

1. Zanghelini, G. M., De Souza, Jr. H. R., Kulay, L., Cherubini, E., Ribeiro, P. T., & Soares, S. R., (2016). A bibliometric overview of Brazilian LCA research. *The International Journal of Life Cycle Assessment, 21*(12), 1759–1775.
2. Hou, Q., Mao, G., Zhao, L., Du, H., & Zuo, J., (2015). Mapping the scientific research on life cycle assessment: A bibliometric analysis. *The International Journal of Life Cycle Assessment, 20*(4), 541–555.
3. Chen, H., Yang, Y., Yang, Y., Jiang, W., & Zhou, J., (2014). A bibliometric investigation of life cycle assessment research in the web of science databases. *The International Journal of Life Cycle Assessment, 19*(10), 1674–1685.
4. Geng, S., Wang, Y., Zuo, J., Zhou, Z., Du, H., & Mao, G., (2017). Building life cycle assessment research: A review by bibliometric analysis. *Renewable and Sustainable Energy Reviews, 76*, 176–184.
5. Huertas-Valdivia, I., Ferrari, A. M., Settembre-Blundo, D., & García-Muiña, F. E., (2020). Social life-cycle assessment: A review by bibliometric analysis. *Sustainability, 12*(15), 6211.
6. Zeng, R., & Chini, A., (2017). A review of research on embodied energy of buildings using bibliometric analysis. *Energy and Buildings, 155*, 172–184.
7. Yu, C., Davis, C., & Dijkema, G. P., (2014). Understanding the evolution of industrial symbiosis research: A bibliometric and network analysis (1997–2012). *Journal of Industrial Ecology, 18*(2), 280–293.
8. Udomsap, A. D., & Hallinger, P., (2020). A bibliometric review of research on sustainable construction, 1994–2018. *Journal of Cleaner Production, 254*, 120073.
9. Chen, H., Jiang, W., Yang, Y., Yang, Y., & Man, X., (2017). State of the art on food waste research: A bibliometric study from 1997 to 2014 *Journal of Cleaner Production, 140*, 840–846.

10. Chen, W., Liu, W., Geng, Y., Brown, M. T., Gao, C., & Wu, R., (2017). Recent progress on emergy research: A bibliometric analysis. *Renewable and Sustainable Energy Reviews, 73*, 1051–1060.
11. Zhao, X., Zuo, J., Wu, G., & Huang, C., (2019). A bibliometric review of green building research 2000–2016. *Architectural Science Review, 62*(1), 74–88.
12. Zhang, Y., Huang, K., Yu, Y., & Yang, B., (2017). Mapping of water footprint research: A bibliometric analysis during 2006–2015. *Journal of Cleaner Production, 149*, 70–79.
13. Marvuglia, A., Havinga, L., Heidrich, O., Fonseca, J., Gaitani, N., & Reckien, D., (2020). Advances and challenges in assessing urban sustainability: An advanced bibliometric review. *Renewable and Sustainable Energy Reviews, 124*, 109788.
14. Huang, M., Wang, Z., & Chen, T., (2019). Analysis on the theory and practice of industrial symbiosis based on bibliometrics and social network analysis. *Journal of Cleaner Production, 213*, 956–967.
15. Ranjan, P., Agrawal, R., & Jain, J. K., (2021). Life cycle assessment in sustainable manufacturing: A review and further direction. *Operations Management and Systems Engineering,* 191–203.
16. Jamwal, A., Agrawal, R., Gupta, S., Dangayach, G. S., Sharma, M., & Sohag, M. A. Z., (2020). Modeling of sustainable manufacturing barriers in pharmaceutical industries of Himachal Pradesh: An ISM-fuzzy approach. In: *Proceedings of International Conference in Mechanical and Energy Technology* (pp. 157–167). Springer, Singapore.
17. Verma, V., Jain, J. K., & Agrawal, R., (2021). Sustainability assessment of organization performance: A review and case study. *Operations Management and Systems Engineering,* 205–219.
18. Rao, H. S., Reddy, D. S. K., Sharma, C., Gupta, S., Jamwal, A., & Agrawal, R., (2021). Assessment of key barriers of sustainable additive manufacturing in Indian automotive company. *Advances in Industrial and Production Engineering: Select Proceedings of Flame 2020*, 245.
19. Jamwal, A., Agrawal, R., Sharma, M., & Kumar, V., (2021). Review on multi-criteria decision analysis in sustainable manufacturing decision making. *International Journal of Sustainable Engineering,* 1–24.
20. Jamwal, A., Agrawal, R., Sharma, M., Dangayach, G. S., & Gupta, S., (2020). Application of optimization techniques in metal cutting operations: A bibliometric analysis. *Materials Today: Proceedings*.
21. Jamwal, A., Agrawal, R., Manupati, V. K., Sharma, M., Varela, L., & Machado, J., (2020). Development of cyber-physical system based manufacturing system design for process optimization. In: *IOP Conference Series: Materials Science and Engineering* (Vol. 997, No. 1, p. 012048). IOP Publishing.

CHAPTER 6

PROCESS AUTOMATION TOOL FOR DESIGN AND COST OPTIMIZATION USING CAD TOOL IN COMBINATION WITH VBA

ABHIJEET VARMA and J. S. KARAJAGIKAR

Department of Production Engineering and Industrial Management, College of Engineering (COEP), Pune, Maharashtra, India,
E-mail: varma21abhi@gmail.com (A. Varma)

ABSTRACT

In this challenging situation, product launch time to market for the automotive products becomes short duration with good product design quality. Design and modeling of the part-time is largely 50–60% of the total time of the product development cycle. The design stage has huge potential in which nonvalue added time can be saved. In design, lots of repetitive activities are performed by the design engineer, including various tasks such as performing quality checks, filing of checklist, e-mailing the status of design, etc., and these tasks are again confirmed by the product design expert. This chapter focuses on how we can save time by reducing the number of mouse clicks and giving assurance of design quality by automating the task and tracking its results which will give good CAD quality with minimum design time and CAD completion percentage. So, using CAD tool in combination with VBA, a macro has been built which does all these repetitive activities performed in single click, which saved 50–60%

Optimization Methods for Engineering Problems. Dilbagh Panchal, Prasenjit Chatterjee, Mohit Tyagi, Ravi Pratap Singh (Eds.)

of time for repetitive task. By using all the results and tracked data, the project CAD completion percentage can be calculated. This results in design time optimization and cost optimization as well.

6.1 INTRODUCTION

Increased global competition in the Automotive sector and in such scenario less product design time to product launch time creates a huge difference in profit. At automotive research and development center, many operations are required to be performed. Many of these activities are repetitive in nature and can be automated with certain logics and conditions applied. Some of such activities which are non-value-added activities performed and can be minimized or eliminated up to a certain extent are considered in this chapter. A major portion of time getting vanish in repetitive as well as non-value added activities such as Cleaning, Correcting, and Quality checking of design and drawings. This repetitive has been captured and standardized so that those will be done in quickly manner without missing any quality aspect and using the data to calculate the CAD completion percentage.

6.2 RESEARCH BACKGROUND

There are many challenges before the design and development department can be shown in Figure 6.1.

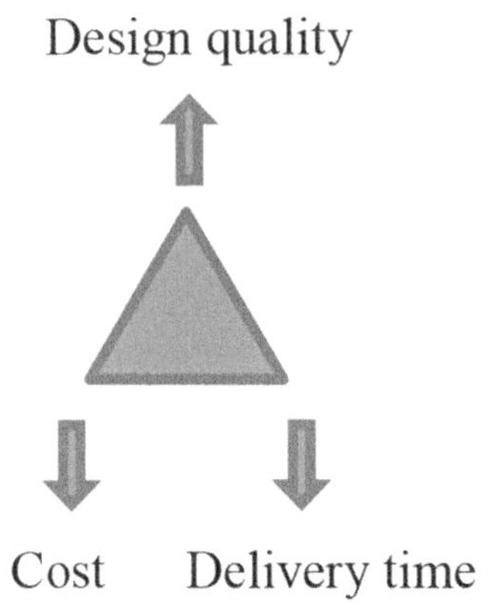

FIGURE 6.1 Challenges in design and development.

To overcome this challenge, less delivery time can be achieved by reducing nonvalue added time. Whereas, the as cost is directly proportional to time. As the design quality can be improved by increasing value-added time. Whereas, repetitive activities include quality checks and various daily activities can be automated by using customized tools in CAD tool.

6.3 METHODOLOGY

6.3.1 IDENTIFICATION OF ACTIVITIES

In this section, the methodology has been briefly discussed by the time saving and cost reduction approach by identifying various repetitive activities performed by design engineer. Many times, quality of design parts can be degraded if all the checkpoints are not check thoroughly which increase manager efforts to closely look into all designed parts follows parametric modeling, standard, and best practices defined by the design center, customer, and OEM specific requirements to perform certain quality checks, etc. So, some of the activities are identified for automation are follows:

- The data records must meet the customer specific requirements before delivery and so after everyday task performed the design engineer needs to give updates to the manager or team regarding the update work done through e-mail.
- The design engineer needs to fill the checklist in excel and attach Q-checker reports, DMU reports, etc.
- Manager needs to find out design completion percentage based on engineering functional milestone for project.
- If any modification done which needs to be rechecked with part comparison tool in Catia and comparison ppt to be shared with all customer/colleagues/manager.

6.3.2 MACRO

These repeated tasks require lot of time, efforts, and involve some cost. Using macros, i.e., set of instructions given to software for any customized repetitive number of commands with certain sequence and logics to be followed. Macros can be built in the scripting language as CATScript (VB), CATVBA (Visual Basic for Applications), and can be automated via CATIA V5 interface. CATIA V5 can be linked with applications of MS Excel, and PowerPoint can be commanded. These macros may be useful for creating, analyzing, measuring, modifying, translating, optimizing surfaces, solids, wireframes and more. Many standard features of CATIA V5 extended by macros are more powerful and can help speed up design procedures. A good example of automation is one can extract the values from CATIA V5, and these values can be used for filling the checklist and confirming the designs are up to set design standards with the said work the time required for the documentation of design process is saved by using such systems. The purpose of writing macros can be text processing, file management, drawing generation, user interface automation, launching programs, report generation, etc.

6.3.3 OBJECT LINKING AND EMBEDDING (OLE) AUTOMATION

CATIA communicates with MS Excel through the OLE Automation. The CATIA has details about the database and application programming interface (API). This enables Visual Basic for Applications (VBA) programs to interact with the application. At the same time, the multiple applications can be automated with the help of one host. This is achieved by creating application objects within the VBA code. Also, the connection to the various libraries must also be provided within the VBA application.

6.3.4 PROCESS FLOW

The methodology followed in this undertaking project is as per the following:

- Design task will be allocated to design engineer by design specialist/manager;
- Once the task is performed by design engineer as per requirement, design engineer needs to run the macro;
- Macro will perform various quality checks based on OEM specific requirement and task specific requirements;

- After final e-mail will be autogenerated with unique user ID and time stamp combination;
- Design specialist/manager needs to approve the work via voting option in terms of percentage;
- Tracker file placed at common location which generates records received from design engineer macro and reply received from the specialist/manager (Figure 6.2).

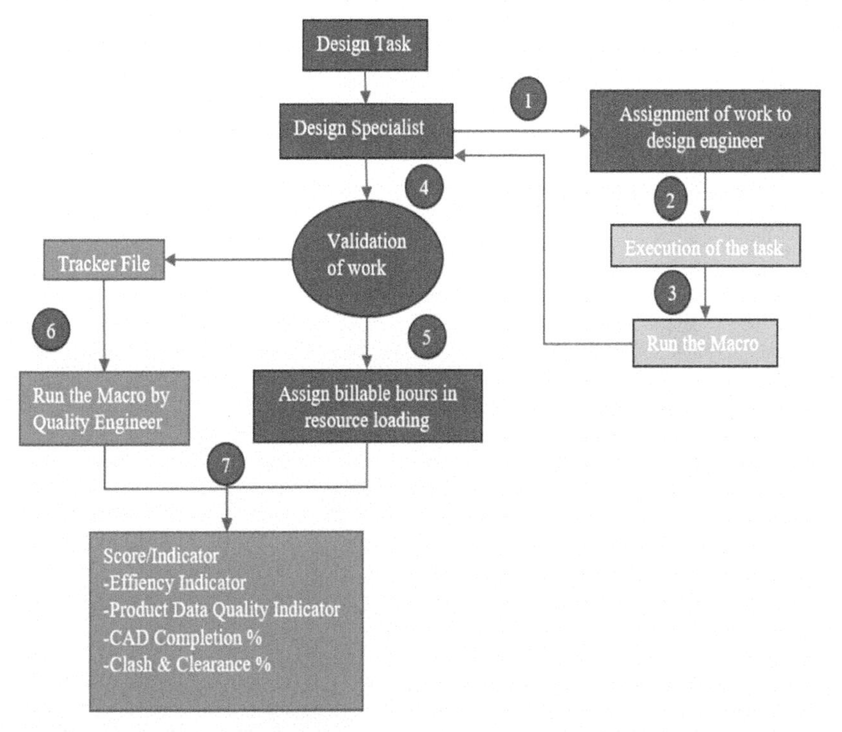

FIGURE 6.2 Process automation tool procedure.

6.4 CASE STUDY

In the real scenario, there are various task category are possible to work upon by design engineer which may include 2D Drawing activities, 3D part design activities, etc., for the sake for time and cost study we will select part involving all the quality checkpoints required, for case study purpose Car seat side cover part is considered. After 3D work of design and modification, the user clicks on the icon as Q check tool in the toolbar (Figure 6.3).

FIGURE 6.3 Q check tool icon in toolbar.

After which some prerequisite of tools will pop up and ask the designer to close any MS office application to be closed which will be automatically opened through macro for various tasks (Figure 6.4).

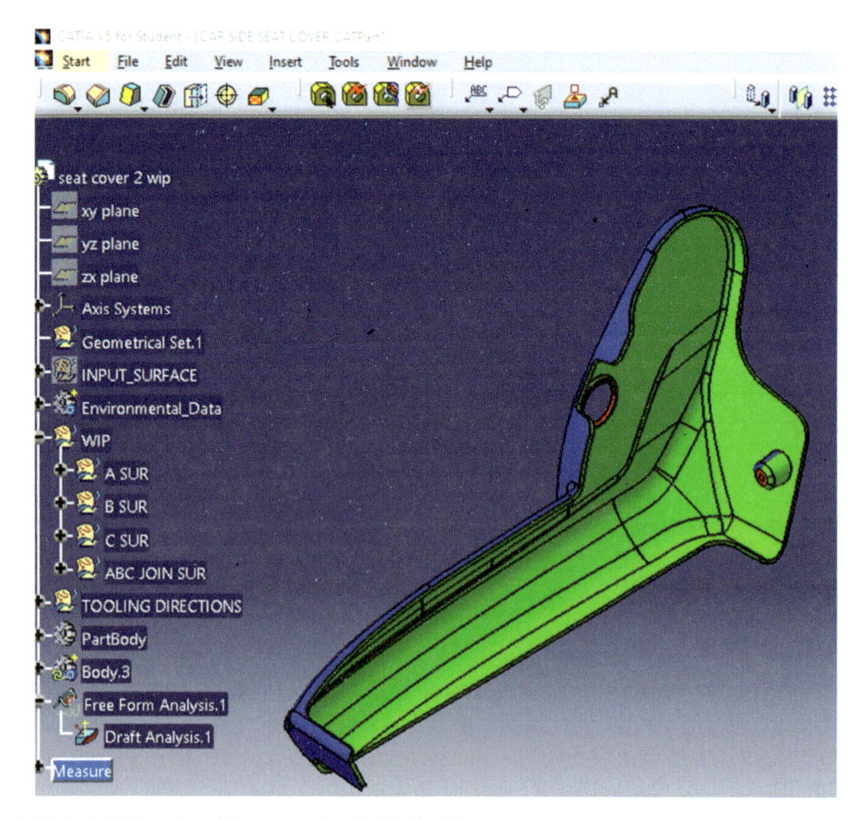

FIGURE 6.4 Seat side cover view in Catia V5.

As this tool can work for 2D work as well as 3D so user needs to select the work performed and checkpoints to be followed, and the designer gets the following window to select the type of work (Figure 6.5).

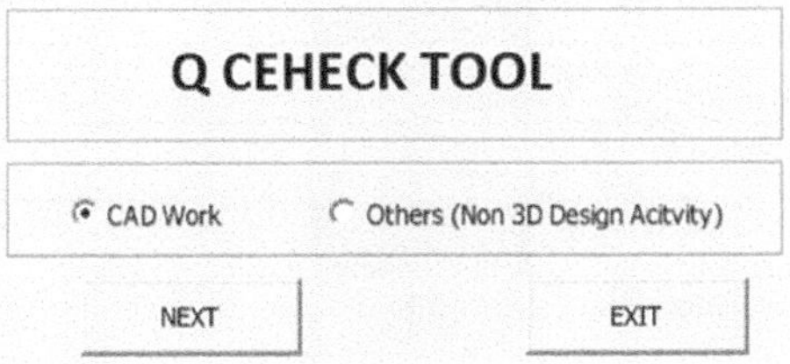

FIGURE 6.5 Work selection window.

This tool is also needing various predefined inputs for further data tracking and analysis work, in which the designer needs to give project name, work status can be in process or completed (Figure 6.6).

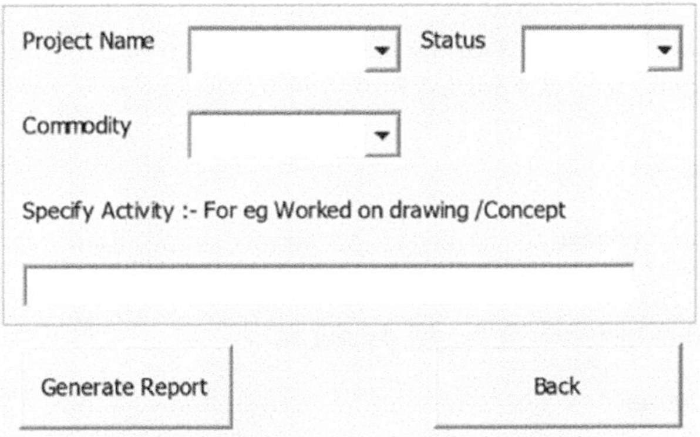

FIGURE 6.6 Initial tool input window.

Based on the requirement and selection of work, all the details are saved to one tracker file over the networked location and which works as database for the dashboard for the tool. After completion of all Quality checks, an e-mail with an attachment has been generated, which contains automated generated clash and clearance analysis, Q Checker reports, etc. (Figure 6.7).

	SR.NO	METHODOLOGY RULES	Methodology Followed	Description
5				
6	1	UPDATE of data model organize	NO	
7	2	No deactivated elements in structure tree	YES	
8	3	All Sketches dimensioned like a drawing and complete constrained	NO	There are sketches which are NOT ISO-CONSTRAIN.
9	4	No usage of Red, Yellow, Magenta, Orange and Cyan colors for geometry	YES	
10	5	Use of publications	YES	
11	6	Publication have same name as element and in upper case	NO	The Publications names must be in UPPER CASE.
12	7	No use of Ordered Geometrical Sets	YES	
13	8	Is the fill function used for topological modifications on historical surface	NA	Fill operation allowed for Frames
14	9	Forbidden characteres (file and entity names)	YES	

Common Check | Common

FIGURE 6.7 Final automated checklist.

E-mail also generated of various details of input details filled and Unique ID created based on the time stamp, whereas clash and clearance data has been given in Figure 6.8.

Hello Team,

Today I've worked on
Please find the attached compressed presentation report for review

Files are at below location Directory path of working part

Program Name V
Status WIP
Draft OK Undercuts __ Removal expected on __
Clash Clearance OK
Total Interferences Env 2 Clash = 2 Clearance = 0 Contacts = 0 Removal expected on __
Total Interferences Int 1 Clash = 1 Clearance = 0 Contacts = 0 Removal expected on __
No of Env components 3
No of Int components 1

Open Points Are attached
Checklist

Thanks and Regards,

FIGURE 6.8 E-mail generated in outlook.

6.5 RESULT DISCUSSION

For this case, a time study has been performed in which the average design user has been considered vs. the tool time, as shown in Table 6.1.

TABLE 6.1 Time Information

Checkpoint	Average User	Using Tool
Part design quality and best practices	10	6
Draft analysis or tooling feasibility	10	6
Part comparison study	4	3
Clash and clearance	8	5
Assembly feasibility checks	5	3
CAD methodology	5	2
Filling of checklist	5	1
E-mailing the status update	3	1
PPT presentation	10	3
Total Time in Minutes	**60**	**30**

So, from the above results, 50% of efficiency can be improved by usage for automated macros at research and development centers and cost savings can be done. Based on general data and some assumptions we can conclude about the percentage of money saved will be huge.

- Total time saved = 30 Min.
- Time invested in checking the design = 60 min.
- Time saved in for one design engineer = 50% = 0.5 Hrs.
- Hours saved for one DE/month = 10 Times X 0.5 Hrs. = 5 hrs./ Month (considering at least 10 times quality check performed).
- Hours saved for all projects/month =500 DE X 5 = 2,500 hrs./ month (considering 500 design at R&D center).
- Cost savings per month = 2,500 × 10$ = 25,000 $ (considering per hrs. cost = 10$).
- Cost savings per year = 25,000 $ × 12 = 300,000 $.

6.6 CONCLUSION

With the presented work, the user can reduce non-productive time, and the user can automate various tasks in the design process. Hence a lot of time and cost involved in the design process is saved. Also, the quality of the design process is improved and maintained. Automotive companies can use customized automation for quality checks, automate repetitive tasks, automatically generate complex geometries, and accelerate design procedures, through which Cost and time optimization is possible. With the use of artificial intelligence (AI) and machine learning in the future with this tracker file, huge data can be studied, and Major problems and hurdles to achieving good CAD quality can be eliminated.

KEYWORDS

- **application programming interface**
- **automotive sector**
- **CAD completion percentage**
- **CAD quality**
- **cost optimization**
- **design time optimization**
- **process automation**
- **VBA macro**

REFERENCES

1. Bala, M. G., Deepak, B. B. V. L., Bijaya, K. K., & Bibhuti, B. B., (2019). CAD-based automatic clash analysis for robotic assembly. *International Journal of Mathematical, Engineering and Management Sciences, 4*(2), 432–441. ISSN: 2455-7749.
2. Carmen Gonzalez-Lluch, Pedro, C., Manuel, C., Jorge, D. C., & Raquel, P., (2017). A Survey on 3D CAD model quality assurance and testing tools. *Computer-Aided Design, 83*, 64–79. ISSN 0010-4485.
3. Yogesh, H. S., & Akshaykumar, K., (2016). Assembly of horizontal screw conveyor in CATIA V5 using VBA. *International Research Journal of Multidisciplinary Studies, 2*(1). ISSN (Online): 2454-8499.
4. Rajesh, P., & Manoj, P., (2016). Macros in Catia to read values from MS excel-a tool for automation. *International Journal of Innovative and Emerging Research in Engineering, 3*(1). ICSTSD.
5. Bei, C., Yan, C., Jiang, D., & Miaomiao, Z., (2016). Research on 3D parametric modeling technology of template parts based on CATIA. *Advances in Computer*

Science Research, 4th International Conference on Information Systems and Computing Technology (ISCT 2016) (Vol. 64).

6. Akshaykumar, V. K., & Nimbalkar, U. M., (2015). Automatic assembly modeling for product variants using parametric modeling concept. *International Journal of Engineering Research & Technology (IJERT), 4*(4). ISSN: 2278-0181.
7. Del Rio-Cidoncha, M. G., Martinez-Palacios, J., & Ortuno-Ortiz, F., (2007). Task automation for modeling solids with Catia V5. *Aircraft Engineering and Aerospace Technology, 79*(1), 53–59.

CHAPTER 7

OPTIMIZING AND FABRICATION OF 3D PRINTED PROSTHETIC FINGER

HARISH KUMAR BANGA, PUNEET KUMAR, AYUSH PUROHIT, and HAREESH KUMAR

Department of Mechanical Engineering,
Guru Nanak Dev Engineering College, Ludhiana, Punjab, India,
E-mail: drhkbanga@gmail.com (H. K. Banga)

ABSTRACT

The aim of the research is to develop a 3D printable body-fueled prosthetic finger for the grown-up in our nation that permits a parametric plan. The upside of the parametric structure is that it tends to be customized for each client, and each grown-up can be fitted with a prosthesis that almost looks like the size of his/her sound hand. A factual examination has been directed to comprehend which parameters are the best decision for parametric structure. The plan was done in Solidwork, and it was associated with an outer record that permits us to change and adjust the structure without requirements to open the CAD document. Acknowledgment of these gadgets is dependent upon the solace of the client, which relies intensely upon the size, weight, and in general tasteful of the gadget. As found in various applications, parametric displaying can be used to create clinical gadgets that are explicit to the patient's needs. Nonetheless, current 3D printed upper appendage prosthetics utilize uniform scaling to fit the prostheses to various us.

Optimization Methods for Engineering Problems. Dilbagh Panchal, Prasenjit Chatterjee, Mohit Tyagi, Ravi Pratap Singh (Eds.)

7.1 INTRODUCTION

Appendage misfortune because of removal is relied upon to arrive at about 3.6 million constantly 2050, which will have drastically expanded from the current 1.6 million of every 2005. Most of these removals are viewed as minor removals, as these people are losing just little extremities, for example, fingers or toes. Removal of the fingers in the upper appendages is a typical event and has huge ramifications on people generally speaking capacity, coordination, and personal satisfaction. Loss of these extremities can lessen useful capacity, bringing about troubles performing exercises of day-by-day living (ADL). The utilization of prostheses has been appeared to improve fruition of ADLs, notwithstanding improving psychosocial confidence, self-perception, interlimb coordination with the contralateral appendage and body evenness. In spite of this, earlier writing found that almost 70% of upper appendage prosthetic clients were unsatisfied with their prosthesis while finishing ADLs. Likewise, it has been demonstrated that about 52% of upper appendage amputees relinquish their prosthetic gadgets due to the utilitarian, tasteful or different impediments. Rather than the announced figures of gadget surrender, reasonable dismissal rates and non-utilization have been assessed to be significantly more prominent because of the absence of correspondence among facilities and prosthetic non-clients. To lessen the enormous level of gadget deserting, it is suggested that prosthetic gadget fitting happen promptly or as fast as conceivable after a careful removal, which may expand the acknowledgment pace of these gadgets. Conventional prosthesis creation is an extensive procedure that requires an ensured prosthetist to make numerous castings of the influenced appendage utilizing mortar, which can be both work and material concentrated. As customary creation strategies may not meet the rate at which prostheses must be made, the requirement for a quickened strategy for creation introduces itself. Present day propels in added substance fabricating (i.e., 3D Printing) have made it feasible for the clump creation of ease, altered upper-appendage 3D prostheses utilizing Fused Deposition Modeling (FDM), where the creation limit is constrained to the size, type, and the all-out number of 3D printers accessible. To diminish the time and error of attachment creation, 3D checking has been recently used to filter the influenced appendage to consider fast prototyping of clinical prostheses by delivering precise stereolithographic (STL) models, which are brought into PC supported plan (CAD) frameworks. Attachment creation utilizing

CAD techniques have been demonstrated to be solid when combined with computerized documents (e.g., STL's) and added substance producing (for example FDM) lessening the measure of time expected to create prosthetic attachments. Moreover, CAD frameworks have been demonstrated to be a reasonable option for the manufacture of useful 3D printable transitional prostheses with exceptionally modified attachments comparative with quiet explicit anthropometrics. Transitional prostheses are alluded to as "brief prosthesis" or "prompt postoperative prosthesis," and have been recently explored for maintenance and rebuilding of strong quality and scope of movement. In this way, the reason for the current investigation is twofold to portray the advancement of a transitional 3D printed prosthesis for fractional finger amputees.

7.2 RESEARCH BACKGROUND

While prosthetic gadgets have kept on progressing from the beginning of time, 3D printing innovation has risen as a progressive method to improve this clinical gadget by maintaining a strategic distance from the customary negative effects that are regularly connected with current prosthetics through the most touchy and extreme customization conceivable. Probably the most punctual recorded employments of prosthetic appendages are found in Ancient Egyptian history, in any case, it was not until 1,536 when French Army hairdresser and specialist Ambroise Paré designed the principal prostheses for both upper-and lower-furthest point amputees. While these unique prostheses were commonly made of overwhelming iron and wood, present day prostheses are currently a lot lighter, as they are regularly included plastic, aluminum, and other composite materials.

The plastic pieces of the prosthesis, of which remember the liners or cushioning found for the gadget, can be framed by utilizing customary plastic shaping techniques, for example, infusing shaping, which powers dissolved plastic into a shape that is then cooled and expelled, or through vacuum shaping strategies. So also, the arch of the prosthesis is framed by compelling the fluid metal, regardless of whether that be titanium or aluminum, into a steel bite the dust that is of the ideal shape. Prosthetist professionals at that point gather the full appendage, which is then fitted explicitly to every individual patient's needs. While current prosthetic gadgets furnish amputees with a more prominent personal satisfaction as they improve their capacity to live gainful lives, there are huge difficulties that are related to the utilization of prosthetics. With an end goal to address a portion of these harmful impacts that can be related with the utilization of prosthetic appendages, three-dimensional (3D) printing innovations have offered a progressive method to customize prostheses without trading off the soundness of the patient. 3D printing, a kind of added substance producing, has just discovered its fruitful application in a wide assortment of enterprises including engineering, nourishment, mechanics, air transportation, ramble innovation, apply autonomy, car, gadgets, medication, and some more.

The plastic pieces of the prosthesis, of which remember the liners or cushioning found for the gadget, can be framed by utilizing customary plastic shaping strategies, for example, infusing forming, which powers liquefied plastic into a shape that is then cooled and expelled, or through vacuum framing techniques. Correspondingly, the arch of the prosthesis is framed by compelling the fluid metal, regardless of whether that be titanium or aluminum, into a steel pass on that is of the ideal shape. Prosthetist specialists at that point amass the full appendage, which is then fitted explicitly to every individual patient's needs. While present day prosthetic gadgets give amputees a more noteworthy personal satisfaction as they improve their capacity to live beneficial lives, there are critical difficulties that are related to the utilization of prosthetics. With an end goal to address a portion of these harmful impacts that can be related with the utilization of prosthetic appendages, three-dimensional (3D) printing advances have offered a progressive method to customize prostheses without trading off the wellbeing of the patient. 3D printing, a sort of added substance producing, has just discovered its effective application in a wide assortment of ventures, including design, nourishment, mechanics, air transportation, ramble innovation, apply autonomy, car, gadgets, medication, and some more. 3D printing is a particular field of added substance fabricating. 3D printing spins around an automated contraption that controls the area of a stream that discharges any kind of condensed, yet quickly setting material, in a particular example. The 3D printing innovation is unfathomably flexible both in structure and material. For all intents and purposes, anything can be 3D printed, as long as the best possible document is produced, whatever can be envisioned can be 3D printed. This innovation is most normally utilized with plastic polymers; however, metal, cells, and sugars are different materials that the 3D printing innovation is as of now good with. The 2010s have just demonstrated to be a period for checked development in the accessibility, flexibility, and common sense of the

3D printing innovation. The rest of the decade will probably proceed or surpass the pattern of the main portion of the decade. At present, the most well-known material that is utilized in 3D printing, both in industry and in close to home use are melded testimony displaying (FDM) thermoplastics. FDM thermoplastics are a general class of numerous kinds of plastics, all of which, for the most part, are entirely tough, and are fundamentally the same as in solidarity to infusion formed adaptations of a similar material. Since these materials have fundamentally the same as qualities to their infusion shaped partners when 3D printed, these materials cause printing a feasible option in contrast to conventional assembling techniques (To emerge 2015). On the bleeding edge of basic 3D printing materials, is the FDM thermoplastic ULTEM 9085. This plastic, notwithstanding being solid and lightweight like the other FDM thermoplastics, has been appraised as fire resistant (Materialize, 2015). This material, on account of its fire-resistant attributes, can promote the adaptability of 3D printed materials to incorporate development and parts for lab testing (Materialize, 2015). These materials have a wide scope of utilizations, from home printing of little new parts to huge mechanical assembling pieces. At present, the cost of 3D printers is easing back the development and improvement of the innovation all in all. Generally accessible to the buyer is the MakerBot printer line. The organization offers three items, the small replicator costing $1,375, the standard replicator costing $2,899, and the replicator Z18 costing $6,499. These printers are additionally constrained in the size of the item that they can print [13]. The smaller than normal replicator can just print inside a space of 10×10×12.5 cm and the biggest of the monetarily accessible MakerBot printers is bound to a 30×30.5×45.7 cm volume [13]. These impediments in the industrially accessible items are as yet a significant detour for some intrigued by 3D printing. Yet, costs have fallen significantly since the commercialization of 3D printers from $20,000 to around $2,000 [11]. 3D printer costs may possibly drop significantly further to around $100 on account of the Peachy Printer, a result of a free creator [1]. 3D printing is a continually advancing field that is developing at an exponential rate. With new printable materials being grown quickly, and enthusiasm for the innovation develops, 3D printing is surprising the assembling scene. The biggest restricting element is the cost of the apparatus, and over the long haul, the costs are falling, and new techniques for 3D printing are being created, 3D printing might be the most helpful and generally utilized innovation since refrigeration.

7.3 RESEARCH METHODS

Throws and finger braces are regularly cumbersome, overwhelming, rancid, and awkward. In the light of the ongoing advancement of the Osteoid cast, 3D sweeps of the tissue over a messed up or harmed bone can be utilized to create increasingly agreeable and better fitting cast or brace. The item will be centered around finger braces as opposed to full broken arms or wrists. Finger braces are basically a metal board that has a froth pad between the metal and the finger. These are massive, awkward, and do not react well to being wet. A 3D examined and printed option can be intended to be increasingly streamlined, and agreeable just as being water verification. The support will be structured likewise to the Osteoid cast. From a 3D output of the influenced finger, a packaging of the finger will be produced. These sweeps will be produced utilizing the 3D sense economically accessible convenient scanner and the 123D find 3D filtering android application. Naturally formed breathing openings will be structured into the finger cast to permit the support to be breathable and dry quickly. The finger opening will likewise have an extended, stretched-out base sense of taste so as to keep the finger immobilized as a brace should. The support will be printed as a solitary piece, and will be custom fitted to the person. The methodology followed for customization of Finger splint is shown in Figure 7.1.

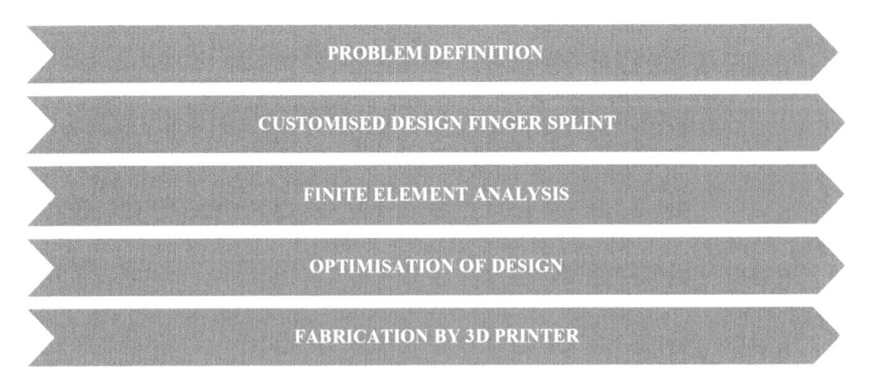

FIGURE 7.1 Methodology for customization of finger splint.

The material should be either clinical evaluation plastic or hypoallergenic material. The structure of the brace will not be without trouble. Creating an exact output of a little structure will be hard for two reasons. The examining innovation accessible experiences issues identifying little structures and the limits of the produced 3D structure some of the times are not fresh. Guaranteeing that the materials will likewise be agreeable and hypoallergenic is another significant worry in the plan of the brace. Finally, tying down the support to the patient may likewise end up being tricky. The least complex answer for fastening the brace to the hand is the utilization of clinical tape.

Clinical tape, in any case, can Figure 7.2: Design of finger splint thrown plan. This structure was the motivation for the regular finger support. A UMass HONORS 397A (2015) Project – "Undertakings in 3D Printing" with A. Schreyer – Page 4 be awkward. On the off chance that the support can be planned tight enough to stay on the finger, however free enough to permit blood to flow, the issue of fixing the brace to the finger will never again be an issue.

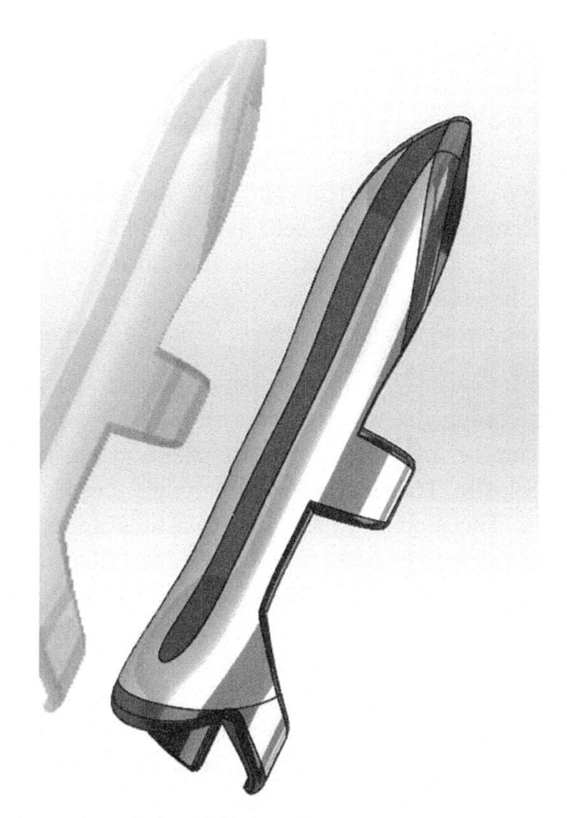

FIGURE 7.2 Customized design of finger splint.

7.4 FINITE ELEMENT ANALYSIS

Finite Element Analysis (FEA) can be used to demonstrate joint supplements to develop a predominant perception of their mechanical lead. Also, FEA gives numerical assessment to the streamlining of the install geometry. FEA involves three phases: preprocessing, course of action, and postprocessing.

- **Preprocessing Stage Involves:** Creation of the geometry, and its subsequent division into center points and parts. In doing accordingly, it is acknowledged that the material properties are uniform for the whole of the segments. Headway of conditions for each part. These conditions address the forces following up on the center points, segments' solidness, and the nodal dislodging. Nodal conditions for each segment are made in the structure:

  $$[\text{Stiffness, K}] \times [\text{Displacement, U}] = [\text{Force, F}]$$

  Get together of the parts together to address the entire geometry of the issue. The overall strength matrix is worked from the conditions made for each center. Use of starting conditions, limit conditions, and loads identifying with the geometry.
- **Plan Arrange Includes:** Settling the arithmetical conditions for all the center points at the same time in order to procure nodal results, like removing.
- **Postprocessing Stage Involves:** Procuring suitable information from the 'got' model.

ANSYS is equipped with a couple of rate-self-ruling, hyperplastic models that can be used to show the additions using the material properties of Silastic. These hyperplastic models join Neo-Hookean and Mooney-Rivlin. The models expect that Silastic's response is isotropic, isothermal, and adaptable. Also, it acknowledges that material is totally incompressible.

There are a couple of wellsprings of error that can add to wrong results during FEA. These consolidate using incorrect material properties or geometries during the preprocessing stage. The assurance of the part type during preprocessing is also truly irreplaceable for showing geometry adequately. Each part type has different degrees of chance and various stacking conditions for which it is suitable. Consequently, care must be taken to ensure the loads following up on the geometry can be managed by the picked segment. Moreover, the assurance of the right work size and segment extraordinarily influences the accuracy of the course of action. Gathering tests should be accomplished after each work refinement. Also, the utilization of exact and proportionate breaking point conditions and loads on the restricted segment model ensures that the model accurately addresses this current reality issue.

The limited component models were utilized to decide the pressure field at the embed, particularly in the stem-pivot interface. The bowing firmness for each point of revolution will be determined to decide the scope of movement of the embed. The lattice of the modified finger brace appears in Figure 7.3.

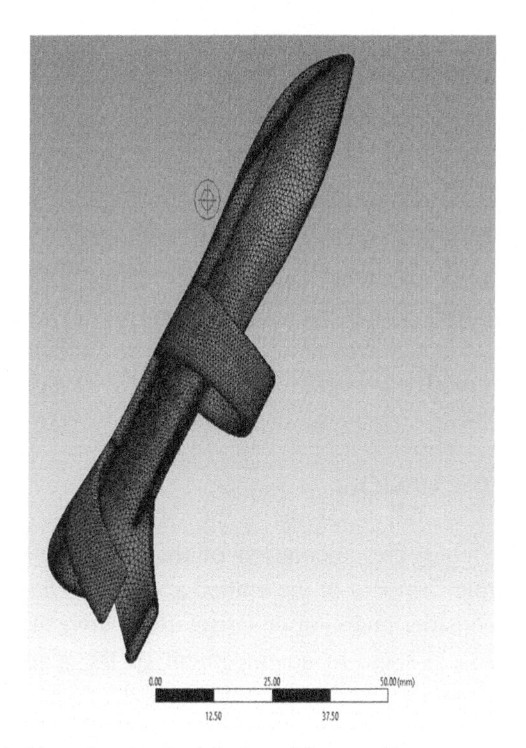

FIGURE 7.3 Meshing of customized design of finger splint.

The deformation of customized design of finger splint by using PLA Material in 3D printer is shown in Figure 7.4.

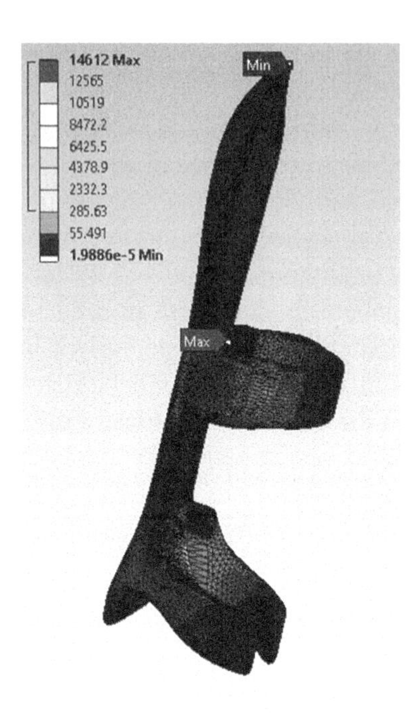

FIGURE 7.4 Deformation of customized design of finger splint.

During the analysis part, three material of 3D printer is identified, which is commonly used in the Prosthetic and Orthotics industry. The analysis values and graph of all material is shown in Table 7.1 and Figure 7.5.

7.5 RESULT DISCUSSION

From the result table, the calculation of the anxieties acting inside an embed and twisting firmness of the embed at different degrees of flexion. These qualities consider quantitative correlation between different embed structures and give a stage to advancement of the plans. The pressure conveyance got from the examination portrays the areas of high burdens, therefore, the structure of these locales can be improved to decrease the measure of stresses experienced. Bowing solidness is a proportion existing apart from everything else required to twist the embed during flexion. In this research, by using a different materials for a 3D printer in the customized design of Prosthetic finger, carbon fiber shows a better result as compared to PLA and nylon polyamide.

TABLE 7.1 Values of Carbon Fiber, Nylon Poly-Amide, and PLA Materials Used in 3D Printer

Graph Between Displacement v/s Stress			
Displacement in mm	**Stress in N/m^2**		
	Carbon Fiber	**Nylon 6**	**PLA**
0.5	3.6	0.957	0.1988
1	6.56	2.27	0.554
1.5	16.86	18.75	2.85
2	69.31	114.94	23.32
2.5	136.96	211.13	43.78
3	204.56	307.32	64.25
3.5	272.18	403.5	84.72
4	339.8	499.69	105.19
4.5	407.43	595.87	125.65
5	475.05	692.06	146.12

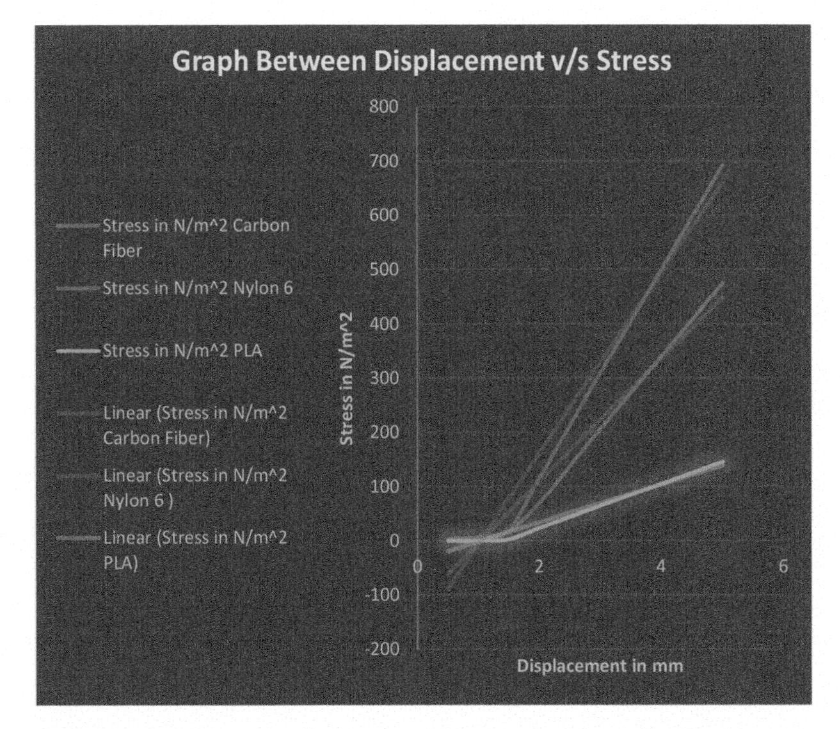

FIGURE 7.5 Graph between displacement v/s stress materials used in 3D printer.

7.6 CONCLUSION

This research work represented that the equivalent geometric models cannot be utilized to direct limited component investigation for hyper-elastic and flexible material models. The geometry of the limited component model must be altered to represent the adjustments in material properties. Accurate geometry was not utilized right now; the area for greatest pressure may change from the real world. Consequently, future examinations exploring the legitimacy of the direct material supposition for silicon elastomers ought to use precise embed geometries. Moreover, the impacts of adduction and snatching on the life span of the inserts must be examined.

KEYWORDS

- **additive manufacturing**
- **finite element analysis**
- **finite element modeling**
- **fused deposition modeling**
- **geometric models**
- **mass customization**
- **prosthetic finger**

REFERENCES

1. Allen, B., (2013). *3D Printer by Sask. Man gets Record Crowdsourced Cash.* CBC News, Web. 5 Feb. 2015.
2. ANSYS, (2008). *Materials Technology*. https://www.ansys.com/en-in/advantage-magazine/volume-xvi-issue-1-2022/increasing-additive-manufacturing-build-success-with-machine-learning. [Online] (accessed on 15 June 2022).
3. ANSYS Contact Technology Guide. (2022). *ANSYS Release 9.0.* [Online] https://www.ansys.com/academic/educators/education-resources/booklet-material-property-data-for-engineering-materials (accessed on 15 June 2022).

4. Biddis, E. A., Bogoch, E. R., & Meguid, S. A., (2004). Three-dimensional finite element analysis of prosthetic finger joint implants. *International Journal of Mechanics and Materials in Design, 1*(4), 317–328.
5. Banga, H. K., Kalra, P., Belokar, R. M., & Kumar, R., (2020). Customized design and additive manufacturing of kids' ankle foot orthosis. *Rapid Prototyping Journal, 26*(10). https://doi.org/10.1108/RPJ-07-2019-0194.
6. Banga, H. K., Kalra, P., Belokar, R. M., & Kumar, R., (2020). *Design and Fabrication of Prosthetic and Orthotic Product by 3D Printing* [Online First]. IntechOpen. doi: 10.5772/intechopen.94846. Available from: https://www.intechopen.com/online-first/design-and-fabrication-of-prosthetic-and-orthotic-product-by-3d-printing (accessed on 15 June 2022).
7. Banga, H. K., Kalra, P., Belokar, R. M., & Kumar, R., (2020). Effect of 3D-printed ankle foot orthosis during walking of foot deformities patients. In: Kumar, H., & Jain, P., (eds.), *Recent Advances in Mechanical Engineering. Lecture Notes in Mechanical Engineering* (pp 275–288). Springer, Singapore.
8. Banga, H. K., Kalra, P., Belokar, R. M., & Kumar, R., (2020). Role of finite element analysis in customized design of kid's orthotic product. In: Singh, S., Prakash, C., & Singh, R., (eds.), *Characterization, Testing, Measurement, and Metrology* (pp. 139–159). CRC Press Taylor & Francis, USA.
9. Banga, H. K., Kalra, P., Belokar, R. M., & Kumar, R., (2020). Improvement of human gait in foot deformities patients by 3D printed ankle–foot orthosis. In: Singh, S., Prakash, C., & Singh, R., (eds.), *3D Printing in Biomedical Engineering, Materials Horizons: From Nature to Nanomaterials*. Springer, Singapore.
10. Banga, H. K., Kalra, P., Belokar, R. M., & Kumar, R., (2018). Fabrication and stress analysis of ankle foot orthosis with additive manufacturing. *Rapid Prototyping Journal* (Vol. 24, No. 2, pp. 300–312). Emerald Publishing.
11. Bilton, N., (2013). *Disruptions: On the Fast Track to Routine 3-D Printing*. Bits. The New York Times.
12. Karasahin, D., (2013). *Osteoid Medical Cast, Attachable Bone Stimulator*. A'Design Award and competition.
13. MakerBot. (2015). *Compare MakerBot Replicator 3D Printers*. MakerBot.
14. Materialise, (2015). *FDM: Materials & Datasheets.* Materialise.
15. Tong, J., Jin, Z., Ligang, L., Zhigeng, P., & Hao, Y., (2012). *IEEE Transactions on Visualization and Computer Graphics, 18*, 643–650.
16. Rosenbaum, P., Paneth, N., Leviton, A., Goldstein, M., Bax, M., Damiano, D., & Jacobsson, B., (2007). A report: The definition and classification of cerebral palsy. *Development Children & Child Neurology, 49*(109), 8–14.
17. Uning, R., Abu, O. N. A., & Abdul, R. R. B., (2008). *3D Finite Element Analysis of Ankle-Foot Orthosis on Patients with Unilateral Foot Drop: A Preliminary Study, 1*, 160–172.
18. Chin, R., Hsiao-Wecksler, E. T., Loth, E., Kogler, G., Manwaring, S. D., Tyson, S. N., & Gilmer, J. N., (2009). A pneumatic power harvesting ankle-foot orthosis to prevent foot-drop. *Journal of Neuro Engineering and Rehabilitation, 6*(1), 1–11.

19. Choi, T. Y., Jin, S., & Lee, J. J., (2006). Implementation of a robot actuated by artificial pneumatic muscles. In: *SICE-ICASE International Joint Conference* (pp. 4733–4737). Busan, South Korea,.
20. Park, Y. I., Chen, B., Young, D., Stirling, L., Wood, R. J., Goldfiled, E., & Nagpal, R., (2011). Bio-inspired active soft orthotic device for ankle foot pathologies. In: *IEEE/RSJ International Conference on Intelligent Robots and Systems*. San Francisco, USA.
21. Stéphanie Dameron, Jane K. Lê, & Curtis Le Baron, (2015). Materializing Strategy and Strategizing Material: Why Matter Matters, *British Journal of Management, 26*, S1–S12, https://doi.org/10.1111/1467-8551.12084.

CHAPTER 8

COMPARATIVE STUDY OF TRUTH INFERENCES ALGORITHMS IN CROWDSOURCING

HIMANSHU SUYAL and AVTAR SINGH

Department of Computer Science and Engineering,
Dr. B. R. Ambedkar National Institute of Technology, Jalandhar,
Punjab, India, E-mail: suyal.himanshu@gmail.com (H. Suyal)

ABSTRACT

In the world, the popularity of artificial intelligence (AI) enlarged the researcher trying to incorporate human behavior into the machine. Due to this, it led to too many problems that machines cannot solve alone. So, the researcher thinks that machines and humans can act together, which led to the new field called crowdsourcing. Crowdsourcing is used to address problems that are very hard to solve by the machine independently and require human intelligence. However, the openness of crowdsourcing increased the requirements of the crowd (called workers), which led to creating the low quality of data and redundant data due to the availability of low-quality workers. To solve this problem, many redundancy-based algorithms can be used in which they assign each task to the worker to find the correctness of the answer called truth, and this fundamental problem is known as truth inference, which decides how effectively they infer the truth. In this chapter, we compare some of the existing truth inference algorithms and make a comparative study of the algorithms in real-time datasets.

Optimization Methods for Engineering Problems. Dilbagh Panchal, Prasenjit Chatterjee, Mohit Tyagi, Ravi Pratap Singh (Eds.)

8.1 INTRODUCTION

With the enhancement of artificial intelligence (AI) amplified lot of problems that came into the depiction. Some of the problems which cannot be solved by the machine self-sufficiently and require human intelligence like sentiment analysis [1], entity resolution [2], etc. So crowdsourcing solution is cast-off to address those machine learning problems which is hard for the machine to solve by incorporating human wisdom. Crowdsourcing solution is very effective and able to find a solution to many hard problems, which led to many public crowdsourcing platforms like Amazon Mechanical Turk [3], Crowd-Flower [4], and Microworkers [5], etc.

Due to the ingenuousness of the crowdsourcing, availability of the crowdsourced data became easy, and several crowdsource databases like CrowdDB [6], deco [6], quark [7], and Crowdforge [8] are easily available, which motivates the researcher to show great attention on these data to find out the efficiency of the data. Several crowdsourcing platforms motivate the crowd (workers) to solve the problem because of the replacement they got paid back. Knowingly or unknowingly, the crowd may yield low quality or even noisy and redundant data. It is a very tedious task to find the quality of the workers, so it becomes more momentous to control the quality of the crowd. To discourse this problem, most of the researchers used the fundamental strategy called redundancy based, in which each task can be given to multiple crowds and try to find the cumulative answer of the different workers. This well-known problem is known as truth inferences which study by the different crowdsourcing solutions [9–14]. One of the simple ways to aggregate the answers and find out the truth by finding the majority answer given by the user is called the MV (Majority voting) approach, which uses the simple approach by selecting the majority answer as truth. The biggest problem with this approach is the equality in which all the workers seem to be equal, but in reality, all the workers do not have the same quality. Some of the workers might be spam and can give faulty data; thus, it became important to find out the quality of the workers. To find out the quality of the workers' someone can give the qualification test to find out the trustworthy workers, workers who qualify for the test seems to be treated as quality workers. On the other end, workers who do not qualify for the test cannot be treated as trustworthy workers. This approach is simple but has two major problems; one of the major problems is it might be possible that works can qualify the test for the money, and after that, they may produce faulty data. The second problem with this approach is that workers do not want to solve the task without pay. The solution to this problem is to take the hidden test, in which we mix the task which ground truth is already known to check the quality of the workers, but this approach has some cons which needs to pay for the extra work which is the waste of money.

Considering these limitations, the database community [15–17] is very keen to study the problems independently [9, 19] and has shown great interest in this problem. This chapter compares several truth inference algorithms with the different diverse databases under the same platform. The rest of the chapter can be summarized in such a way that Section 8.2 will discuss about the Truth Inferences algorithms; Section 8.3 has the experiments study, and Section 8.4 contain the conclusion.

8.2 TRUTH INFERENCE ALGORITHMS

Since crowdsourcing is become prevalent and exposed the marketplace for the crowd, the quality of the crowd has become the most challenging task. The fundamental problem is known as truth inference, which decides how effectively they infer the truth. Truth inferences algorithms got a considerable amount of attention in the machine learning research field due to the direct effect on the prediction model. Truth inference refers to the process of valuing the true label of each task from the multiple truth label set.

Previous all work [9, 11, 13–15, 20, 21] can be summarized into three categories: direct calculation method [6, 23], Optimization method [13, 14, 17, 20], and probabilistic propagative methods [9–11, 15, 21], these methods discuss briefly below.

Table 8.1 summarize some of the truth inference algorithms with their technique and task type. It is noted that single-choice task is those tasks in which a worker need to select one choice out of several choices; for example, a worker needs to select the sentiment of the movie ('very good,' 'good,' 'bad,' 'worst') from the given movie review. On the other binary, a class task is a specific type of single-choice task in which the user has only two choices, either 'yes' or 'no.'

TABLE 8.1 Comparison of Different Methods

Method	Technique	Task Type
MV	Direct calculation method	Binary class, single choice
Mean	Direct calculation method	Numeric
Median	Direct calculation method	Numeric
Minmax	Optimization	Binary class, single choice
CATD	Optimization	Binary class, single choice, numeric
PM	Optimization	Binary class, single choice, numeric
GLAD	PPM	Binary class, single choice
ZC	PPM	Binary class, single choice
EM	PPM	Binary class, single choice

8.2.1 DIRECT CALCULATION METHOD

This method unswervingly estimates the truth. Majority voting (MV) is the simple approach under this category in which the majority answer seems to be ground truth. For the numeric glitches, mean and median methods are used by taking the mean and median of the workers answers, respectively, to find the ground truth.

8.2.2 OPTIMIZATION METHOD

The main objective of this method is to find out the relation between workers quality and the ground truth by regulation the self-define optimization function. These optimization functions can be used to imprisonment the relation between two parameters: first one is worker probability, PM [13] used this approach in which each worker quality can be treated as a single value and optimization function can be referred to the formula 1.

$$min_{\{Q^W\},\{V_i\}} f\left(\{Q^W\},\{V_i\}\right) = \sum_{W \in W^*} Q^W \cdot \sum_{T_i \in \tau} w\, d\left(V_i^w, V_i\right) \tag{1}$$

where; $\{Q^W\}$ represents the set of all workers quality; and $\{V_i\}$ signify the set of all truth; $d(V_i^w, V_i)$ is denoted the distance between the worker answer V_i^w; and the ground truth V_i where $d(V_i^w, V_i)$ should not be negative.

Lower the value of the $d(V_i^W, V_i)$ means worker answer is close to the ground truth, eventually we have to minimize the $f(\{Q^W\}, \{V_i\})$.

Second one is the worker confidence; CATD [17] used the both parameter worker probability and worker confidence. Worker confidence is a function which determines the confidence of the worker more the answer given by the worker more the confidence of the worker.

MiniMax [20] uses the idea of miniMax entropy [21]. Specifically, this approach is based on the probability distribution in which each task can T_i can be assumed that the answer given by W for the different task can be focus on the single label task with the c choices and generated by the probability distribution $\pi_i^W = [\pi_{i1}^W, \pi_{i,2}^W \ldots\ldots\ldots\ldots \pi_{i,c}^W]$. Where; $\pi_{i,j}^W$ is the probability of that worker W answer the task T_i with the choice of j. Considering this in mind, two oblige can be enforced for the task and workers. First impose is for the task T_i number of answers can be the sum of the corresponding probabilities; for a worker W, and second one that, for the worker W among all the task which is answered by W and its already given the truth is j^{th} choice number of answers collected for the kth choice is equal to the sum of the corresponding generated probability.

8.2.3 PROBABILISTIC PROPAGATIVE METHOD (PPM)

Probabilistic propagative model having the concrete base which is based on the very renowned approach Dawid and Skene's model. One of the most important mainstreams in the PPM is the Expectation-Maximization (EM) algorithm [11], which extensively used to estimate the latent variable. RY [22] is a binary estimation algorithm which use two parameters sensitivity and specificity where sensitivity parameter denotes the worker bias towards the positive and specificity towards negative. GLAD [10] introduces the two phenomena instance difficulty and expertise into their model. However, due to the spars of the data accuracy of the algorithm may be compromised, so the researcher introduces ZC [9], which uses only one parameter reliability of the worker with two elements {good, bad}. The main advantage of using one parameter is it ignores the problem of large estimation of the deviation due to the sparsity of the data. ZC is a slightly more simple method compared to EM. If nothing is known, it assumes the reliability is 0.5 based on the maximum entropy principle.

KOS [11] familiarizes the simple model to imprisonment the presence of the spammer, which is known as the spammer hammer model, in which it introduces the new parameter intention of the worker, which is used to distinguish spammer worker from the normal ones.

The primary focus of the PPM is to solve several machine learning problems such as sampling, EM algorithms, convex optimization, etc.

8.3 EXPERIMENTS

In this section, we first introduce the experimental setup, which contains the datasets and performance criteria and the next we discuss the comparative study of each algorithm on different datasets.

8.3.1 EXPERIMENT SETUP

We implement all experiments in Python 3.0 on a server with CPU 3.02 GHz and 6 GB RAM.

- **Data Sets:** There are lots of crowdsourced data available publicly [18] form which we take 5 datasets with different domain which are given below: Table 8.2 summarize the datasets with their description. It is noted that we used all the datasets which ground truth already available.
- **Performance Criteria:** We use two methods to check the performance of the above algorithms which is described below:
- **Accuracy:** It is the way to measure the fraction of task who inferred correctly. For given method if $V_i^{*\sim}$ is the inferred truth for the task T_i then accuracy can be defined as:

$$\text{Accuracy} = \sum_{i=1}^{i=n} \|_{(V_i^{\sim} = V_i^{*\sim})/n} \tag{2}$$

Where || (.) returns 1 if it is true; otherwise, it returns 0.

- **F1 Score:** F1 score is defined as the harmonic mean of the precision and recall.

$$\text{F1 Score} = 2 \Big/ \frac{1}{precision} + \frac{1}{recall} = 2 \times \frac{\sum_{i=1}^{i=n} \|_{\{V_i^{\sim} = T\}} \cdot \|_{\{V_i^{*\sim} = T\}}}{\sum_{i=1}^{i=n} \|_{\{V_i^{\sim} = T\}} + \|_{\{V_i^{*\sim} = T\}}} \tag{3}$$

For numeric datasets, we use two methods which are given below:

- **MAE:** Mean Absolute Error can be defined below:

$$\text{MAE} = \frac{\sum_{i=1}^{i=n} |V_i^{\sim} - V_i^{*\sim}|}{n} \quad (4)$$

- **RMSE:** Root Mean Square Error can be defined below:

$$\text{RMSE} = \sqrt{\frac{\sum_{i=1}^{i=n} (V_i^{\sim} - V_i^{*\sim})^2}{n}} \quad (5)$$

Accuracy and F1 Score having the value in the interval [0,1], having more value more the performance of the algorithm on the other end MAE and RMSE gives the error and having the value in the interval $[-\infty, \infty]$, having less the value more the performance.

TABLE 8.2 Datasets and Their Description

SL. No.	Database Name	Size (Question, Answer)	Questions with Ground Truth	Category	Description
1.	HIT spam detection	5380,42762	110	Binary class	The task is to find weather the HIT is spam or not.
2.	Duck identification	108,4212	108	Binary class	Task to find out whether the image contains duck or not.
3.	Face sentiment	584,5256	584	Multi-class	The task is to identify the sentiment on whether it is neutral(0), happy(1), sad(2) or angry(3) for a given face image.
4.	Weather sentiment	300,6000	300	Multi-class	The task is used to find out the sentiment of the tweet (positive, negative, neutral, and cannot say) for weather.
5.	Emotion	700,7000	700	Multi-class and numeric	The task is to give the emotion rate, having 7 emotions rate from –100 to 100

8.3.2 RESULT AND DISCUSSION

This section discusses the comparative study of the various algorithms on different datasets. For each evaluation criteria ↓ means the lowest the value having the better performance ↑ means the highest value has the better performance.

Table 8.3 shows the comparative study of the algorithms, as we know, MV is the simplest method but always not give poor results always. Expectation-Maximization gives the best results in most of the databases. If we compare ZC and GLAD, ZC gives promising results.

TABLE 8.3 Accuracy and F1 Score of the Datasets ↑

Method	HIT Spam Detection		Duck Identification		Face Sentiment	Weather Sentiment
	Accuracy	F1 Score	Accuracy	F1 Score	Accuracy	Accuracy
MV	0.867	0.7967	0.842	0.7857	0.6245	0.8766
Minmax	0.9745	0.9712	0.9605	0.96	0.7157	0.9666
GLAD	0.8834	0.8625	0.8703	0.8571	0.6678	0.8790
ZC	0.9554	0.9431	0.9259	0.9230	0.6917	0.8809
CATD	0.9338	0.9158	0.9074	0.9019	0.6952	0.8866
PM	0.8967	0.8735	0.88	0.88	0.6404	0.8671
EM	0.9767	0.9761	0.9629	0.9629	0.7123	0.88

Figure 8.1 shows a comparison of the various algorithms of truth inferences algorithms on the HIT spam detection data. Minmax gives the most accuracy for that dataset, and EM algorithm also gives a promising result. Due to the simplicity of the MV algorithm, the MV algorithm also gives good results.

Figure 8.2 shows a comparison of the various algorithms of truth inferences algorithms on the Duck identification data. For the duck identification dataset, Minmax gives the most accuracy in that dataset, EM algorithm also gives a promising result.

Figure 8.3 showing comparison of the various algorithms of truth inferences algorithms on the face sentiment data. Minmax and EM giving the best results, although ZC and CATD also giving the promising results.

Figure 8.4 showing comparison of the various algorithms of truth inferences algorithms on the weather sentiment analysis data. Minmax is giving the most accurate results. Apart from that, majority voting (MV) is also giving good results.

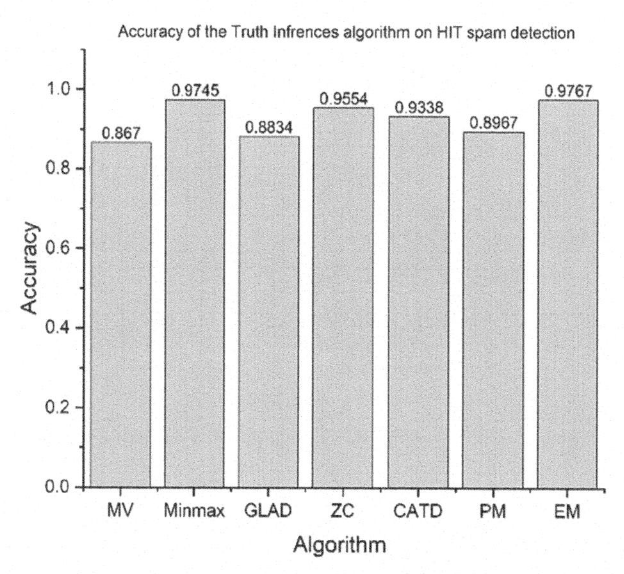

FIGURE 8.1 Accuracy of the truth inferences algorithm on HIT spam detection.

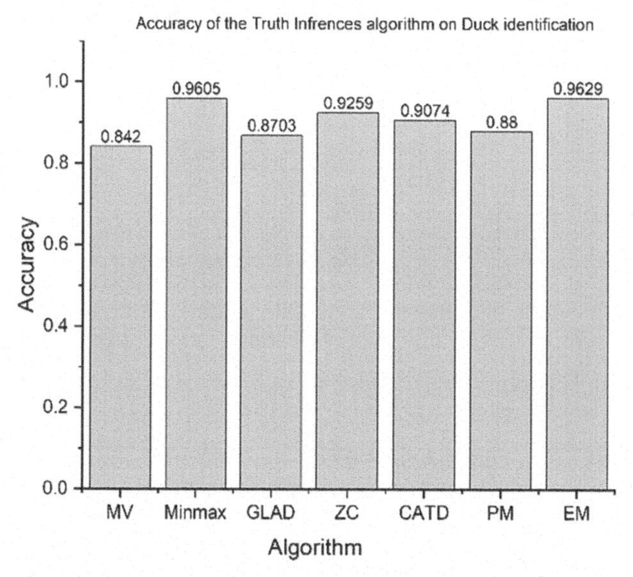

FIGURE 8.2 Accuracy of the truth inferences algorithm on duck identification.

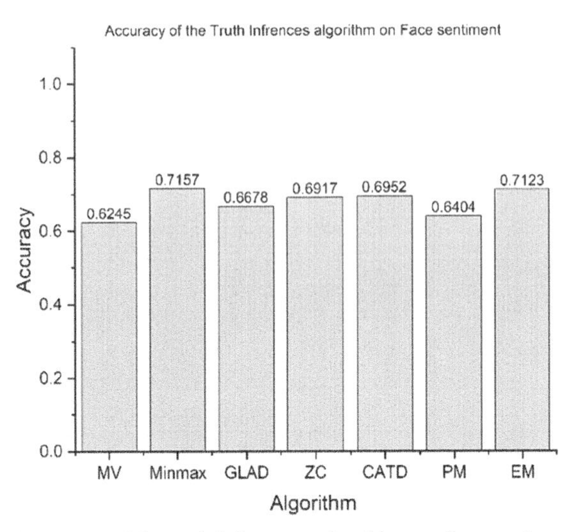

FIGURE 8.3 Accuracy of the truth inferences algorithm on face sentiment.

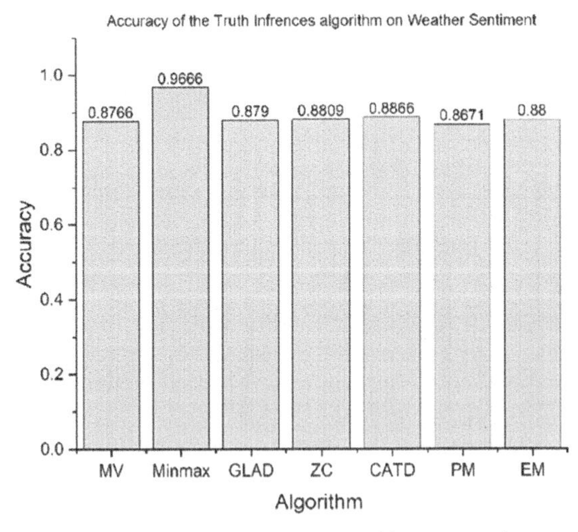

FIGURE 8.4 Accuracy of the truth inferences algorithm on weather sentiment.

Table 8.4 shows the comparative study for the numeric data set. For numeric data sets, CATD and PM are not giving good results because these methods are based on probability and might not well study the numeric datasets, so mean and median give the best results for the numeric datasets.

TABLE 8.4 MAE and RMSE for the Numeric Dataset ↓

Dataset	CATD		PM		Mean		Median	
	MAE	RMSE	MAE	RMSE	MAE	RMSE	MAE	RMSE
Emotion	16.36	25.94	13.91	21.96	12.02	17.84	13.53	21.26

Figure 8.5 shows the error rate comparison of the numerical data. As shown in the figure mean and median give the most promising results for the numeric dataset.

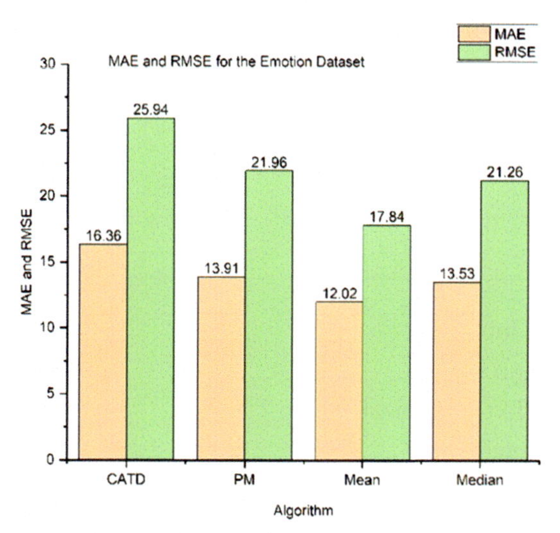

FIGURE 8.5 MAE and RMSE for the emotion data set.

8.4 CONCLUSION AND PROSPECT

As the popularity machine of machine learning increased it emerge the new paradigm of crowdsourcing, which is used to address those problems which hard for the machine to solve. Due to the development of the crowdsourcing system, which led to producing several crowd source data, so it became more important for the researcher to find out the correctness of the crowdsourced data and introduce the terminology truth inferences.

This chapter summarized several truth inference algorithms, and based on the analysis, we can conclude that for the binary class, one can use the simplest method MV if datasets have the answer by sufficient workers, and someone wants to get a little complex method by taking little overhead to get more promising results it can go for the PPM methods to use GLAD, ZC, and Minmax methods. For multi-class datasets, we study several algorithms, and MV is not giving poor results always; GLAD and ZC give good results and Minmax gives the most promising results. For numeric datasets, we strongly recommend the baseline method mean algorithm. In the past 10 years, a lot of research has busted out and came up with a lot of new methods still crowdsourcing having a young age and having more potential. Problems that are more worthy for the future are that most of the algorithms focus on binary label datasets, although very few algorithms claim that they are suitable for multi-class datasets. So, more study needs on the multi-label datasets and how to reduce the negative impact of sparsity generated by the multi-label datasets.

KEYWORDS

- **artificial intelligence**
- **crowdsourcing**
- **Expectation-Maximization**
- **human intelligence**
- **probabilistic propagative method**
- **truth inference**

REFERENCES

1. Liu, B., (2012). Sentiment analysis and opinion mining. *Synthesis Lectures on Human Language Technologies, 5*(1), 1–167.
2. Brizan, D. G., & Tansel, A. U., (2015). A. survey of entity resolution and record linkage methodologies. *Communications of the IIMA, 6*(3), 5.
3. Amazon mechanical Turk. https://www.mturk.com/ (accessed on 14 June 2022).
4. CrowdFlower. https://visit.crowdflower.com/ (accessed on 14 June 2022).
5. CrowdDb https://dbdb.io/db/crowddb (accessed on 14 June 2022).
6. Parameswaran, A. G., Park, H., Garcia-Molina, H., Polyzotis, N., & Widom, J., (2012). *Deco: Declarative Crowdsourcing* (pp. 1203–1212). In CIKM.

7. Marcus, A., Wu, E., Madden, S., & Miller, R. C., (2011). *Crowdsourced Databases: Query Processing with People* (pp. 211–214). In CIDR.
8. Kittur, A., Smus, B., Khamkar, S., & Kraut, R. E., (2011). CrowdForge: Crowdsourcing complex work. In: *Proceedings of the 24th Annual ACM Symposium on User Interface Software and Technology, ACM* (pp. 43–52).
9. Demartini, G., Difallah, D. E., & Cudr´e-Mauroux, P., (2012). *Zencrowd: Leveraging Probabilistic Reasoning and Crowdsourcing Techniques for Large-Scale Entity Linking* (pp. 469–478). In WWW.
10. Welinder, P., Branson, S., Perona, P., & Belongie, S. J., (2010). *The Multidimensional Wisdom of Crowds* (pp. 2424–2432). In NIPS.
11. Karger, D. R., Oh, S., & Shah, D., (2011). *Iterative Learning for Reliable Crowdsourcing Systems* (pp. 1953–1961). In NIPS.
12. Kim, H. C., & Ghahramani, Z., (2012). *Bayesian Classifier Combination* (pp. 619–627). In AISTATS.
13. Aydin, B. I., Yilmaz, Y. S., Li, Y., Li, Q., Gao, J., & Demirbas, M., (2014). *Crowdsourcing for Multiple-Choice Question Answering* (pp. 2946–2953). In AAAI.
14. Fan, J., Li, G., Ooi, B. C., Tan, K. L., & Feng, J., (2015). *iCrowd: An Adaptive Crowdsourcing Framework* (pp. 1015–1030). In SIGMOD.
15. Liu, X., Lu, M., Ooi, B. C., Shen, Y., Wu, S., & Zhang, M., (2012). CDAS: A crowdsourcing data analytics system. *PVLDB, 5*(10), 1040–1051.
16. Ma, F., Li, Y., Li, Q., Qiu, M., Gao, J., Zhi, S., Su, L., Zhao, B., Ji, H., & Han, J., (2015). *Faitcrowd: Fine-Grained Truth Discovery for Crowdsourced Data Aggregation* (pp. 745–754). In KDD.
17. Li, Q., Li, Y., Gao, J., Su, L., Zhao, B., Demirbas, M., Fan, W., & Han, J., (2014). A confidence-aware approach for truth discovery on long-tail data. *PVLDB, 8*(4), 425–436.
18. *Crowdsourcing Datasets*. http://dbgroup.cs.tsinghua.edu.cn/ligl/crowddata/ (accessed on 14 June 2022).
19. Whitehill, J., Wu, T. F., Bergsma, J., Movellan, J. R., & Ruvolo, P. L., (2009). *Whose Vote Should Count More: Optimal Integration of Labels from Labelers of Unknown Expertise* (pp. 2035–2043). In NIPS.
20. Zhou, D., Basu, S., Mao, Y., & Platt, J. C., (2012). *Learning from the Wisdom of Crowds by Minimax Entropy* (pp. 2195–2203). In NIPS.
21. Raykar, V. C., Yu, S., Zhao, L. H., Valadez, G. H., Florin, C., Bogoni, L., & Moy, L., (2010). Learning from crowds. *JMLR, 11*, 1297–1322.
22. Raykar, V. C., Yu, S., Zhao, L. H., Valadez, G. H., Florin, C., Bogoni, L., & Moy, L., (2010). Learning from crowds. *Journal of Machine Learning Research, 11*, 1297–1322.
23. Franklin, M. J., Kossmann, D., Kraska, T., Ramesh, S., & Xin, R., (2011). CrowdDB: Answering queries with crowdsourcing. In: *SIGMOD* (pp. 61–72).

CHAPTER 9

DESIGN OF THE VENTILATION SYSTEM OF CAR PARKS USING CFD SIMULATIONS

PAWANDEEP SINGH MATHAROO and GIAN BHUSHAN

Department of Mechanical Engineering,
National Institute of Technology, Kurukshetra–136119, Haryana, India, E-mail: pawandeep_31806203@nitkkr.ac.in (P. S. Matharoo)

ABSTRACT

Automobile parking either be completely enclosed, or it can be partially open. Completely closed parking is generally underground and needs a ventilation system. Partially open car park garages are often above decks that have an open side. Natural and Mechanical ventilation, along with their consolidated use, can provide an ajar car park. The aim of the present work is to examine the designed ventilation system of the car parks using CFD Simulations. The specific aim is to check the ventilation of the car parking garages for the predefined positions of Inlet and Outlets such that it meets requirements of CO exposure limit, which is set by health and safety regulations. It involves 3D modeling of the car park area, meshing into finite volumes, and carrying out the simulation using CFD tools to have a quick analysis of the CO concentration and velocity profile of the flow in the domain. The results emphasize that the designed ventilation system was able to extract CO to maintain its safe level in the car park.

Optimization Methods for Engineering Problems. Dilbagh Panchal, Prasenjit Chatterjee, Mohit Tyagi, Ravi Pratap Singh (Eds.)

9.1 INTRODUCTION

Two main concerns are presented by the automobiles in the car parking. The important one is the emission of CO (Carbon monoxide). The other emission is in the form of oil and gasoline fumes, which can cause various illnesses like nausea and headache, and constitutes a potential fire hazard. Also, NOx and haze smoke emission from the engines needs to be considered for safety reasons. However, it is enough to reduce the CO contaminant concentration within the acceptable limits, and it will automatically take care of the other contaminants and harmful emissions concentration to a safe limit.

There are various models like ASHRAE Standardized 62.1 [1], and also it has its predecessors that allow a flat exhaust rate of 1.5 CFM/ft2 in enclosed car parking and garages. Over the years, the emissions from vehicles have been reduced. A study sponsored by ASHRAE found the ventilation rate, which is required to control and reduce the level of contaminants in the enclosed parking garages [2]. It was found in that study that in some cases, much less than 1.5 CFM/ft2 of ventilation is satisfactory for the proper removal of the contaminants. However, ANSI/ASHRAE [3] and the Council of international code, and the International Mechanical Code® [4, 5] allowed 0.75 CFM/ft2 ventilation, whereas NFPA Standard 88A [6] recommends a minimum of 1.0 CFM/ft2, so it must be properly understood by the engineer about these specific codes and standards that applied in the design. It may be required by the engineer to request variation, or waiver, from the authorities which have the needed jurisdiction before implementing any changes like lesser ventilation system design. If the fans used are large in order to achieve code requirements, they may not necessarily increase overall power consumption, as if we use them in a properly designed CO level monitoring and ventilation system, the fans used to run only for a shorter time period to maintain acceptable CO level. Nowadays, it is also important to reduce energy consumption, and much focus is done on it as well; in order to achieve energy savings, several CO-based ventilation system controls are invented and used that provide enough cost savings during the operation of the parking garages. Some standards suggest providing ventilation of 3–12 ACH (air changes per hour).

9.1.1 VENTILATION REQUIREMENT AND DESIGNING

ASHRAE finds design ventilation rate required in parking depends on these four aspects:

- Admissible amount of contaminants in the parking area;
- Number of vehicles in operation at peak condition;
- Operating hours of vehicles in the warehouse; and
- Discharge ratio of a conventional vehicle in diverse conditions.

9.1.2 CO CONCENTRATION GRADE STANDARD

ACGIH [7] advices CO gas to restrict at 25 ppm for 8 h exposures, and U.S. EPA [8] determines that CO concentration of 35 ppm for 1 h is allowed. So, the important criteria for car garages are the carbon monoxide level which needs to be always below a certain safe limit value. Currently, the rate of ventilation required in car parks is such that it maintains safe level of CO at the average human height of 1.7 m which includes limits as (35 ppm) for 1 hour exposure, CO level of (65 ppm) if exposure is 30 minutes and if the exposure is only 15 minutes, then the maximum limit should be (120 ppm) only.

9.1.3 NUMBER OF VEHICLES IN OPERATION

Based on the type of facility which is served by the car parking, a number of vehicles which are in operation varies. For example, the parking areas of the apartments and shopping complexes have continued and a distributed usage; hence, the vehicles in operation is around 4% to 5% of the total capacity of the car park lot. For areas like airports, stadiums, and other heavy usage parking, the total vehicles in operation attains about 15–20% of the total capacity.

9.1.4 OPERATING TIME OF VEHICLES IN THE GARAGE

This time is the time for which the vehicle is switched on and being operated while inside the parking area. It depends upon various aspects like the size and the outline of the car park area. It also depends upon the total count of the vehicles which area going to enter or exit from the garage at a particular given time. Based on the various factors, it is found that the time for which vehicles operate generally ranges from 60–180 secs but in some cases, it can go as high as 600 secs also.

9.1.5 VEHICLE EMISSION RATES

The rate at which the emissions are exhaust from the vehicle depends upon the factors like whether it is entering the car parking or going out of the car park or is standing still and the engine is switched on. Since the vehicles travel at low speed only in the car parking so they are most in low gear and hence consume rich fuels and hence emission is also higher than normal operation. If the vehicle is entering in the car park lot, then it is most likely that the engine is hot, and in case the vehicle is moving out of the car park then it is cold start and hence the engine is cold. So, the emissions from the vehicles are different in both cases, and it is important to consider it for the determination of the ventilation rates.

In this chapter, we are designing and analyzing the car park for normal/ pollution ventilation mode or CO simulation only.

9.1.6 THEORY OF CFD (COMPUTATIONAL–FLUID–DYNAMICS)

It uses the technique of numeric method and data structures in order to analyze and solve the problem which involve fluid flows. So, with the help of a computer, we can predict the fluid flow based on the various governing laws which are used to design the CFD software.

Various critical aspects like airflow, heat transfer rate, transport of the contaminants, temperature variation with time, etc., can be easily predicted with the help of Computational-Fluid-Dynamics technique. Different basic physical equations are used to build a CFD model which is used for the analysis. It involves energy equation; fluid flow equations and other assumptions are made for the simplification. The derived equations are in the differential form which governs the fluid flow. It can be used to achieve steady state or transient solutions based on the requirement.

The important uses of CFD include various applications that were difficult earlier. Complex models involving intricate geometries can now be easily analyzed that too along with complex flows over it and just using the basic physics and mechanical laws in CFD tools. Once developed and evolved, the technology has been used in almost every engineering field that involves automobiles, aerospace, turbo machines, nuclear power engineering, hydropower engineering, medical field, civil engineering, fire hazard mechanics and a lot more.

However, while designing a particular CFD software, we need to make several assumptions, and also different approximations are to be included in order to simply the tool and the storage and cost requirement also. This depends upon the application of the software and varies for different engineering areas. Apart from these assumptions we also do compromise on the number of iterations or run-times that we do in CFD in order to get the result faster as it can be very time-consuming based on the complexity of the problem. All these factors can lead to deviation in the simulation and the results obtained and it can affect the reliability of the results. So, it is very important that all these aspects are considered while the designing of the CFD tool.

9.1.7 KEY BENEFITS OF CFD

Complex geometries involving complex flows can be easily solved and represented by it. Also, it is difficult to obtain the results by experiments because they can be time-consuming, costly, and sometimes fire and smoke-related experiments can be dangerous as well.

9.2 METHODOLOGY

9.2.1 PROCESSES INVOLVED IN CFD SIMULATION OF A SIMPLE CAR PARK

Several steps are taken before and during the application of the CFD software used for the analysis of the designed car park that includes both flow and thermal analysis. These are as follows:

- Defining the geometry and domain of the car park area using the 2D layout drawings of the car park and the proposed ventilation system.
- Selecting physical sub-models (Laminar or Turbulence models) that varies with the simulation method used. Here in our simulation, we are using standard (k-ε) turbulence model, which comes under the category of Reynolds-averaged Navier-Stokes equations.
- Boundary conditions are specified for the domain involved. This involves specifying the walls, exhaust fans, jet fans, ramp inlets, openings, supply fans.

- The mathematical equations are discretized; it means a mesh model is created in order to divide the domain into small finite volumes and then the numerical sub-model is selected based on the application.
- The iterative solution process is continuously monitored.
- The solution obtained is then analyzed.
- Highlighting the different uncertainties that may have occurred at any of the above steps and they are noted for the correction.
- Visualizing the obtained solution.

9.2.2 GOVERNING EQUATIONS

The governing equations are basically in the form of differential equations. These equations are the mathematical form of the laws that govern the phenomenon of heat transfer, flow of fluid and other processes that are related to the numerical simulation in computational dynamics.

A certain conservation principle is expressed by each individual differential equation that is being used. A physical quantity is employed by every equation as its dependent variable and a balance between the different factors that affect the variable is ensured by it. Generally, a specific property is used as the dependent variable in these governing equations.

Ruling computations are of two sorts: Compressible and In-compressible. For creating a mathematical model of fluid zone and solving the 3-D Navier-Stokes equations numerically over a flow field which is discretized, we will be using in-compressible Navier stokes equations; which includes species transport equations, energy equation, momentum equation, and conservation of mass.

9.2.3 MASS CONSERVATION LAW

The average mass conservation of the same mass means that the rate at which mass is stored in a given fixed amount of control volume, due to variable density, is proportional to the rate at which mass arrives. In the case of a continuous flow condition, the mass conservation principle gives a mass equation which means that the flow inside must flow continuously. The sum is described as below:

$$\frac{\partial \rho}{\partial t} + \nabla \cdot \rho u = 0$$

In this equation, the change in the density with respect to time is described by the first term and the convection of mass by the second term. The vector describing the velocities in the x, y, and z directions is shown as 'u.'

9.2.4 SPECIES CONSERVATION

Let Y denote the mass fraction of a chemical species. By this vector u, the conservation of mass fraction 'Y' of a test item 'i' is proposed by:

$$\frac{\partial}{\partial t}(\rho Y_i) + \nabla \cdot \rho Y_i u = \nabla \cdot \rho D_i \nabla Y_i + \dot{m}_i'''$$

The terms of the above equation are explained as–the mass of the contaminant species which is changing with time is represented by the former term of the left-hand side equation and due to convection, the species inflow and outflow from the control volume is represented by the latter term of left side. As a result of dissemination, the species move in and out of the control volume, which is described by the former term of the right side, and the latter term describes the rate at which the particular species operate inside the control volume due to chemical changes.

9.2.5 MOMENTUM CONSERVATION

The differential formula for the governing of momentum conservation for a given direction of a Newtonian fluid could be described along similar lines. Based on Newton's second law of motion, we can write the numerical formula for the momentum conservation, according to which the rate of pressure of a substance is proportional to the amount of energy acting on it. The sum is written in the form as:

$$\rho\left(\frac{\partial}{\partial t} + (u \cdot \nabla)u\right) = -\nabla p + \nabla \cdot \tau + \rho g + f$$

In the above equation left hand side symbolizes the increase of the forces of superior forces and inertia, while the left side contains the forces that are acting on it. There are several types of forces that are included in it such as force due to pressure p, gravity force, the external state of the f (symbolizing the drag associated with the spray droplets inside the domain) and the rate at which the viscous pressure is reduced which is holding the fluid inside the domain or the control volume. The most important force among these is the gravity force as it describes the impact of the flowing force.

9.2.6 *ENERGY CONSERVATION*

It is based on the first law of thermodynamics which states 'Heat and work are two forms of energy and they are always conserved as per the conservation of energy principle according to which energy can neither be created nor be destroyed but can only be transformed from one form to another. The governing equation based on this principle is as follows:

$$\left(\frac{\partial}{\partial t}(\rho E)+\nabla\cdot\left(\vec{v}\left(\rho E+p\right)\right)\right)=\nabla\cdot\left(k_{eff}\nabla T-\sum_{j}h_{j}\vec{J}_{j}+\left(\bar{\bar{\tau}}_{eff}\cdot\vec{v}\right)\right)+S_{h}$$

In the above equation, k_{eff} is called active thermal efficiency and J_j is defined as the occurrence of species variation 'j.' The energy transfer occurring in the control volume as a result of conduction, species introduction and viscous dissolution is described by the first 3 terms of the right-hand side equation. The temperature of the chemical reaction and other volumetric heating sources is represented by the term S_h. The target is over there:

$$E=h-\frac{p}{\rho}+\frac{v^{2}}{2}$$

where; 'h' is the sensible enthalpy defined for an ideal gas as:

$$h=\sum_{j}Y_{j}h_{j}$$

And for in-compressible flow as:

$$h=\sum_{j}Y_{j}h_{j}+\frac{p}{\rho}$$

and mass fraction of species *j* is represented by Y_j and,

$$h_j = \int_{T_{ref}}^{T} C_{pj}\, dT$$

9.2.7 TURBULENCE MODELING

Turbulence is defined as a fluid flow which is which is characterized as unsteady flow, completely irregular with respect to space and time co-ordinates, a 3-Dimensional flow, rotational, dissipative flow which dissipates the energy of the flow and also it is diffusive in nature at such high Reynolds number values. There is a very small magnitude of fluctuations in the velocity, temperature, and the pressure which are caused as a result of the divergence in the turbulent flows. The study and use of turbulence models is important as it is used to determine the equations that are further used to find the time-averaged temperature, flow velocity and the pressure fields easily and hence there is no need to completely define the turbulent flow pattern. This is achieved by using Reynolds Averaged Navier Stokes Equations (RANS).

In this simulation we are using the standard k-ε Turbulence model which is one of type of RANS models as RANS requires minimal hardware, time required for the iterations as well as human effort.

Turbulent kinetic energy is denoted by *k*. It is the measure used to determine the kinetic energy contained in the fluctuations.

ε denotes turbulent kinetic energy dissipation. It is a measure of the rate at which kinetic energy is dissipated. It determines the scale of turbulence.

There are many unknowns and immeasurable terms in actual k-ε equations. Thus, we use the standard k-ε model for a much more practical aspect. Thus, the unknown terms are minimized and a set of equations is obtained which can be used in numerous turbulent applications.

For the equation of turbulent kinetic energy (k):

$$\frac{\partial(pk)}{\partial t} + \frac{\partial(pku_i)}{\partial x_i} = \frac{\partial}{\partial x_j}\left[\frac{\mu_t}{\sigma_k}\frac{\partial k}{\partial x_j}\right] + 2\mu_t E_{ij} E_{ij} - \rho\varepsilon$$

For the dissipation of kinetic energy (ε):

$$\frac{\partial(\rho\varepsilon)}{\partial t} + \frac{\partial(\rho\varepsilon u_i)}{\partial x_i} = \frac{\partial}{\partial x_j}\left[\frac{\mu_t}{\sigma_\varepsilon}\frac{\partial \varepsilon}{\partial x_j}\right] + C_{1\varepsilon}\frac{\varepsilon}{k}2\mu_t E_{ij} E_{ij} - C_{2\varepsilon}\rho\frac{\varepsilon^2}{k}$$

The above equation is described as:

Rate of change of (k or ε) in time + Transport of (k or ε)
by advection = Transport of (k or ε) by diffusion +
Rate of production of (k or ε) – Rate of destruction of (k or ε)

where; u_i is used to define velocity component in the co-ordinates directions; E_{ij} is used for defining the component of deformation rate; μ_t is used for eddy viscosity.

$$\mu_t = \rho C_\mu \frac{k^2}{\varepsilon}$$

Some adjustable constants like σ_k, σ_ε, $C_{1\varepsilon}$ and $C_{2\varepsilon}$ are also used in the above equations. So, a wide range of turbulent flows have been analyzed and after several numerical iteration of data fitting, the values of the constants mentioned above is found as follows:

$C_\mu = 0.09$, $\sigma_k = 1.00$
$\sigma_\varepsilon = 1.30$, $C_{1\varepsilon} = 1.44$
$C_{2\varepsilon} = 1.92$

9.2.8 FEATURES OF k-ε MODEL

- It has good convergence and less memory requirements;
- Can be used for both compressible and in-compressible flows, external flow interactions with complex geometry.

9.2.9 CFD SIMULATION SETUP AND BOUNDARY CONDITIONS

As per the details in the form of 2D layout drawings and information of ventilation system; the 3D geometrical model for Car park Project was prepared.

9.2.10 POLLUTION MODE/NORMAL VENTILATION MODE CO SIMULATION MODELING

Few simplifying assumptions are used in the creating a mathematic model of given car park for the simulation analysis and in order to meet the safety requirement proper care is taken with the assumptions used in the design. The CO emissions generated within the car park area are calculated as per ASHRAE standards and the calculated CO emission value was distributed in the whole car park area.

The car park analyzed is having an area of 811.24 m^2 and the clearance height is 4.26 m. Per vehicle CO emitted is 186.165 mg/sec. A total number of cars in the concerned geometry is 22. Thus, total CO emitted is 204.781 mg/sec (assuming that 5% of total vehicles shall be in active condition and same was considered in the simulations as per ASHRAE standard). To extract CO from the car park, we are using two IV Smart EC induction fans with each fan having a flow rate of 1,322 CFM (thrust 12N).

The 3D model of the Car Park area is shown below depicting the domain in which simulation is done, and it also shows the exhaust and supply air openings (Figure 9.1).

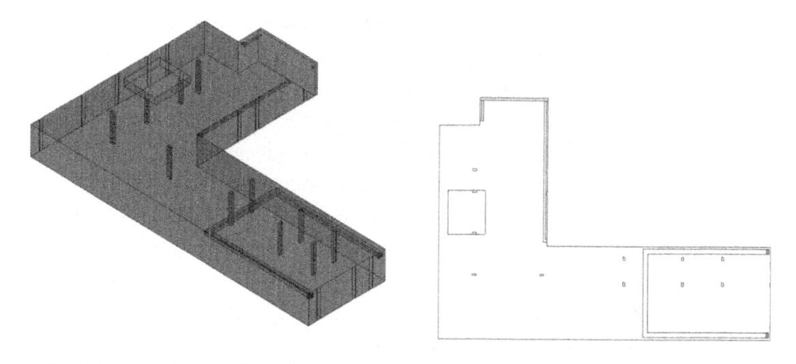

The 3-D Model of the Car Park area is shown below depicting the domain in which simulation is done and it also shows the exhaust and supply air openings.

FIGURE 9.1 Position of exhaust air fan openings and ramp opening.

As shown in the images, the *green object* represents the ramp opening (fresh air supply-natural supply) and the *red object* represents the exhaust air openings. Fresh air supply is natural supply and exhaust air supply is mechanical supply with flow rate of 3,300 CFM.

➢ **Exhaust Air Fan Specifications:**

Exhaust Air Fan Openings	Total Volume Flow Rate in CFM
Exhaust air fan openings	3,300 CFM

- **Induction Fan Specifications:**

Type	Number of Fans in Final Configuration	Flow Rate of Each Fan in CFM (Thrust in Newtons)
IV Smart EC	02	1,322 CFM (12N)

- **Meshing:** This was done using the same open-source software Salome in which 3-D model was prepared and tetrahedral and hexagonal mesh with mesh sizes of 0.05 m to 0.3 m were created and total mesh cells of 1.0 million were obtained (Figure 9.2).

FIGURE 9.2 Mesh model prepared in Salome.

9.3 CO SIMULATION RESULTS

CO simulation results without using induction fans in the car park area are shown in Figures 9.3 and 9.4.

We check CO concentration at 1.7 m from the ground as we take 1.7 m as an average human head height and thus, we need to make sure that at this height, the CO concentration is within the safe acceptable limit. It can be seen clearly from the above image of the contour plot that the maximum CO concentration is 693.1 ppm which is much more than the safe acceptable limit, i.e., 120 ppm for 15 minutes exposure.

Thus, in order to make the car park area safe for breathing, we need to design the ventilation system by placing the induction fans at appropriate places in order to induce flow in the stagnant areas having high CO concentration. Thus, we have used 2 induction fans and carried out CFD simulation again with the fans. Following are the results obtained with jet fans.

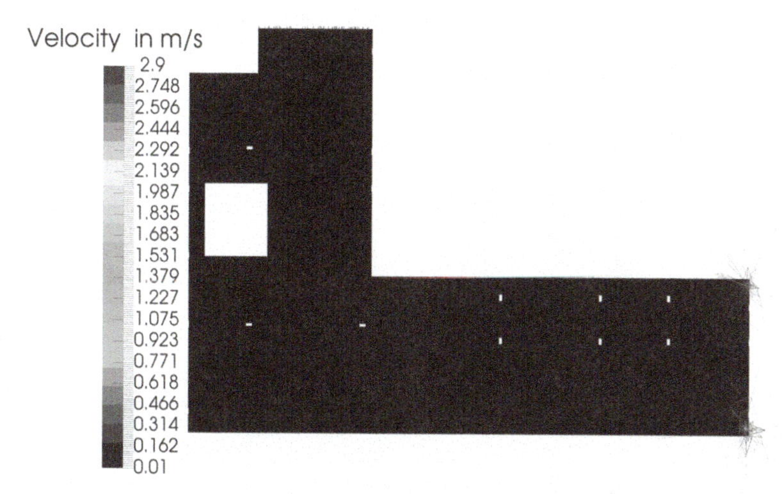

FIGURE 9.3 Vector plot of velocity magnitude at jet fan height from ground.

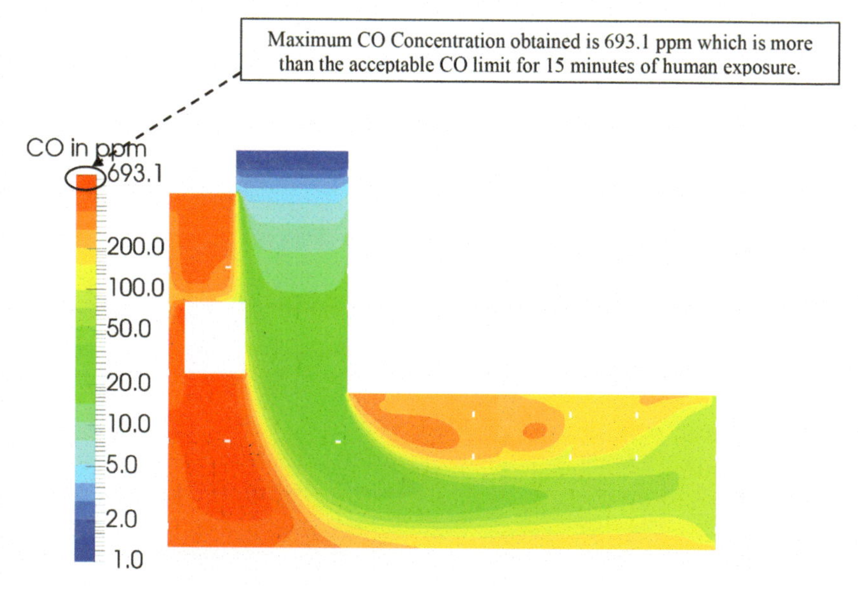

FIGURE 9.4 Contour plot of mass fraction of CO at 1.7 m from ground.

9.3.1 CO SIMULATION RESULTS WITH INDUCTION FANS IN THE CAR PARK AREA

Thus, from the above plot, we can see that by using the two jet fans in the car park at the locations mentioned, the maximum CO concentration has been reduced successfully, and hence design is safe for the human exposure (Figures 9.5–9.7).

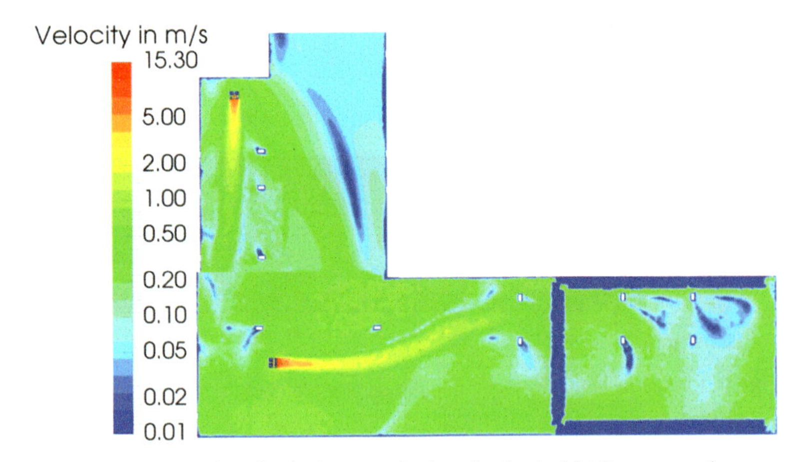

FIGURE 9.5 Vector plot of velocity magnitude at jet fan height from ground.

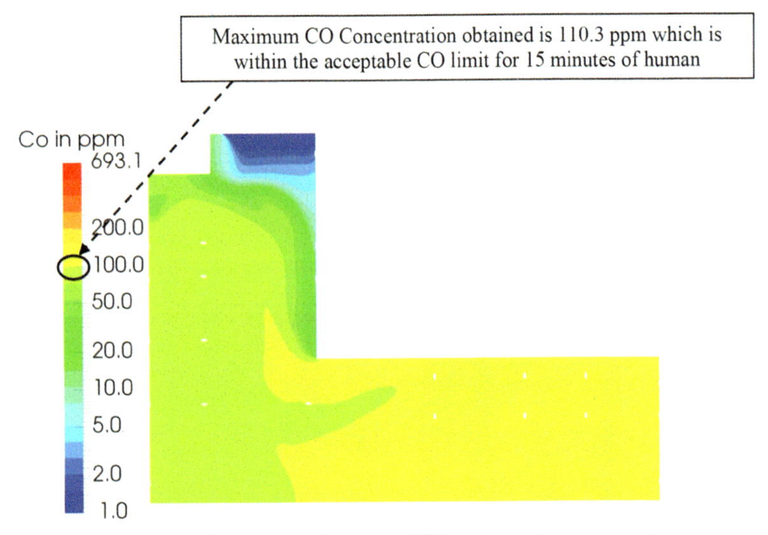

FIGURE 9.6 Contour plot of mass fraction of CO at 1.7 m from ground.

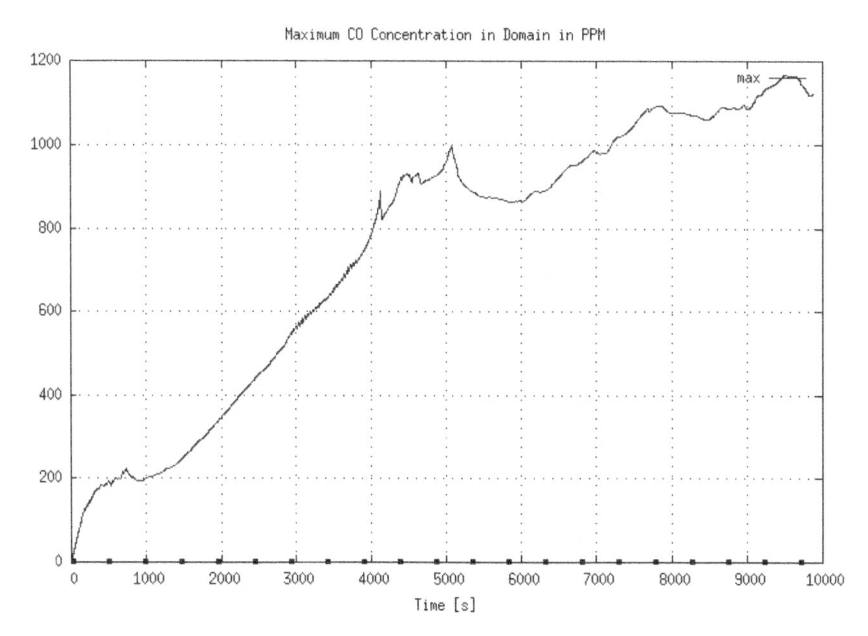

FIGURE 9.7 This graph talks about the CO concentration variation in the entire domain of the car park.

The CO variation in ppm is shown over the iterations which are performed during the simulation in over 10,000 iterations. The high CO concentration is shown by this curve as it depicts about the variation of CO in the entire domain of the car park at any point or location in it. But we are mainly concerned for the human height which is on average 1.7 m. Thus, this curve is not of enough use (Figure 9.8).

The residuals are shown in the above graph are the error terms which are produced after each iteration and in order to have stable solution the fluctuation in the graph must be minimum as the iterations are performed and we can see that the graph is almost stable for each residual term.

9.4 CONCLUSION

From the CFD simulation of induction fan ventilation system, we can draw the following conclusions for the undertaken car park area:

- Ventilation System in the car park area has been designed to reduce the CO level within the acceptable limits. This has been done with the help of CFD tools.
- All assumptions have been taken according to Industrial Standards.
- The CO concentrations from the CFD simulations, without using induction fans has been obtained and it is found that the maximum CO concentration in the car park area at average human height (1.7 m) is 693.1 ppm (831.72 mg/m^3), which is more than the acceptable CO limits for 15 minutes of human exposure.
- After conducting CFD analysis with placing the induction fans in the car park area, maximum CO concentration obtained is 110.3 ppm (132.36 mg/m^3) which is within the acceptable CO limits for 15 minutes of human exposure.

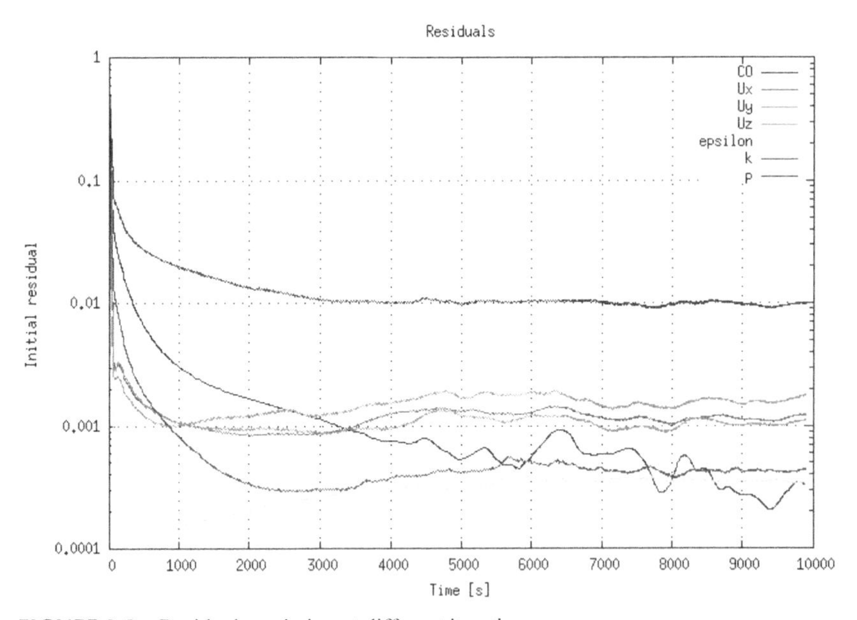

FIGURE 9.8 Residuals variation at different iterations.

Thus, we found that based on the Induction ventilation system we have used with two induction fans, their position, direction, and the specifications are appropriate and it is able to extract the CO contaminant from the car park area to maintain it safe level for breathing in that area (Figure 9.9).

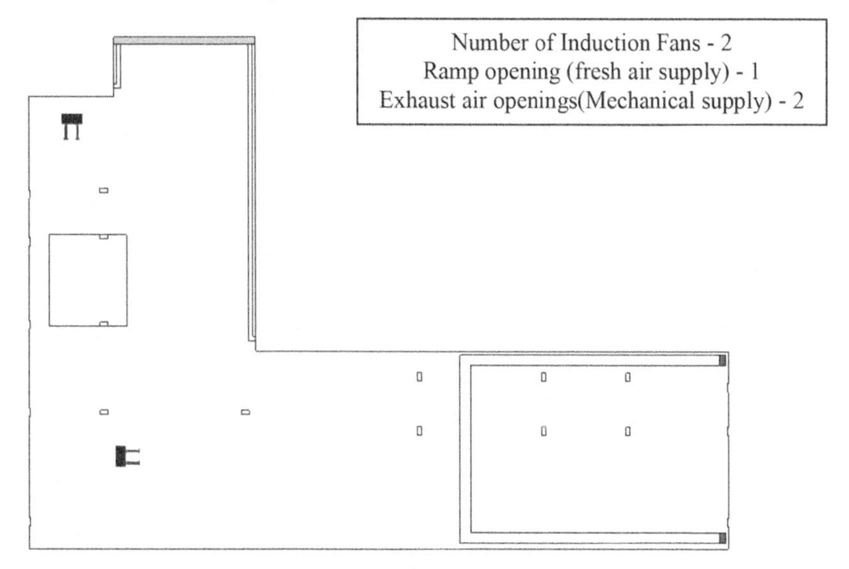

FIGURE 9.9 Final position, direction, and number of the induction fans and openings (2D view).

KEYWORDS

- **automobiles**
- **carbon dioxide**
- **carbon monoxide**
- **computational-fluid-dynamics**
- **thermodynamics**
- **ventilation system**

REFERENCES

1. ASHRAE, (2004). *Ventilation for Acceptable Indoor Air Quality.* ASHRAE Standard 62.1-2004.
2. Krarti, M., & Ayari, A., (1998). Overview of existing regulations for ventilation requirements of enclosed vehicular parking facilities (RP-945). *ASHRAE Transactions, 105*(2), 18–26.

3. ANSI/ASHRAE, (1999). *Laboratory Methods of Testing Fans for Rating*. Standard 51-1999 (AMCA Standard 210-99).
4. ICC, (2009a). *International Mechanical Code®*. International Code Council, Country Club Hills, IL.
5. ICC, (2009d). *International Fuel Gas Code®*. International Code Council, Country Club Hills, IL.
6. NFPA, (2011). *Standard for Parking Structures*. Standard 88A. National Fire Protection Association, Quincy, MA.
7. ACGIH, (1998). *Industrial Ventilation: A Manual of Recommended Practice* (23rd edn.). Appendix A. American conference of governmental industrial hygienists, Cincinnati, OH.
8. EPA, (2000). *Air Quality Criteria for Carbon Monoxide*. EPA/600/P-99/001F.U.S. Environmental Protection Agency, Research Triangle Park, NC.

CHAPTER 10

COMPARATIVE STUDY OF SINGLE-, DOUBLE-, AND FOUR-INLET CONDITIONS IN A ROCKET NOZZLE IMPLYING CFD

SAIF AHMAD, SAMEER MISHRA, VYASMUNI PRAJAPATI, SUMIT SINGH, and SANDEEP CHHABRA

Department of Mechanical Engineering, KIET Group of Institutions, Ghaziabad, Uttar Pradesh, India,
E-mail: saifahmad1602@gmail.com (S. Ahmad)

ABSTRACT

Rocket nozzle is a mechanical device designed in such a way as to control the flow rate, direction, velocity, and pressure of the exhaust. In this research paper, Bell nozzle is considered, and the inlet condition to the combustion chamber is varied keeping the design parameters to be the same in each case. Single, double, and four inlets in the combustion chamber is analyzed to estimate the best results in these three cases, and ultimately, a comparative study is done to predict that the single inlet is a better inlet condition for an efficient flow. Emphasis is given on computational fluid dynamics to solve the problem and ANSYS fluent software package is used for the purpose.

Optimization Methods for Engineering Problems. Dilbagh Panchal, Prasenjit Chatterjee, Mohit Tyagi, Ravi Pratap Singh (Eds.)

10.1 INTRODUCTION

The propulsion system in a rocket is the main component that provides a sufficient thrust to beat the gravitational influence of the planet. Since, the starting of the rocket ages or the space era, various developments have been done which has marked the importance of the rockets and its systems. The rocket nozzle plays a very vital role in the working of a whole propulsion system. Earlier conical nozzles were used, which had a very straight geometry and did not provide the required efficiency but, in this chapter, the most used and tested bell nozzle geometry is considered. The inlet in a nozzle through the combustion chamber can vary and it could be a deciding factor for the selection of better nozzle geometry [1]. In this chapter, the single-, double- [4, 5], and four-inlet conditions are tested computationally by the use of computational fluid dynamics to make a comparison between the three. The important factor is that the geometry of the nozzle is the same in all three cases except the variation in the number of the inlets.

The selection of a suitable geometry depends on various factors like the pressure, the velocity of the fluid flow, temperature, and the Mach number at convergent, divergent, and the throat section [1–7].

10.2 DESIGN AND GEOMETRY OF THE NOZZLE

The geometry and design are kept the same for all the inlet conditions and just the number of inlet value is varied to provide stable geometrical conditions so that whatever the result comes out can only be a consequence of the inlet variation and not the geometry. The use of similar geometrical parameters for the single, double, and four inlet eradicates any error occurring due to the geometry. Table 10.1 shows the dimensional parameters for the section of the nozzle along with the value [8].

TABLE 10.1 Nozzle Parameters and Related Values

Parameters	Values (m)
Convergent diameter	0.433
Divergent diameter	1.85
Combustion chamber length	0.27155
Throttle diameter	0.331
Nozzle divergence inlet radial	0.182
Nozzle divergence outlet radial	1.848
Nozzle divergence length	1.824

Now, the design is made using the above-given parameters and values, the design for the nozzle has been kept simple for easy understanding.

10.2.1 SINGLE INLET DESIGN (Figure 10.1)

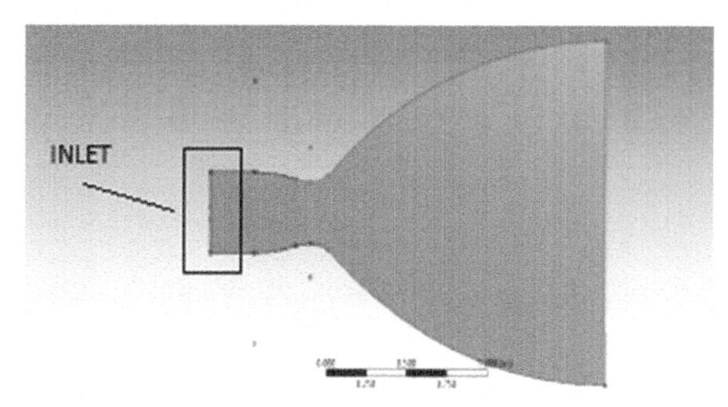

FIGURE 10.1 Geometry of a single inlet design.

10.2.2 DOUBLE INLET DESIGN (Figure 10.2)

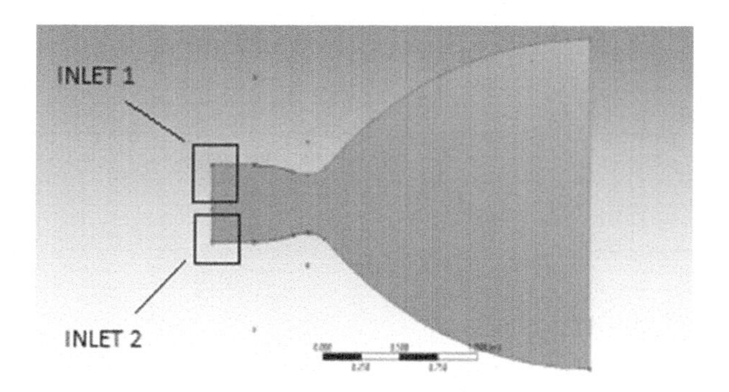

FIGURE 10.2 Geometry of a double inlet design.

10.2.3 FOUR INLET DESIGN

A 2-dimensional design is prepared using the ANSYS Design Modeler to work on the analysis, the design is straight and simple to understand and is purely made according to the dimensional parameters as discussed above in the table. Only the inlet value is changed from one to two and then four otherwise the overall design is the same (Figure 10.3).

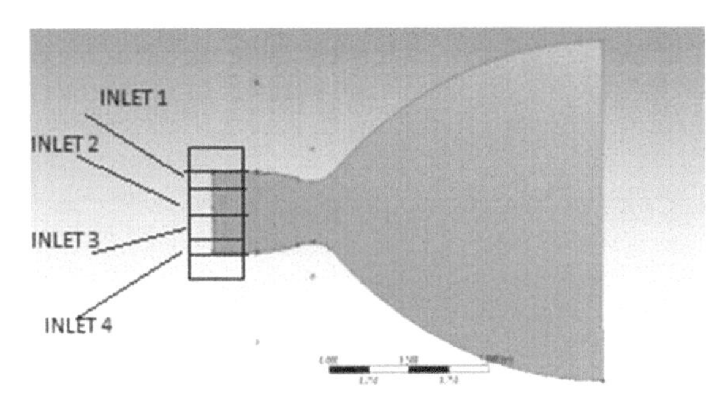

FIGURE 10.3 Geometry of a four-inlet design.

10.3 ANALYSIS OF THE NOZZLE GEOMETRY

Now the analysis is done on the geometry prepared, and for the analysis, ANSYS Fluent software package is used. The analysis is done to find out various parameters like Velocity, pressure, and temperature in all the three nozzle cases. Each nozzle analysis is done separately, and stated accordingly.

10.3.1 SINGLE INLET ANALYSIS

The same analysis pattern is followed for all the nozzle geometry and therefore, firstly the single nozzle is discussed below along with the analysis of various parameters.

10.3.1.1 TEMPERATURE

The nozzle here is analyzed based on the temperature variation, and it can be observed that the usual CD nozzle phenomenon of decreasing temperature is happening as the flow is moving from the converging section to the diverging section. This variation is very important for thermodynamic stability. A graph is made out to clearly understand the phenomenon. The temperature at inlet values 3,500 K and gradually decreases throughout the geometry, and at the exit, the temperature becomes less than 1,000 K. This temperature variation supports the pressure changes as they are directly related to each other (Figures 10.4 and 10.5). The governing equation for temperature is:

$$T = T_0\left(1+\frac{\gamma-1}{2}M^2\right)^{-1} \tag{1}$$

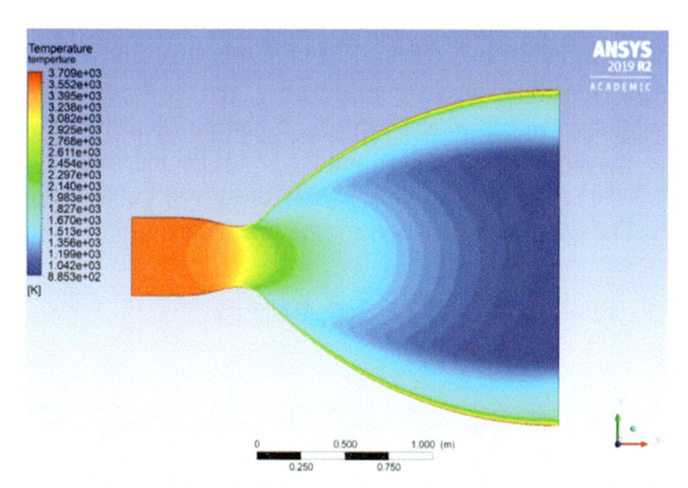

FIGURE 10.4 Analysis of temperature of single inlet design.

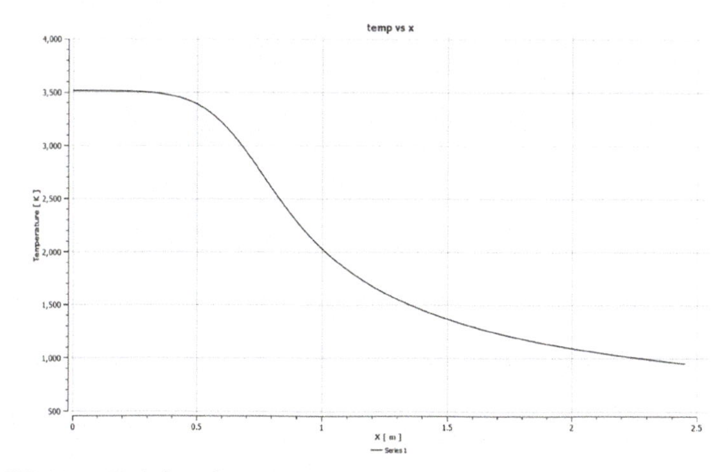

FIGURE 10.5 Variation of temperature along the X-axis.

10.3.1.2 PRESSURE

Here the nozzle is analyzed on the basis of pressure variation, and as the graph and analysis, image suggests the pressure is decreasing throughout the nozzle with the maximum in the converging section and minimum in the diverging section. A sudden decrease in the slope is observed at the throat section which is not as steep in the temperature graph. Now so as to meet the choking condition at the throat, we need to have a pressure ratio $P/P_0 = 0.528$ at the nozzle throat. The chamber pressure should be *1.89* times the pressure at the throat therefore only then the supersonic solutions can be obtained. And from the analysis it can be seen that the pressure inside the converging section is around 2.1×10^7 Pa and in the throat, it is around 1.3×10^7 Pa which roughly equals the given value (Figures 10.6 and 10.7). The governing equation for the pressure is:

$$P = P_0 \left(1 + \frac{\gamma - 1}{2} M^2\right)^{-\gamma/(\gamma+1)} \tag{2}$$

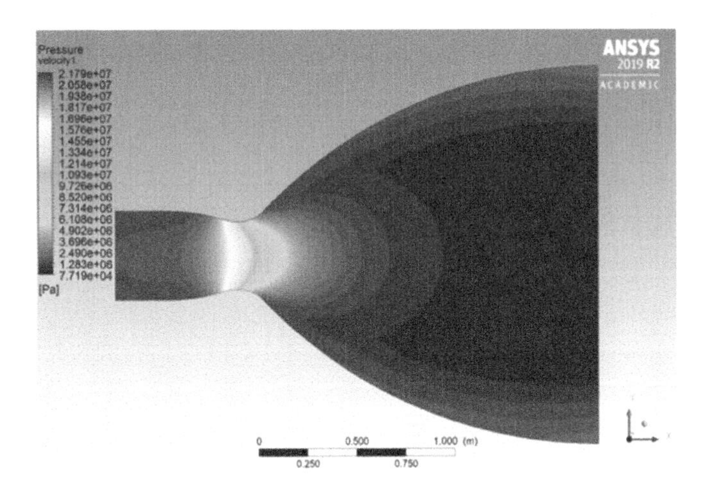

FIGURE 10.6 Analysis of pressure of single inlet design.

10.3.1.3 VELOCITY

The main concept of the nozzle is to obtain maximum velocity possible at the exit of the nozzle, and this can be done when the pressure in that area is less. The velocity here has a minimum value of around 0 ms^{-1} at the inlet point and reaches a maximum value of 2,300 ms^{-1} (Mach number: 6.76) at the exit of the divergence section. Both the Mach number and velocity correlates with each other. The throat section has a velocity of around 1

Mach (M) which is sonic, and therefore the exit point has M>1 which is supersonic (Figures 10.8 and 10.9).

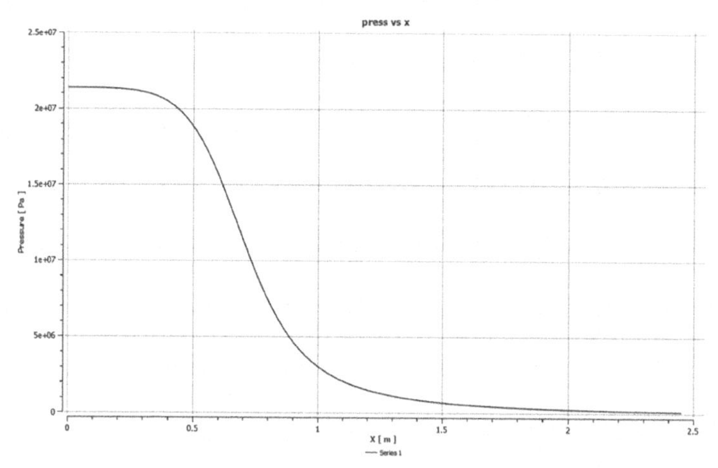

FIGURE 10.7 Variation of pressure along the X-axis.

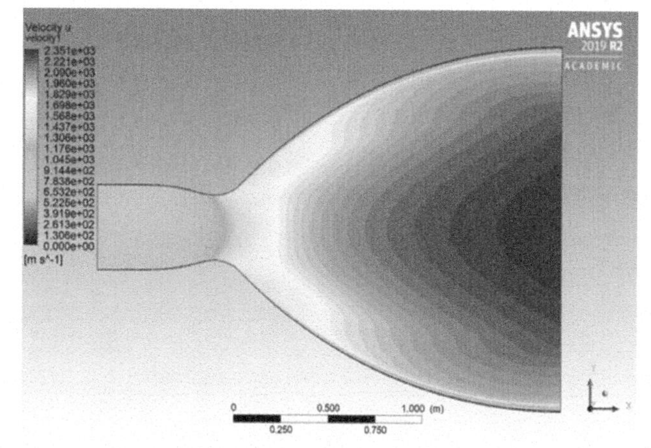

FIGURE 10.8 Analysis of velocity of single inlet design.

10.3.2 DOUBLE INLET ANALYSIS

Since the parameters of evaluation are the same and just the inlet condition has been varied so the results coming out have been analyzed on the basis of them only. The parameters are Temperature, Pressure, and Velocity.

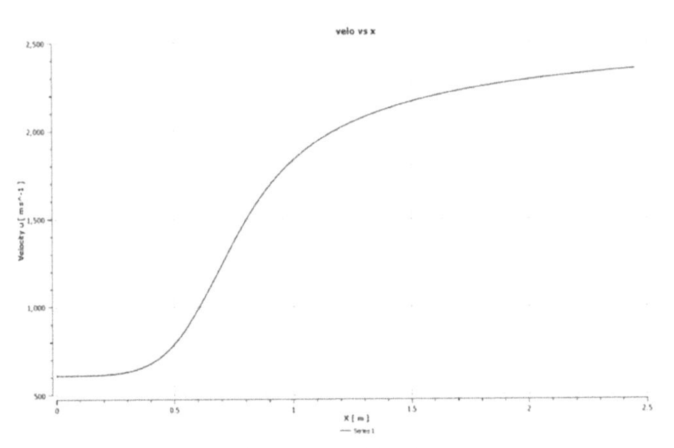

FIGURE 10.9 Variation of velocity along the X-axis.

10.3.2.1 TEMPERATURE

The temperature variation, in this case, has an inlet value of around 3,600 K and decreases to 900 K at the exit. The two inlets allow the flow to have separate paths and the junction can be seen to have a bit higher temperature of around 3,800 K due to the interface of both the inlet flows (Figures 10.10 and 10.11).

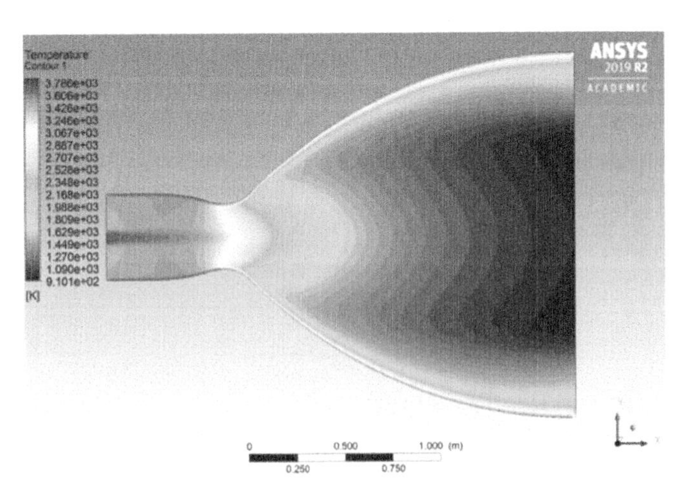

FIGURE 10.10 Analysis of temperature of double inlet design.

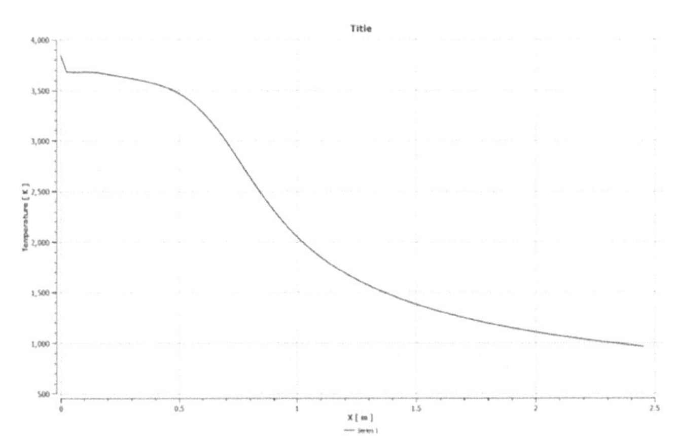

FIGURE 10.11 Variation of temperature along the X-axis.

10.3.2.2 PRESSURE

The pressure variation, in this case, has an inlet value of around 1.9×10^7 Pa and decreases to a value of 9.6×10^4 Pa at the exit. The two inlet helps in lowering down the pressure as the surrounding is not confined and opens with two path flows. The interface even has the lower pressure. Though the inlet shows a bit lower pressure but the supersonic condition is still achievable at the exit (Figures 10.12 and 10.13).

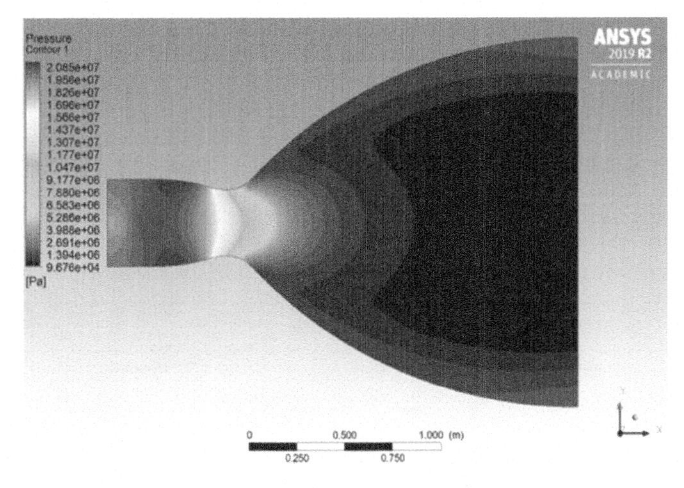

FIGURE 10.12 Analysis of pressure of double inlet design.

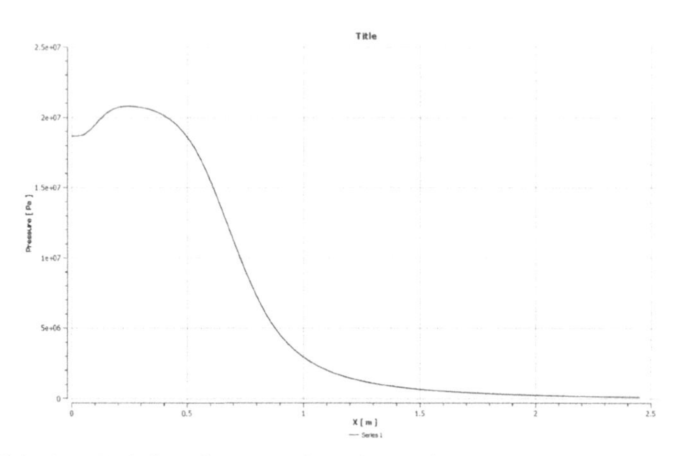

FIGURE 10.13 Variation of pressure along the X-axis.

10.3.2.3 VELOCITY

The velocity achieved in two inlet conditions at the inlet is around 660 ms^{-1} and at the exit is around 2,290 ms^{-1} (Mach number: 6.73). The values are right with the nozzle perspective and the sonic condition of M=1 is satisfied and the M>1 is also satisfied for the supersonic conditions (Figures 10.14 and 10.15).

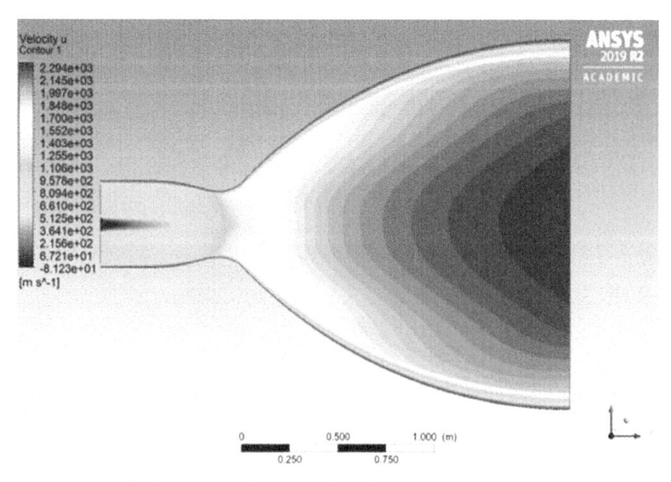

FIGURE 10.14 Analysis of velocity of a double inlet design.

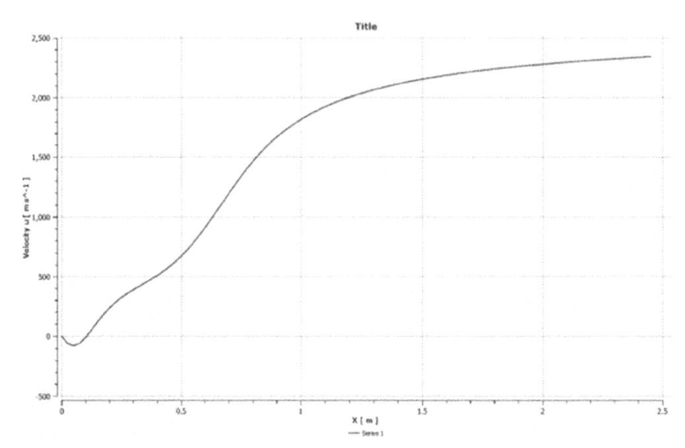

FIGURE 10.15 Variation of velocity along the X-axis.

10.3.3 FOUR INLET ANALYSIS

This is the final condition among the three to be analyzed and then the comparison is done for the prediction. The analysis of four inlet condition is similar as the above two conditions. Here the velocity, temperature, and pressure parameters are analyzed based on the four inlet conditions.

10.3.3.1 TEMPERATURE

The temperature variation if look closely than the maximum temperature of approximately 4,100 K is there and then there is a steep lowering of the temperature of around 3,800 K and then prevails up to the 0.4 m length and then it reaches to value of 3,600 K at 0.5 m length which is 100 K more as obtained as maximum in single inlet but after that it starts decreasing and lowers down up to 1,000 K at 2.45 m length of the nozzle. This is the general trend of temperature variation for a Converging Diverging nozzle that has been observed for this case. Here X-axis is considered because the length of the nozzle is along X-axis and any variation happening can be monitored easily because of the length of the nozzle in the same axis of the flow (Figures 10.16 and 10.17).

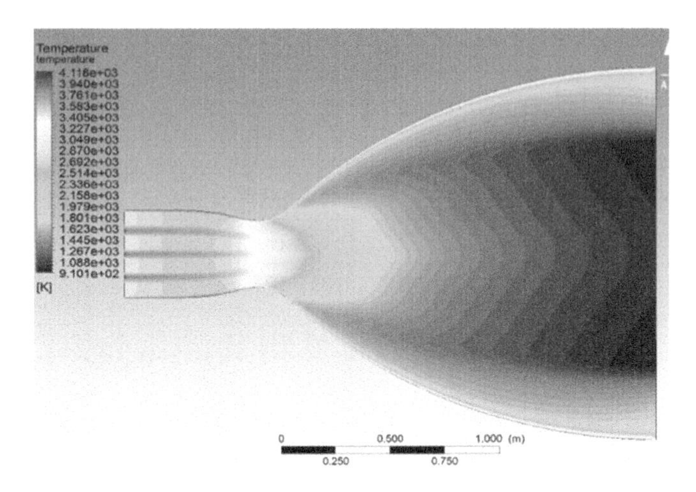

FIGURE 10.16 Analysis of temperature of four inlet design.

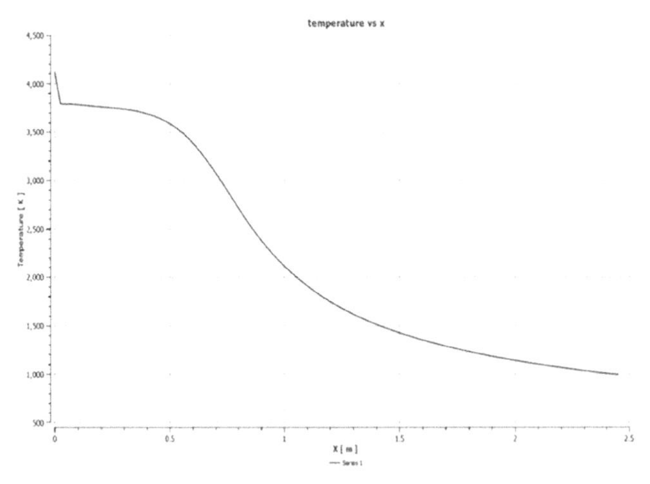

FIGURE 10.17 Variation of temperature along the X-axis.

10.3.3.2 PRESSURE

The pressure variation if look closely than the maximum pressure of approximately 2×10^7 Pa is up to the 0.4 m length then it starts decreasing and lowers down up to 0 Pa at 2.45 m length of the nozzle. A sudden bump in the pressure can be seen which starts at 1.7×10^7 Pa and overs at 2×10^7

Pa. The maximum pressure here, in this case, is lesser in comparison to the double inlet condition. This is the general trend of pressure variation for a Converging Diverging nozzle in this case. Here X-axis is considered because the length of the nozzle is along X-axis and any variation happening can be monitored easily because of the length of the nozzle in the same axis of the flow (Figures 10.18 and 10.19).

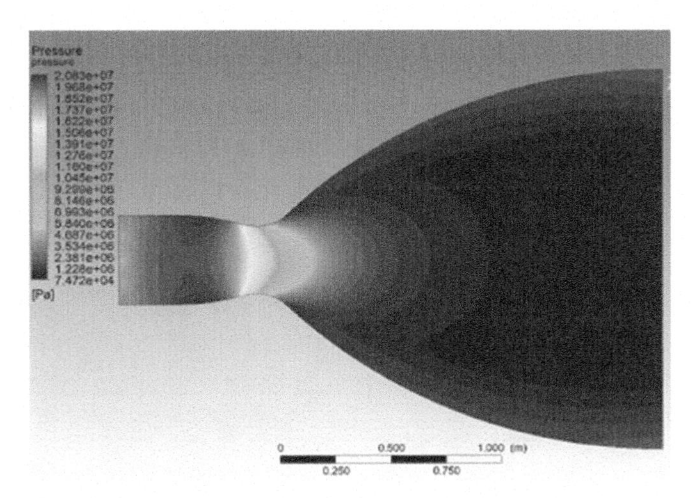

FIGURE 10.18 Analysis of pressure of four inlet design.

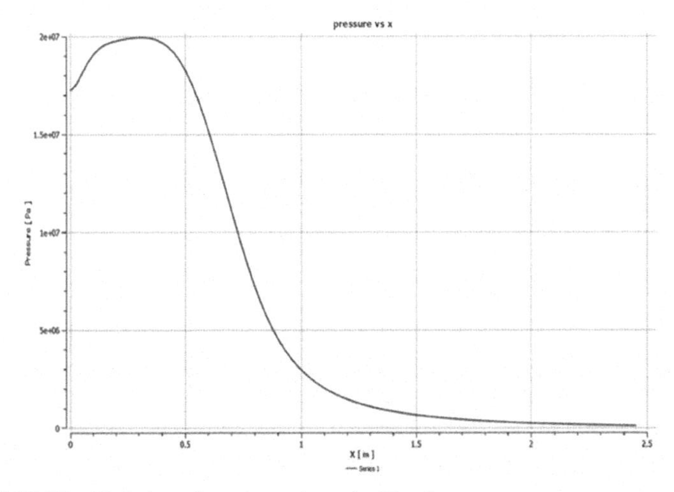

FIGURE 10.19 Variation of pressure along the X-axis.

10.3.3.3 VELOCITY

The values are right with the nozzle perspective and the sonic condition of M=1 is satisfied and the M>1 is also satisfied for the supersonic conditions at the exit of divergence section. Since we haven't worked on the boundary layer phenomenon in this project work but if look closely then the velocity near the wall is around 651 ms^{-1} and is much higher as compared to double inlet having 360 ms^{-1} and for the single inlet the velocity near to the wall is around 390 ms^{-1} which shows that four inlets has a higher velocity range in the divergence section and that's why the final exit velocity is quite high in comparison to the single inlet and double inlet (Figures 10.20 and 10.21).

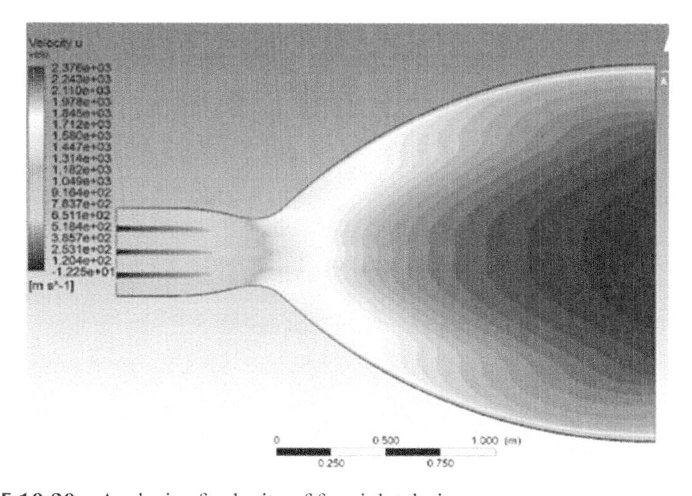

FIGURE 10.20 Analysis of velocity of four inlet design.

10.4 CONCLUSION

A comparative study is done in this chapter between the single, double, and four inlet conditions of a rocket nozzle. Now the comparison is done based on the parameter's temperature, velocity, and pressure.

When the temperature is concerned, then a lesser temperature inside the nozzle will support the material from getting damaged, but it works in accordance with the pressure change. By the observation of temperature, the inlet temperature in single inlet is around 3,700 K and in the double inlet it is around 3,800 K whereas in the four inlet it is approximately 4,100 K which is a higher than the other two and is due to the interface of flows due to four inlets. Similarly at the exit in the single inlet it is around 880 K and in the case of double, and four inlet it is around 910 K. Since the temperature is a little higher in double and four inlet condition. A certain amount of heat energy is lost to the walls as the system is not purely adiabatic.

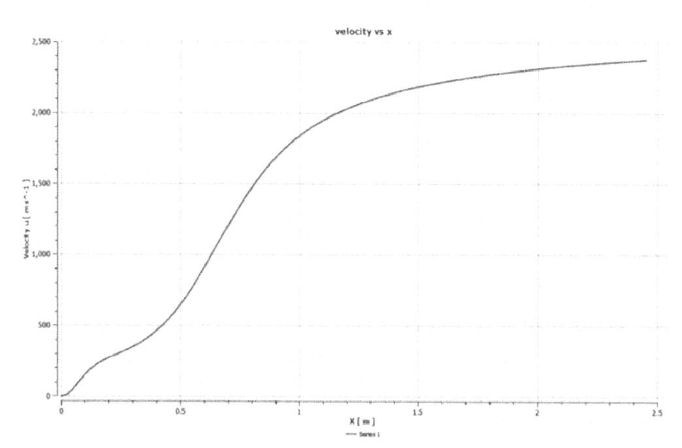

FIGURE 10.21 Variation of velocity along the X-axis.

And when the pressure change is compared, then the pressure at the entry of a single inlet nozzle is around 2.179×10^7 Pa, for a double inlet, it is around 2.085×10^7 Pa and for four inlet it is approximately 2.083×10^7 Pa. The single inlet is confined and does not have an opening, therefore, exert a bit more pressure as compared to the double and four inlets. Similarly, the exit pressure for the single inlet is around 7.719×10^4 Pa which is higher than the double inlet is approximately 9.676×10^4 Pa and the four inlet it is 7.472×10^4 Pa but as the better condition is concerned the exit pressure has to be lower as possible because only then the pressure energy change can produce higher velocity at the exit.

Now when the two parameters are checked, then the velocity is concerned so that maximum exit velocity can be obtained for producing a higher flow rate through the nozzle. Anyway, if the velocity is increased in the converging section than also the ultimate velocity will be less than the sonic speed or $0<M<1$. The choke condition has to be obtained, which means the final velocity in the throat has to be $M=1$ irrespective of the inlet velocity as it will produce no change in the throat section. In this chapter, the exit velocity for the single inlet is approximately 2,300 ms^{-1} and for the double inlet is 2,290 ms^{-1} whereas, for the four inlet is 2,370 ms^{-1} which seems to be a little variation but if the flow around the wall is seen then the double inlet has lower velocity forming zones as compared to the single, and four inlets. The graph for single inlet shows a very gradual change as compared to the double inlet and four inlet which has high degree of variability in the velocity that means some degree of turbulence.

Now by going through the results and analysis, it can be recommended that the four-inlet design is better as compared to the double inlet and single inlet, considering higher exit velocity as the governing parameter with the following values as discussed in this chapter.

KEYWORDS

- **ANSYS fluent**
- **combustion chamber**
- **computational fluid dynamics**
- **inlet conditions**
- **rocket nozzle**
- **thermodynamic stability**

REFERENCES

1. Biju, K. P., & Sanjesh, M. (2013). Optimization of divergent angle of a rocket engine nozzle using computational fluid dynamics. *The International Journal Of Engineering and Sciences, 2*(2), ISSN: 2319-1813.
2. Roy, P., Mondal, A., & Barai, B., (2016). CFD analysis of rocket engine nozzle. *International Journal of Advanced Engineering Research and Sciences (IJAERS), 3*(1), ISSN: 2349-6495.
3. Natta, P., Kumar, V. R., & Rao, Y. V. H., (2012). Flow analysis of rocket nozzle using computational fluid dynamics (CFD). *International Journal of Engineering Research and Application (IJERA), 2*(5), 1226–1235.
4. Pandey, K. M., & Yadav, S. K., (2010). CFD analysis of a rocket nozzle with two inlets at Mach 2.1. *Journal of Environmental Research and Development, 5*(2).
5. Pandey, K. M., & Yadav, S. K., (2010). CFD analysis of a rocket nozzle with four inlets at Mach 2.1. *International Journal of Chemical Engineering and Applications, 1*(4), ISSN: 2010-0221.

6. Thakur, B., & Pegu, I. J., (2017). A review of cryogenic rocket engine. *International Research Journal of Engineering and Technology (IRJET), 04*(08). ISSN: 2395-0056.
7. Prathibha, M., Gupta, M. S., & Naidu, S., (2015). CFD analysis on a different advanced rocket nozzles. *International Journal of Engineering and Advanced Technology (IJEAT), 4*(6), ISSN: 2249-8958.
8. Mahzad, C., Hassan, K. M., & Ehsan, T., (2012). A method of simulation of a cryogenic liquid propellant engine. *Applied Mechanics and Materials.*

CHAPTER 11

EXPERIMENTAL ANALYSIS OF IMPROVED VORTEX TUBE

SUBHASH N. WAGHMARE,[1] KETAN S. MOWADE,[2] SAGAR D. SHELARE,[1] and SANJAY W. MOWADE[3]

[1]*Department of Mechanical Engineering, Priyadarshini College of Engineering, Nagpur, Maharashtra, India, E-mail: Sagmech242@gmail.com*

[2]*Department of Mechanical Engineering, Michigan Technological University, Houghton, USA*

[3]*Department of Mechanical Engineering, Smt. Radhikatai Pandav College of Engineering, Nagpur, Maharashtra, India*

ABSTRACT

The vortex tube is a particular refrigerator and industrial tools which utilized compressed air as a working substance. Vortex tubes can isolate a pressurized gas stream from a colder stream to a hotter stream. Even so, the use has been restricted by the poor quality of the equipment. The main concern of any industry is to keep up environmental safety. This chapter focuses on the investigation and efficiency of vortex tube, which works on the principle of the hot gas stream through one end and a cold gas flow from the further end without any external source of energy. The nozzle design is the prime concern of the research as it will give a greater cooling effect as compared to the inlet and outlet orifice of the vortex tube.

Optimization Methods for Engineering Problems. Dilbagh Panchal, Prasenjit Chatterjee, Mohit Tyagi, Ravi Pratap Singh (Eds.)

The geometrical parameters have been analyzed to get the better and more efficient design of the improved vortex tube.

11.1 INTRODUCTION

Lifestyles have become highly energy-consuming. After industrialization, our reliance on fossil fuels has only risen. As a result, carbon dioxide levels are the maximum they have been and are rising in the past 3 million years. The vortex tube is a cylindrical apparatus in which the hot gas air stream enters from one termination and the further side cold gas stream leaves as shown in Figure 11.1 displays the operational principle for the vortex tube. The tube operates along accepting the compressible fluid tangentially via the nozzles. As the tube is cylindrical, the inlet pressure and the speed result in an internal circular movement. As the tube wall and the center are at different pressure due to wall friction facilitates the energy transfer from the center region to the tube walls. The confined liquid which has cooled now moves against the major flow direction. This liquid exits the tube in a direction opposite to the main flow while the hot liquid exits following the direction of the main flow. The RHVT works for heating as well as cooling purposes.

The existing vortex tube has some drawbacks like vortex tubes have smaller efficiency than conventional air conditioning apparatus; it required turbulent flow and has a limited range of construction material. Therefore, the available design and manufacturing of vortex tubes are discussed below.

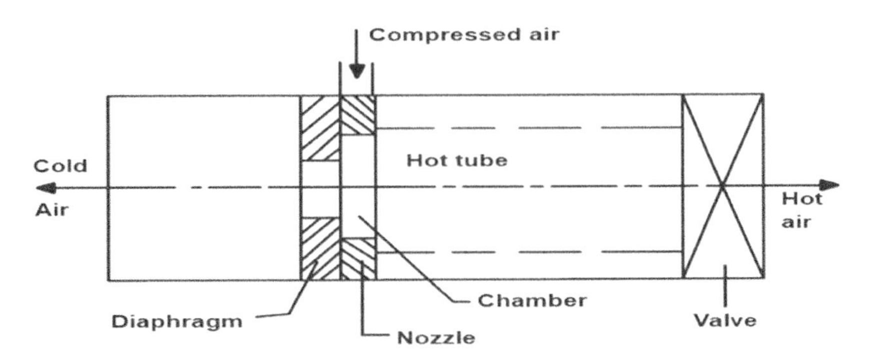

FIGURE 11.1 Vortex tube.

11.2 RESEARCH BACKGROUND

In this research, the authors have provided a brief review of available literature related to vortex tube, various investigators studied hypothetically vortex tubes and attempt to illustrate in what dimensions the top presentation obtained. Thermo physical variables and geometrical variables are necessary variables exert influence on the vortex tube performance George Ranque a French physicist invented the vortex tube in 1928 while working to develop a pump where he noticed an exhaust of cool air from one end and hot air from another.

Rafiee and Sadeghiazad [13] investigated the heat transfer characteristics for variables like nozzle area of inlet compressed air, hot end area of the tube, cross-section area of the hot and cold end, L/D ratio and cold orifice area. Saberi et al. [14] demonstrated experimentally analysis of a vortex tube and optimizing the operational efficiency when put to use for industrial applications like spot cooling, weld cooling, etc. Bidwaik and Dhavale [2] have worked upon a double inlet vortex tube with dc = 5, 6, and 7 mm and eight tangential nozzles having L/D =11.5. This resulted in increased intensity of swirl. The governing equation for the same was obtained using ANSYS FLUENT 15.0 on a 3D model for the fluid domain. New experiments carried out by Purwanta [12] with the nozzle diameter and have been able to state that the nozzle diameter influences the performance and cooling efficiency and has been able to identify that the nozzle diameter can be optimized to give the best performance. Devade and Pise [3] have parametrically reviewed the effects of around parameters that define the performance of the vortex tube with a focus on reviewing works carried on in order to enhance the refrigeration effect. Saidi and Valipour [17] have defined and researched factors effecting vortex tube output as thermophysical variables like type of gas and cold mass fraction, inlet gas pressure, the outlet orifice diameter, inlet gas moisture and width and length of the main tube and (Table 11.1) [5].

11.3 RESEARCH METHODS

11.3.1 OUTLINE OF THE DEVELOPMENT OF VORTEX TUBE

Accordingly following invention relates to a vortex tube that precisely breaks a compressed gas flow into a cold and hot flow without any

TABLE 11.1 List of Abbreviations Used

Abbreviations Nomenclature Unit		
Cp	**Specific Heat of Air**	**Kj/KgK**
M	Mass flow rate	Kg/Sec
T	Temperature	°C
ΔT	Temperature difference	°C
Tc'	Static temperature drop	°C
ΔTrel	Relative temperature drop	°C
μ	Cold mass fraction	–
Q	Cooling or heating rate	Kj/sec
V	Velocity	m/sec
W	Actual work done by the compressor	Kw
ηad	Adiabatic efficiency of vortex tube	%
i	Inlet to vortex tube	–
h	Hot air exit	–
c	Cold air exit	–

chemical response or outside energy supply. Herein compressed air is gone through the nozzle where air expands and obtains huge velocity due to the selective shape of the nozzle. A vortex motion is generated in the chamber and air travels in a spiral-like motion through the edge of the hot side [20]. The valve prevents this movement. As the pressure of the nearby air valve is created rather than outside by partly shutting the valve, the aft axial flow between the center of the hot boundary begins from the high-pressure region to the low-pressure area. Via this process, the heat transfer takes place between the reversed flow and the forward flow. The airflow through the center is then cooled below the inlet temperature of the air in the vortex tunnel, while the airflow in the forward direction is warmed up. The cold flow is going through the diaphragm hole to the cool side, while the heat flow is moving via the gap of the valve. The amount of cold air and its temperature can be varied by controlling the opening of the valve [19]. The vortex tube is a device that employs no devices which need to make some motion to operate and employs compressed air to produce hot and cold air from two of its ends. Lots of research has been carried out to know the heat dynamics of a vortex tube with reference to the variables such as the area of nozzle, cross-sectional area of cold and hot end, hot end region of the tube, cold orifice area, along with the length to diameter ratio. It has been observed that the investigation carried out fails to present the perfect picture and hence physical verification has been carried to come to a defining conclusion.

11.3.2 THEORETICAL ISSUE

As far as the design is concerned the vortex tube can be fabricated in two ways with the first variant being the maximum temperature drop design for generating air with very low temperature and the other variant being a design that dishes out maximum cooling effect essential for air in bulk and moderate. These variants are put to use to maximize the heat convey rate through the forward motion of swirl air and backflow of axial air. These two design variants are employed in work for increasing the heat convey rate while the movement of swirl air axial's air in its reverse movement.

A vortex tube considers the following conditions for its operation:

- Ambient pressure (Pa);
- Pressure of inlet (Pi);
- Temperature of the inlet (Ti);
- Temperature of cooling (Tc);
- Hot body temperature (Th).

For the conditions above vortex tubes can be designed with reference to the below-mentioned parameters:

- Hot tube diameter (DT);
- Hot tube length (L);
- Diameter of a cold orifice (DC);
- Nozzle diameter (DN).

The present study has been focused on maximum temperature drop design for generating air with very low temperatures. Considering Soni and Thomson the nozzle area to tube area proportion of 0.11+ 0.01 was set up for the highest temperature fall and a ratio of 0.084+0.001 to attain the highest efficiency. Similarly, as per their suggestion, the cold orifice area to tube area ratio confirms to 0.08+0.001 fixed to maximize temperature drop and 0.145+0.035 for achieving the highest efficiency. The length of the vortex tube was set up to be 45 times greater than the tube diameter as per their suggestions [18]. The material used for a cold end (inlet cap) is SS 304, and brass for the hot end to harness its excellent thermal conductivity while the remaining assembly is fabricated in stainless steel to keep a tab on the machining and the overall cost [15].

The research is focused on checking the efficiency and application of the proposed apparatus when put to use in industry. The tube is used for instant cooling purposes, where we require dry coolant such that applications do not get damaged thus application where we need cooling for small application, we can use a vortex tube [16].

11.4 DESIGN AND CONSTRUCTIONAL FEATURES

11.4.1 CAD MODEL PART DIAGRAM OF VARIOUS PARTS OF A VORTEX TUBE (Figures 11.2–11.7)

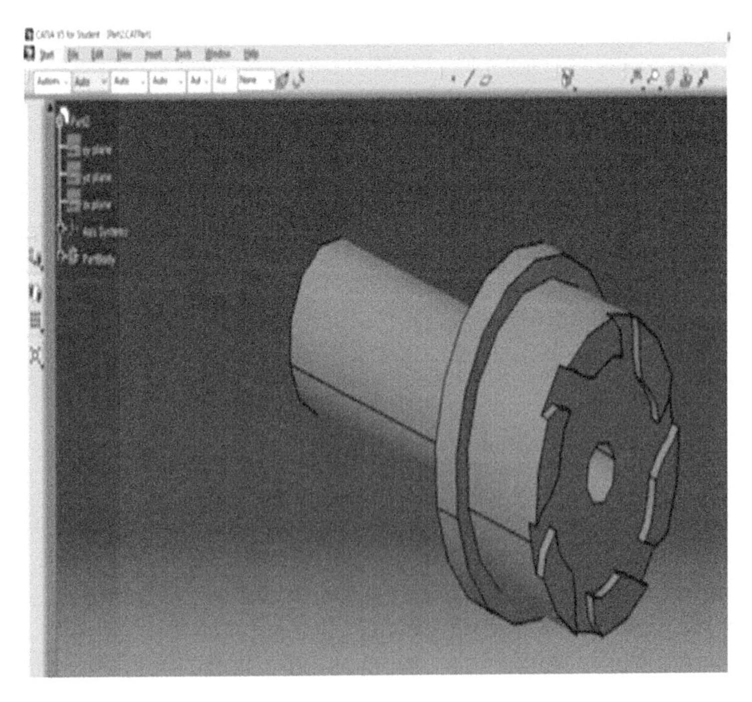

FIGURE 11.2 Vortex generator.

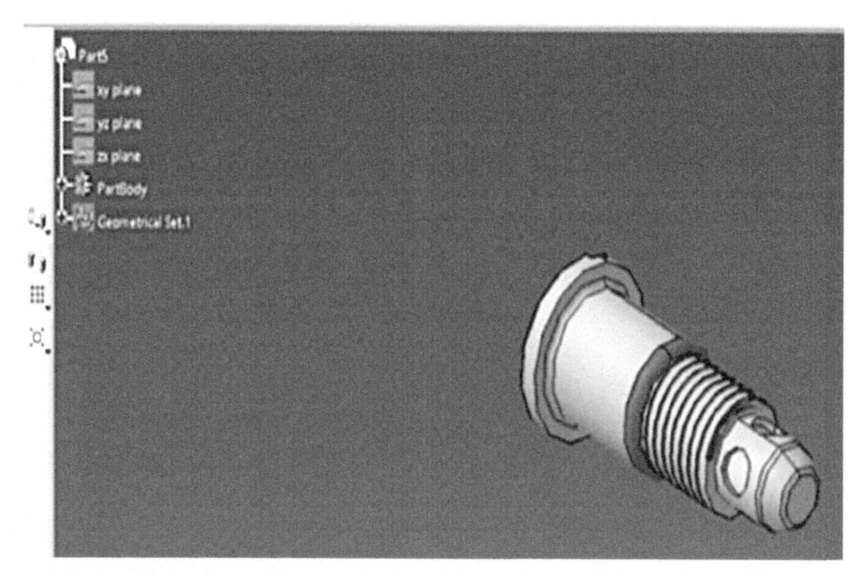

FIGURE 11.3 Conical valve.

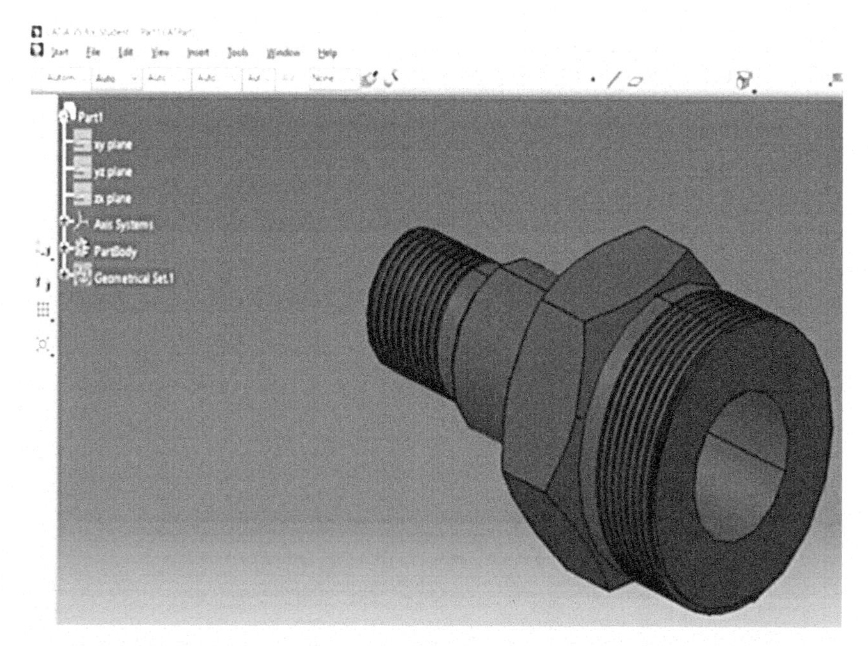

FIGURE 11.4 Hot end part.

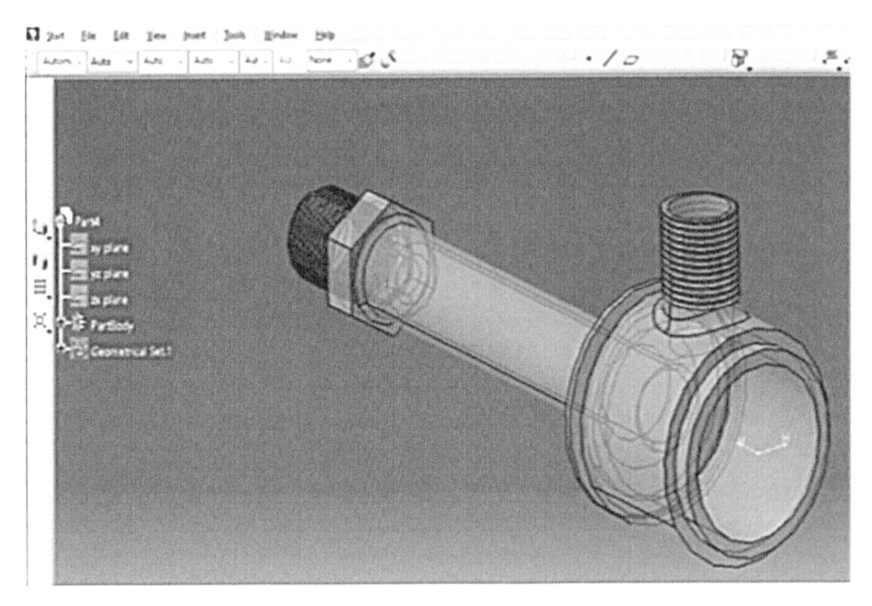

FIGURE 11.5 Cold end part.

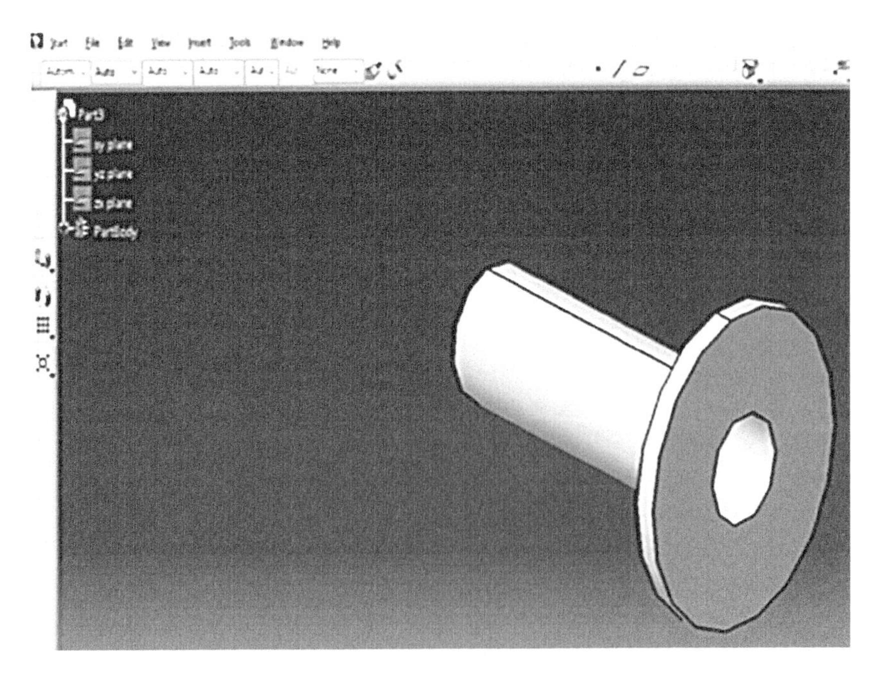

FIGURE 11.6 Hot end sleeve.

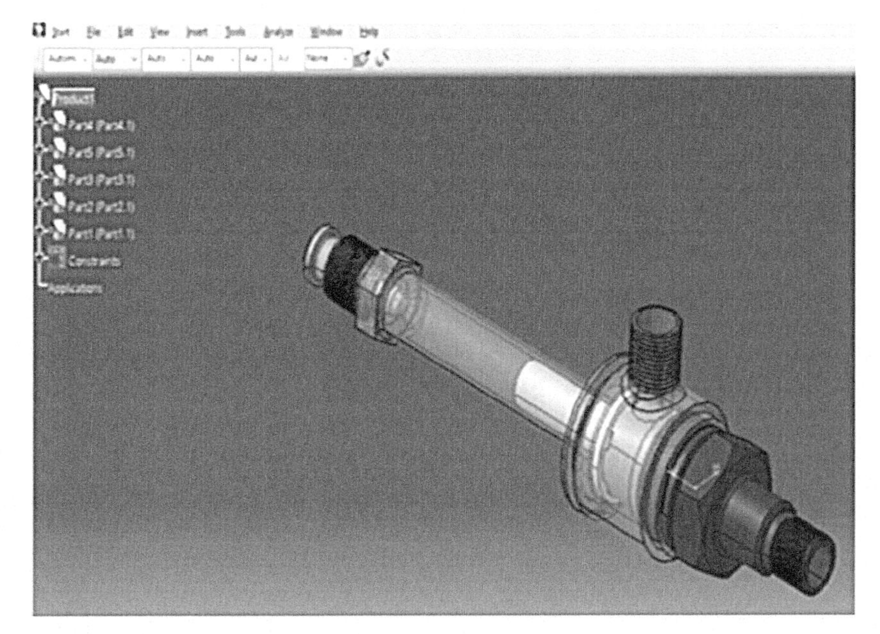

FIGURE 11.7 Vortex tube.

11.4.2 IMAGE DRAFTING WITH A DIMENSION OF VARIOUS PARTS OF VORTEX TUBE

Experimental design for set up – in the following figure, the size of the vortex tube under study (the vortex tube cold end, vortex generator hot end sleeve, conical valve, and hot end) has been shown the vortex tube dimensions under study (Figures 11.8–11.12).

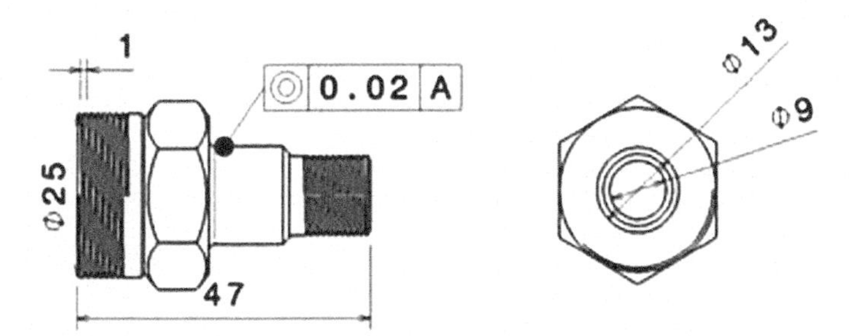

FIGURE 11.8 Vortex tube cold end.

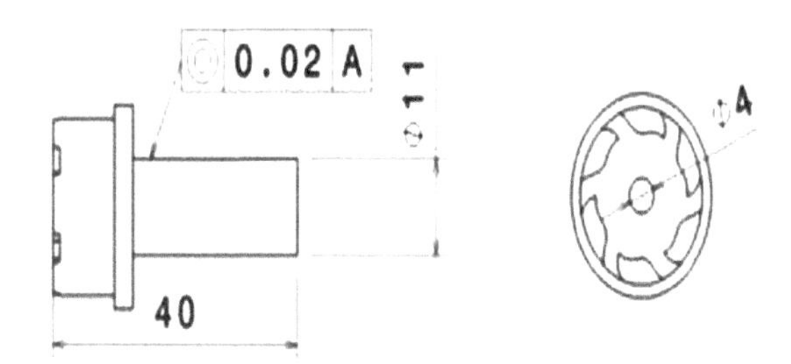

FIGURE 11.9 Vortex generator.

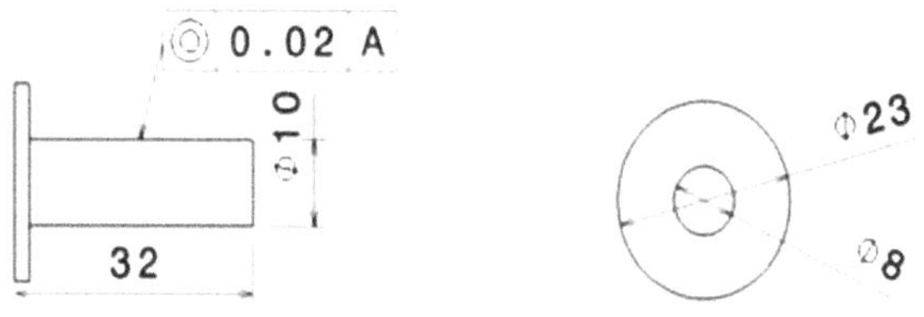

FIGURE 11.10 Hot end sleeve.

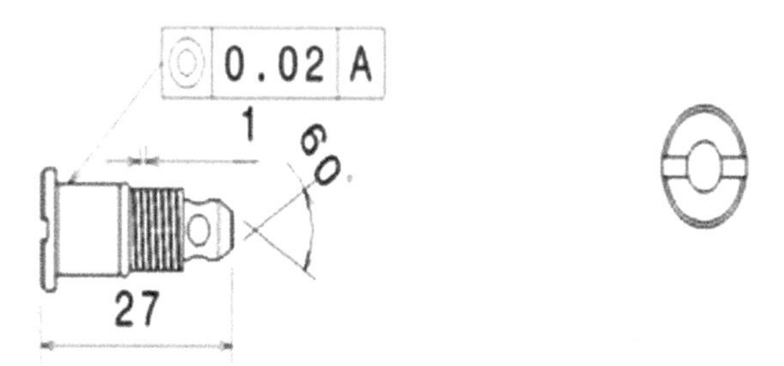

FIGURE 11.11 Conical valve.

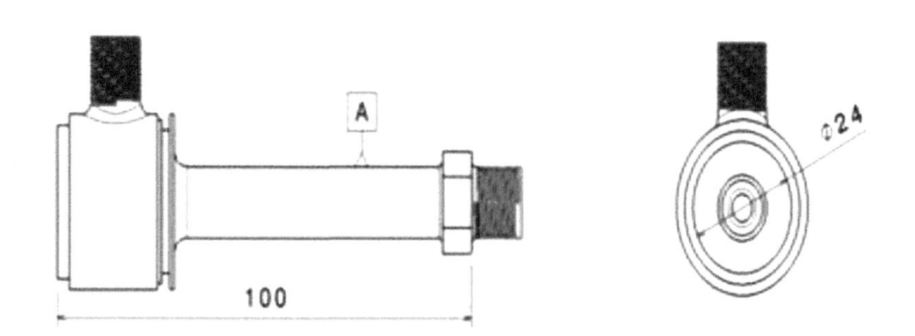

FIGURE 11.12 Hot end.

11.4.3 FINAL FABRICATED PRODUCT OF VARIOUS PARTS OF VORTEX TUBE ASSEMBLY

Designed various parts of vortex tube apparatus (Figures 11.13–11.18).

FIGURE 11.13 Vortex generator.

FIGURE 11.14 Hot end sleeve.

FIGURE 11.15 Conical valve.

FIGURE 11.16 Hot end.

FIGURE 11.17 Assembled vortex tube.

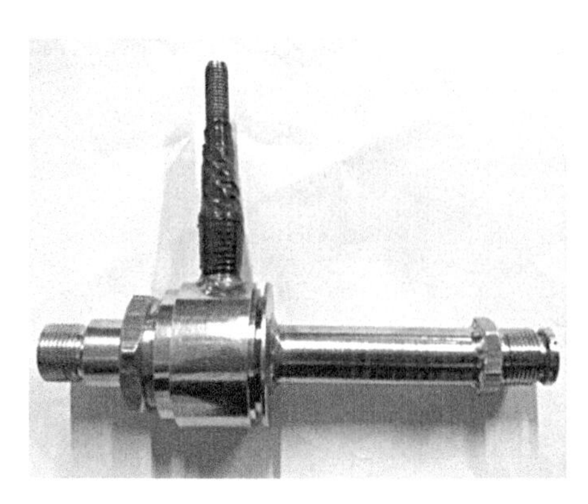

FIGURE 11.18 Final designed manufactured vortex tube.

11.5 OPERATIONAL PROCEDURES

The experimentation is performed with a counterflow vortex tube at different valve positions and varying pressures ranging from 7.2 to 8.3 bars. There are 7 valve positions from 0 to 100% opening at the hot end side to restrict or release the hot mass of air [10]. For every valve position, the temperature rise and drop are recorded. So as to have a correct set of readings 3 measurements are recorded for every valve position and then the average temperature rise and drop are estimated, respectively [8, 9].

The input fluid from the compressor which is the compressed air is passed through the pressure regulator to adjust the inlet pressure and then passed through the generator which is of brass material having 6 aerofoil shaped cut due to which the air vortices are generated inside the vortex tube at very small vortex angle of 6. The rotation of the generator and simultaneously the vortices work in direct proportion to the inlet pressure [22]. The restrictor at the other end regulates the mass flow measurement of the hot air to see its effect on temperature distribution by the ends of the vortex tube [6]. The inlet and cold mass flow rate is calculated with the help of an anemometer, and the hot mass is calculated from the mass conservation principle.

11.5.1 ASSESSMENT

The assessment works upon the following important terms.

The cold mass fraction denoted by μ is equated as the mass of cold air to the mass air ratio [11].

$$\mu = \frac{mc}{mi} \tag{1}$$

where; mc is the cold mass flow rate in kg/s and mi is the inlet mass flow rate in kg/s.

The cold end temperature is figured out as the difference in the temperature of the inlet end and the cold end [20].

$$\Delta\, Tc = (Ti - Tc) \tag{2}$$

with Ti being the inlet temperature; and Tc the cold end temperature.

Temperature drop at the hot end is figured out as the difference in the temperature of the hot end and the inlet end and the hot end [7]. With T_h being the temperature at the hot end and Ti being the inlet temperature [1, 4].

$$\Delta Th = (Th - Ti) \tag{3}$$

At a given case the cooling effect Qc is represented as below:

$$Qc = mc\ Cp\ (Ti - Tc) \tag{4}$$

Heating effect Qh at a given case at hot end is given as:

$$Qh = mh\ Cp\ (Th - Ti) \tag{5}$$

Estimation of the work accomplished by a compressor is equated as below [13].

$$W = \left(n/_{n-1} \right) x\ P1V1 x \left[\left(P2/_{P1} \right)^{\left(n - \frac{1}{n} \right)} - 1 \right] \tag{6}$$

with P1 being the inlet atmospheric pressure accepted by the compressor and P2 being the outlet gauge pressure.

The coefficient of performance is equated as the ratio of cooling effect to the accomplished work of the compressor.

$$COP = Qc/_{W} \tag{7}$$

The static temperature drop is the temperature drop due to adiabatic expansion which is represented by:

$$\Delta\ Tc'\ \Delta\ Tc' = (Ti - Tc') \tag{8}$$

where; $Tc' = Ti \left[1 - \left(Pa/_{Pi} \right)^{(n-1/n)} \right]$

The relative temperature drop is the ratio of temperature drop in a vortex tube to the static temperature drop which is denoted by $\Delta\ Trel$

$$\Delta Trel = \frac{\Delta Tc}{\Delta Tc'} \tag{9}$$

Adiabatic efficiency is the ratio of real cooling gets in the vortex tube to the cooling possible with adiabatic expansion. It is denoted by $\varepsilon\ ad$

$$\varepsilon\ ad = \mu\ x\ \Delta\ Trel \tag{10}$$

11.6 RESULT DISCUSSION

Table 11.2 shows the experimental result of the vortex tube.

TABLE 11.2 Experimental Result of Vortex Tube

SL. No.	Pressure (bar)	Conical Valve (Rotation)	Hot End Temperature (°C)	Cold End Temperature (°C)	Temperature Difference (°C)
1.	8.3 (120 psi)	0.5	42.2	27.7	14.5
		1	47.9	29	18.9
2.	7.6 (110 psi)	0.5	41.7	28	13.7
		1	45.6	29.6	16
3.	7.2 (105 psi)	0.5	41.4	28.5	12.9
		1	38.5	28.7	9.8

Figure 11.19 shows the effect of inlet pressure on ΔTc. The value of Tc increases with the increase in Pi for 0.5 rotation of valve opening with a straight-line relation.

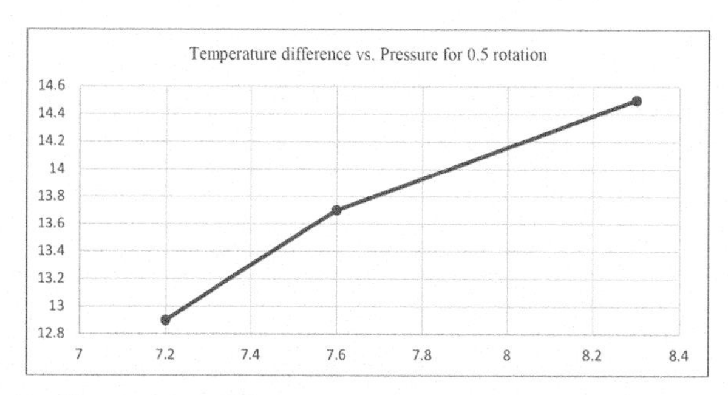

FIGURE 11.19 Temperature difference vs. pressure for 0.5 rotation.

Figure 11.20 displays the bearing of inlet pressure on ΔTc. The value of Tc increases with the increase in Pi for 1 rotation of valve opening with a straight-line relation.

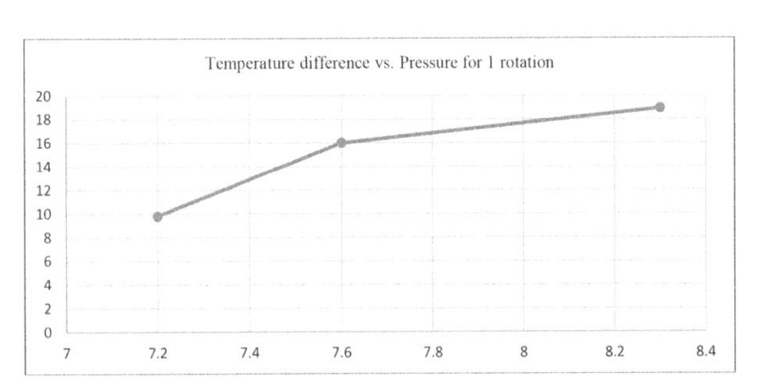

FIGURE 11.20 Temperature difference vs. pressure for 1 rotation.

11.6.1 CALCULATION OF COEFFICIENT OF OPERATION

$$C.O.P = \frac{m * Cp * \Delta T * 10^{-3}}{1 * 746 * 0.75} = 1.79 \times m \times \Delta T$$

1. $\Delta cfm = 6$
   i. For 0.5 rotation:

$$C.F = \frac{(3.5 - 0.5)}{3.5} = 0.857$$

Mass of air coming out of cold end = 0.857 × 6 = 5.1426

$$m = 5.1426,\ cfm = 2.57 \times 10^{-3}\ m^3\ s^{-1}$$

$$\Delta T = 14.5\ °C$$

$$C.O.P. = 0.066$$

   ii. For 1 rotation:

$$C.F = \frac{(3.5 - 1)}{3.5} = 0.71428$$

Mass of air coming out of cold end = 0.71428 *x* 6 = 4.2857

$$m = 4.2858,\ cfm = 2.143\ x\ 10^{-3}\ m^3\ s^{-1}$$

$$\Delta T = 18.9\ °C$$

$$C.O.P = 0.073$$

2.

i. For 0.5 rotation:

$$C.F = \frac{(3.5-0.5)}{3.5} = 0.857$$

$$Mass\ of\ air\ coming\ out\ of\ cold\ end = 0.857\,x\,6 = 5.1426$$

$$m = 5.1426,\ cfm = 2.57\ x\ 10^{-3}\ m^3\ s^{-1}$$

$$\Delta T = 13.7\ °C$$

$$C.O.P = 0.063$$

ii. For 1 rotation:

$$C.F = \frac{(3.5-1)}{3.5} = 0.71428$$

$$Mass\ of\ air\ coming\ out\ of\ cold\ end = 0.71428\,x\,6 = 4.2857$$

$$m = 4.2858, cfm = 2.143\,x\,10^{-3}\ m^3\ s^{-1}$$

$$\Delta T = 16\ °C$$

$$C.O.P = 0.061$$

3.

i. For 0.5 rotation:

$$C.F = \frac{(3.5-0.5)}{3.5} = 0.857$$

$$Mass\ of\ air\ coming\ out\ of\ cold\ end = 0.857\,x\,6 = 5.1426$$

$$m = 5.1426, cfm = 2.57\,x\,10^{-3}\ m^3\ s^{-1}$$

$$\Delta T = 12.9\ °C$$

$$C.O.P = 0.059$$

ii. For 1 rotation:

$$C.F = \frac{(3.5-1)}{3.5} = 0.71428$$

$$Mass\ of\ air\ coming\ out\ of\ cold\ end = 0.71428\,x\,6 = 4.2857$$

$$m = 4.2858, cfm = 2.143\,x\,10^{-3}\ m^3\ s^{-1}$$

$$\Delta T = 9.8\ °C$$

$$C.O.P = 0.038$$

Table 11.3 shows the experimental result of pressure, conical valve, and calculated C.O.P.

TABLE 11.3 Experimental Result of Pressure, Conical Valve and Calculated C.O.P

SL. No.	Pressure	Conical Valve	C.O.P
1.	8.3	0.5	0.066
		1	0.073
2.	7.6	0.5	0.063
		1	0.061
3.	7.2	0.5	0.059
		1	0.038

Figures 11.21–11.23 display the C.O.P vs. pressure for 1 rotation.

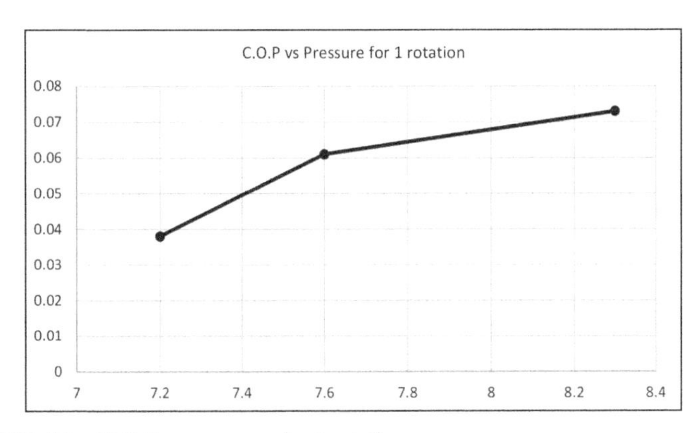

FIGURE 11.21 C.O.P vs. pressure for 1 rotation.

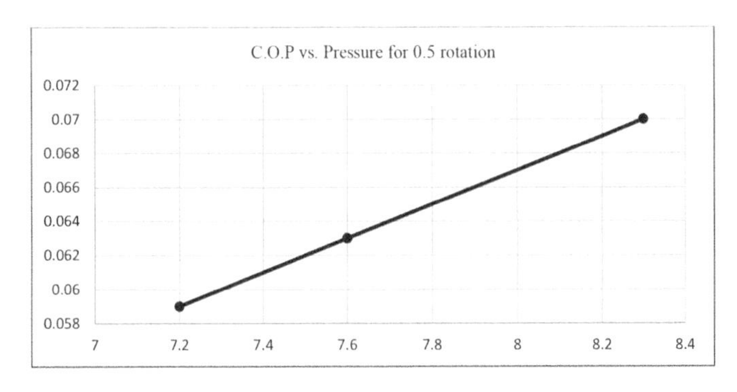

FIGURE 11.22 C.O.P vs. pressure for 0.5 rotations.

11.6.2 RESULT OF A RISE IN PRESSURE ON COP AT DIFFERENT VALVE OPENINGS

The figure displays the bearing of a rise in pressure on COP at different locations of valve opening. We can make an observation that the COP follows a direct proposition with pressure. It follows a straight-line relationship which clearly shows the impact of pressure in producing the cooling effect.

11.7 CONCLUSION

Based on the research, few conclusions can be drawn. Temperature drops and flows work in direct proposition to inlet pressure. The performance of the vortex tube suffers on account of the opposite exerted on the cold air exhaust. Low back pressure measuring till 2 PSIG (0.1 bar), has zero effect on the performance but for pressure to tune of 5 PSIG (0.3 bar) changes the performance by around 5°F (2.8°C).

A Vortex tube has the capacity to drop the air temperature from the temperature when the air was accepted in. High input temperatures will result in a corresponding rise in cold air temperatures. The temperature drops or gains are proportional to the inlet temperature.

During testing two different types of valves were used viz.

i. Needle valve (bolt type); and
ii. Inline conical valve.

It was observed that the Inline conical valve gave the maximum temperature drop. Thus, this valve was recommended for further performance testing of the Vortex Tube.

The nozzle used is altogether different than the one that was used by Prof. Parulekar in his short vortex tube. External U-threads were used for the tangential entry. The air attains the path followed by the threads with small helix angle and this helps producing the free vortex which after facing the restriction in the flow (Hot end control valve), becomes the forced vortex traveling in opposite direction with same angular velocity corresponding to sonic linear velocity.

KEYWORDS

- **cold and hot air**
- **nozzle**
- **nozzle diameter**
- **pressure of inlet**
- **temperature of cooling**
- **valve**
- **vortex tube**

REFERENCES

1. Baghdad, M., Ouadha, A., & Addad, Y., (2018). Effects of kinetic energy and conductive solid walls on the flow and energy separation within a vortex tube. *International Journal of Ambient Energy, 42*(3), 297–313. doi: 10.1080/01430750.2018.1550013.
2. Bidwaik, A., & Dhavale, S., (2015). To study the effects of design parameters on vortex tube with CFD analysis. *International Journal of Engineering Research And, V4*(10). doi: 10.17577/ijertv4is100149.
3. Devade, K. D., & Pise, A., (2017). Parametric review of Ranque-Hilsch vortex tube. American *Journal of Heat and Mass Transfer*. doi: 10.7726/ajhmt.2017.1012.
4. Dhande, H. K., Shelare, S. D., & Khope, P. B., (2020). Developing a mixed solar drier for improved postharvest handling of food grains. *Agricultural Engineering International: CIGR Journal, 22*(4), 166–173.
5. Hu, Z., Li, R., Yang, X., Yang, M., Day, R., & Wu, H., (2020). Energy separation for Ranque-Hilsch vortex tube: A short review. *Thermal Science and Engineering Progress, 19*, 100559. doi: 10.1016/j.tsep.2020.100559.
6. Jawalekar, S. B., & Shelare, S. D., (2020). Development and performance analysis of low cost combined harvester for rabi crops. *Agricultural Engineering International: CIGR Journal, 22*(1), 197–201.
7. Khosravi, M., Sadi, M., Arabkoohsar, A., & Ebrahimi-Moghadam, A., (2020). Numerical investigation of the thermo-hydraulic characteristics for annular vortex tube: A comparison with Ranque–Hilsch vortex tube. *Journal of Energy Resources Technology, 142*(11). doi: 10.1115/1.4047969.
8. Mali, P., Sakhale, C., & Shelare, S., (2015). A literature review on design and development of maize thresher. *International Journal of New Technologies in Science and Engineering, 3*(9), 9–14.
9. Mathew, J. J., Sakhale, C. N., & Shelare, S. D., (2020). Latest trends in sheet metal components and its processes—A literature review. *Algorithms for Intelligent Systems*, 565–574. doi: 10.1007/978-981-15-0222-4_54.

10. McGavin, P., & Pontin, D. I., (2018). Vortex line topology during vortex tube reconnection. *Physical Review Fluids, 3*(5). doi: 10.1103/physrevfluids.3.054701.
11. Mowade, S., Waghmare, S., Shelare, S., & Tembhurkar, C., (2019). Mathematical model for convective heat transfer coefficient during solar drying process of green herbs. *Computing in Engineering and Technology*, 867–877. doi: 10.1007/978-981-32-9515-5_81.
12. Purwanta, J., (2020). Engineering of cement dust catching nozzle diameter optimization on the tools fogging method. In: *2nd International Conference On Earth Science, Mineral, And Energy.* doi: 10.1063/5.0012095.
13. Rafiee, S., & Sadeghiazad, M., (2016). Heat and mass transfer between cold and hot vortex cores inside Ranque-Hilsch vortex tube-optimization of hot tube length. *International Journal of Heat and Technology, 34*(1), 31–38. doi: 10.18280/ijht.340105.
14. Saberi, A., Rahimi, A. R., Parsa, H., Ashrafijou, M., & Rabiei, F., (2016). Improvement of surface grinding process performance of CK45 soft steel by minimum quantity lubrication (MQL) technique using compressed cold air jet from vortex tube. *Journal of Cleaner Production, 131*, 728–738. doi: 10.1016/j.jclepro.2016.04.104.
15. Sadi, M., & Farzaneh-Gord, M., (2014). Introduction of annular vortex tube and experimental comparison with Ranque–Hilsch vortex tube. *International Journal of Refrigeration, 46*, 142–151. doi: 10.1016/j.ijrefrig.2014.07.004.
16. Sahu, P., Shelare, S., & Sakhale, C., (2020). Smart cities waste management and disposal system by smart system: A review. *International Journal of Scientific & Technology Research, 9*(3), 4467–4470.
17. Saidi, M. H., & Valipour, M. S., (2003). Experimental modeling of vortex tube refrigerator. *Applied Thermal Engineering, 23*(15), 1971–1980. doi: 10.1016/s1359-4311 (03)00146-7.
18. Shelare, S. D., Kumar, R., & Khope, P. B., (2020). Formulation of a mathematical model for quantity of deshelled nut in charoli nut deshelling machine. *Advances in Metrology and Measurement of Engineering Surfaces*, 89–97. doi: 10.1007/978-981-15-5151-2_9.
19. Sinha, J., Hauhnar, M., & Dubey, A., (2020). Validation of vortex tube experimentally. In: *8th International Conference on Reliability, Infocom Technologies and Optimization (Trends and Future Directions) (ICRITO)*. doi: 10.1109/icrito48877.2020.9197919.
20. Waghmare, S. N., Sakhale, C. N., Tembhurkar, C. K., & Shelare, S. D., (2019). Assessment of average resistive torque for human-powered stirrup making process. *Computing in Engineering and Technology,* 845–853. doi: 10.1007/978-981-32-9515-5_79.
21. Waghmare, S., Mungle, N., Tembhurkar, C., Shelare, S., Sirsat, P., & Pathare, N., (2019). Design and analysis of power screw for manhole cover lifter. *International Journal of Recent Technology and Engineering, 8*(2), 2782–2786. doi: 10.35940/ijrte.B2628.078219.
22. Waghmare, S. N., Shelare, S. D., Tembhurkar, C. K., & Jawalekar, S. B., (2020). Development of a model for the number of bends during stirrup making process. *Advances in Metrology and Measurement of Engineering Surfaces*, 69–78. doi: 10.1007/978-981-15-5151-2_7.

CHAPTER 12

PERFORMANCE COMPARISON OF DENOISING METHODS FOR FETAL PHONOCARDIOGRAPHY USING FIR FILTER AND EMPIRICAL MODE DECOMPOSITION (EMD)

NIKITA JATIA and KARAN VEER

Instrumentation and Control Department, Dr. B. R. Ambedkar National Institute of Technology, Jalandhar, Punjab, India, E-mails: nikitajatia2604@gmail.com (N. Jatia), veerk@nitj.ac.in (K. Veer)

ABSTRACT

Long-term and frequent checking of a fetal health state is still a difficult task, mainly in high-risk type of pregnancies. This chapter shows a fully non-invasive technique to extract out the fetal heart sound (FHS) from sound signals which are recorded from the surface of mothers abdominal. This chapter mainly relates to fetal phonocardiogram (fPCG) processing of signal. Signal denoising is always a major task after recording the signal, so in this chapter basically, two types of denoising methods that are finite impulse response filter (FIR Filter) and empirical mode decomposition (EMD) methods are used and compared. For testing, or comparing these two, recordings of real dataset were used, and the estimate depends

Optimization Methods for Engineering Problems. Dilbagh Panchal, Prasenjit Chatterjee, Mohit Tyagi, Ravi Pratap Singh (Eds.)

on cognitive observation and signal advancement after performing both methods. The results in this chapter proved that both the methods assisted for improving fetal PCG signal by de-noising the signal. On the ground of the results, in the end we concluded that EMD as a suitable method for denoising and processing the fetal PCG signal.

12.1 INTRODUCTION

Heart which plays vital role in human body, hence whenever there is any effect on heart it is always a major disease or heart disease is major disease for a human body throughout the world. Treatment for these types of diseases can be efficient and easy if the disease is diagnosed at an early stage. Heart diseases which are also called Coronary Diseases are diagnosed successfully and easily using stethoscope as it demands less equipment. Sometimes auscultations can be the only accessible way for examining, such as in PHC (primary health care centers), at these centers or in rural areas such type of high-fi instruments are not easily available for diagnosis and cases of children where alternative facilities like ECG are difficult to perform. Traditional listening of heart sound mainly requires broad training, storage of records for further use and experience which is not possible at every place [1]. To overcome these types of shortcomings, one can use automatic auscultation using electronic stethoscopes. Doctors can observe any problem in heart by just listening heart sound, and heart sound is comprising of four types of components from the two are dominant one which is S1 and S2 and are caused by closure of valves. S3 is the result of the first rapid filling and produce vibration of ventricular valves, S4 take place when the atria valve contract in 2nd phase of the ventricular filling. S1 is lower in pitched and shorter duration and can be listen at mitral auscultation which is caused by closing of tricuspid and mitral valves. S2 is a louder sound and caused by the closing of the aortic valve and pulmonary valve [2].

The Phonocardiogram (PCG) is a plot of the heart sound, which demonstrate the mechanical activity of the heart [15]. These types of sounds are produced by the closing and opening of heart valves and vibration of cardiovascular structures. Adult PCG contains two major and (commonly) distinct components known as S1 and S2, due to the time interval between the closures of atrioventricular and semilunar valves. In fetal PCG, this time interval is much shorter than that of the adult. For this reason, while auscultation, typically one dominant sound can be heard. The average FHR at term is 110 to 160 beats per minute (BPM), and as the fetus matures, the average FHR baseline decreases, while the heart rate variability (HRV) increases [3]. Cardiac auscultation is a process to listen and analyze the sound of the heart using an electronic stethoscope, which gives digital recordings of the heart sound, called phonocardiogram (PCG). This signal contains useful information mainly about the condition and functionality of the heart. Based on that PCG signal, a signal of heart sound is divided in two types of categories, i.e., abnormal and normal. Stages which are involved in this processing the signal and diagnosing heart sound signal are, recording heart sound signal, then elimination of noise, sampling of PCG signal at a particular frequency rate, extraction of features, classification, and training. From the signal handling perspective, the PCG is a sound signal having energy mostly focused between 20 and 200 Hz [1]. It is a highly nonstationary signal with pseudo-periodic time-frequency characteristics. Although there is no deterministic behavior in this signal, there are clear similarities between local energy trends of the PCG successive beats in the time-frequency (TF) domain.

12.2 METHODOLOGY

Database is created using recordings of 109 pregnant females (aged between 16 and 47 years, mean ±SD of 29.3±5.8 with Body Mass Index between 19.5 and 38.9 mean±SD of 29.2±4.0), using digital JABES Electronic stethoscope of GS Technology Co. Ltd., South Korea placing on mother's lower abdominal as explained by Samieinasab and Sameni [4].

Twin's cases, (seven cases) the data is recorded two times by placing the sensor advised by the gynecologist or experts. Audio software is Audacity® cross-platform, used for collecting recording and after that performing editing the signals on a computer. From 109 subjects, 99 subjects were having single signal recording, three subjects having two and seven cases of twins were recorded separately, in total 119 recordings. 90 seconds is the mean time for each recording, with sampling rate of 16,000 Hz and 16-bit quantization.

12.3 METHOD

12.3.1 FINITE IMPULSE RESPONSE (FIR) FILTER

Linear type of filtration is known as the main process for removing unwanted or distorted signal components; here, the most frequent one is harmonic interference. Therefore, this is mainly applied where there is a mixture of valuable signal along with noisy signal is provided as input to the filter. These types of filters are generally frequency selective. Hence, they can remove elements of signal at a particular frequency band.

Fetal PCG signal is non-deterministic and non-stationary. Traditional filtering filtered the data by improving SNR and eliminating noise or signal which is outside the band pass filter, but then to some of the noise remains in the fetal PCG signal [5]. FIR filters are convolutional and non-recursive in nature. This type of filter achieves a linear phase-frequency characteristic. But there is a similar type of delay for the individual harmonic components and they are distortion less. As shown in the equation below Y_n shows the response of linear filter and h_i is impulse characteristics. Waveform of modulus frequency characteristic is one of the basic methods for designing FIR filter [1].

$$Y_n = h_0.x_n + h_1.x_n - 1 + h_n - 1.x_n - (n - 1) \quad (1)$$

12.3.2 EMPIRICAL MODE DISTRIBUTION

Empirical mode decomposition (EMD), recently created by Huang et al. [9] this technique is best used for decomposing any nonlinear signal and nonstationary signal with the oscillating components following some fundamental qualities. The most important key feature of EMD is that it can be used as a spontaneous decomposition and completely data adaptive method. The process of the EMD method is to decompose a given noisy signal into a summation of band-limited functions called intrinsic mode function (IMFs). Each of the IMF fulfills two basic conditions.

- In the given signal it is observed that the number of extrema and zero-crossings in the signal should be differ or same atmost by one; and
- Average values of IMF wave must be zero.

There are many ways to compute the EMD method [5–10]. Below shown, Algorithm is used here to decompose signal into various IMFs components [14].

1. To find Maxima and Minima, Figure 12.1(a) [16].

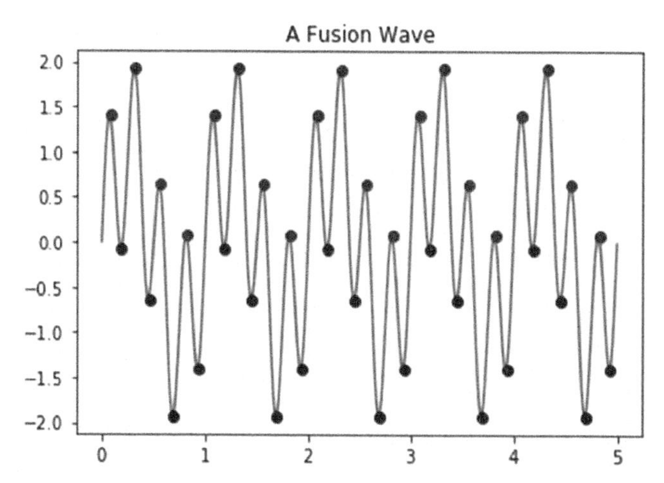

FIGURE 12.1(a) To find maxima and minima.

2. Generate an envelope of maxima and minima from the array of maxima and minima shown in Figure 12.1(b) [16].

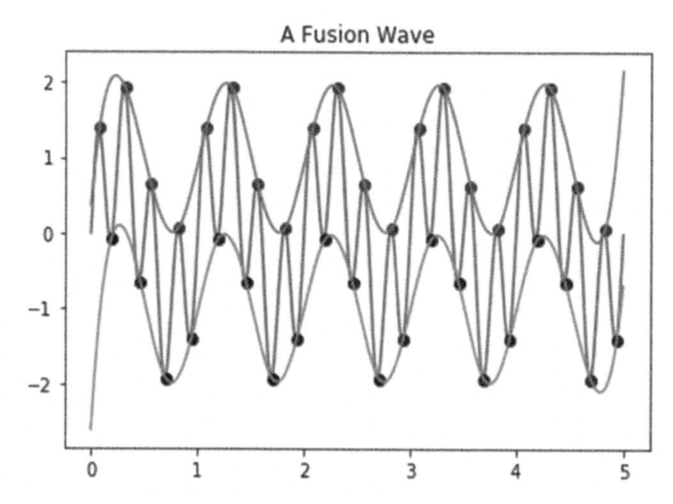

FIGURE 12.1(b) Envelope of maxima and minima.

3. Get the middle value or number from the envelope of maxima and minima as in Figure 12.1(c) [16].

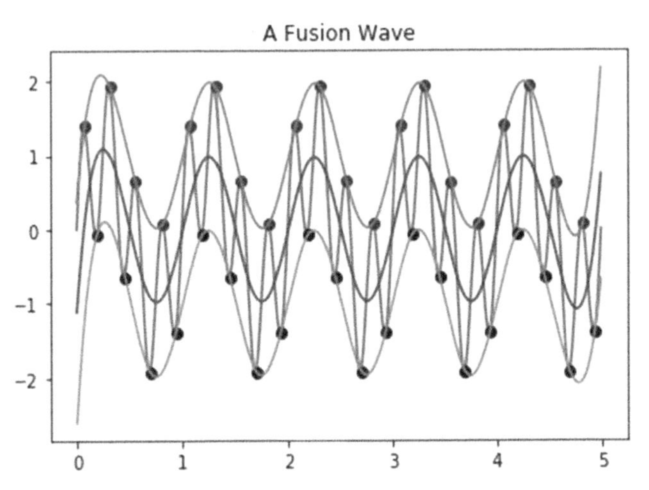

FIGURE 12.1(c) Number from the envelope of maxima and minima.

4. Reduce the number or value of real dataset signal from the mid value of the envelope depicted in Figures 12.1(d) and 12.1(e) [16].

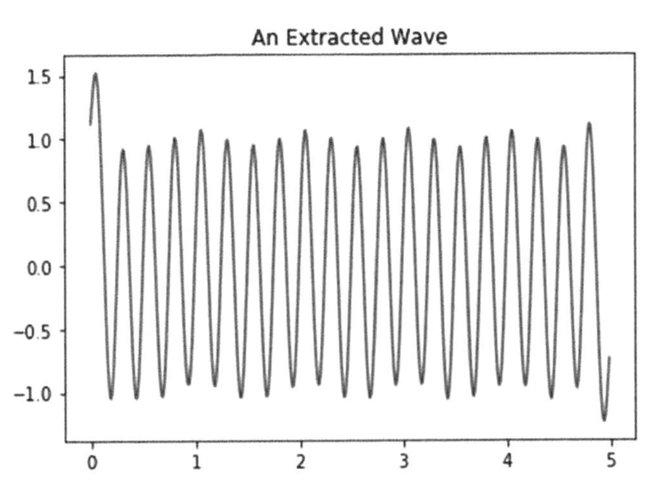

FIGURE 12.1(d) Value (extracted wave) of real dataset signal from the mid value of the envelope.

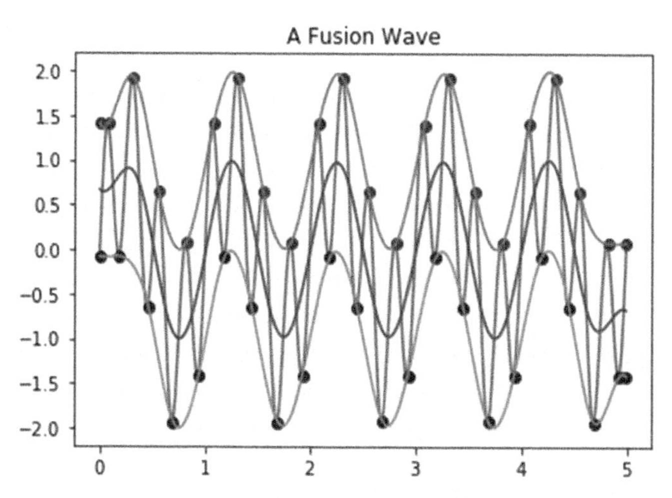

FIGURE 12.1(e) Value (fusion wave) of real dataset signal from the mid value of the envelope.

5. Check that the extracted signal that it is IMF or not [16].
6. Reduce the original dataset signal earlier IMF extracted in Step 5, depicted in Figures 12.1(f) and 12.1(g) [16].

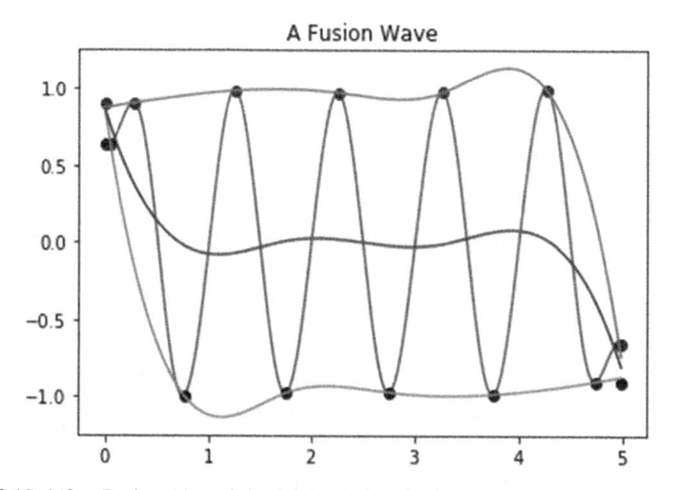

FIGURE 12.1(f) Reduce the original dataset signal.

The main reason for decomposing fetal PCG recordings into the IMFs is to allow the fundamental cardiac sounds in various oscillating components, which increases the monitoring of the system. Besides, the time and frequency depiction of the PCG recordings can deliver important information about the automatic detection of heart sounds and detecting pathologies from patterns of these type of heart sounds in the IMF.

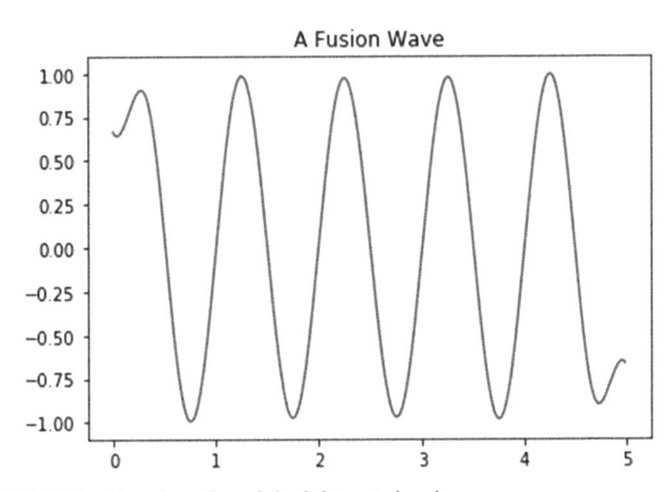

FIGURE 12.1(g) To reduce the original dataset signal.

12.4 RESULT AND DISCUSSION

12.4.1 DATASET

The dataset of fetal PCG signals, which used in my work, data is accessible in Physio Bank archive. This dataset was taken at Hafez Hospital in Shiraz University of Medical Sciences. The dataset was taken by using recordings of 109 pregnant women (whose age is in between 16 and 47, Body Mass Index between 19.5 and 38). These phonograms were taken by digital JABES stethoscope of GS Technology Co. Ltd., South Korea, placed at the lower side of the abdomen of pregnant women. Audacity® software employed in the whole process for recording and editing acoustic recordings on computer. In total, 119 phonograms were obtained. 90 seconds is the mean length of every recording, f_s or sampling frequency is 16,000 Hz with around 16-bit Quantization and various recordings were at f_s of 44,100 Hz. Frequency range is in between 20 Hz and 1 kHz [5].

12.4.2 EXPERIMENT ON REAL DATA

After performing test on real type of data it is found that right use of FIR filter band is in between 20 Hz and 200 Hz with the sampling frequency of 16,000 Hz and filter's order at 1,000. Figure 12.2 represent a real type of recording of the f102m.mat after and before filtering the data using FIR filter. Utilizing vision monitoring, it is observed that the filtration decreases ambient noise of signal.

And the second method is EMD, EMD is not bound to use in frequency spectrum. This is the main key feature of EMD which is generally not present in any other type of filtering method. And the output from EMD is also in time spectrum. There is no requirement of assuming EMD as periodic signal and it is not depending on simple sine wave, it is based on Intrinsic Mode Function (IMF) (Figure 12.3).

12.5 CONCLUSION

To enhance the noise suppression operation of fetal PCG an efficient denoising techniques based on FIR filters and EMD are generally applied by the not-adaptive type of signal processing algorithms. Here is my work, testing to check the performance for extracting a fetal PCG signal from abdominal PCG using real type of data. Testing was performed on each of the individual for checking the efficiency of the signal before and after performing the filtering process. There are so many limitations in the FIR filter then to it has stayed the most powerful technique for signal analysis and processing [14]. Even though it is unsuitable for analyzing signals which are generated by non-linear and non-stationary processes, i.e., real-world, or natural signals. But then comes the savior EMD, and its capability to analyze real-world natural signals The ability of EMD to decompose.

A real-world signal into their true and meaningful components has been widely appreciated in the various fields of science and engineering. And the result which we get after performing both the methods, in my work also validate that EMD achieved better results. The main feature of EMD is that it does not care about the signal is periodic or not and it is not based on the simple sine wave type rather it is based on Intrinsic Mode Function (IMF). It is powerful spontaneous decomposition algorithm for

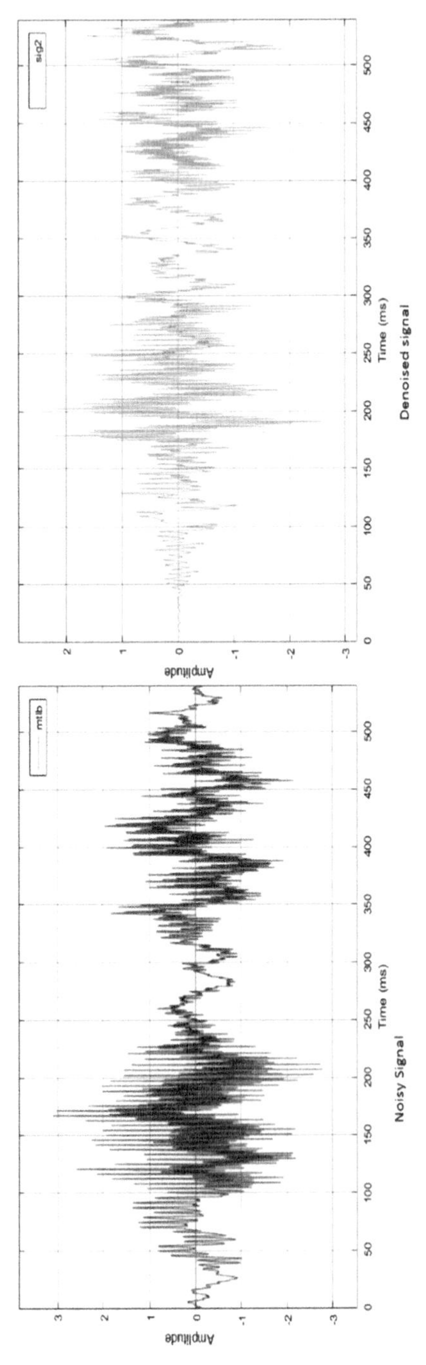

FIGURE 12.2 Noisy signal and filtered signal using FIR filter.

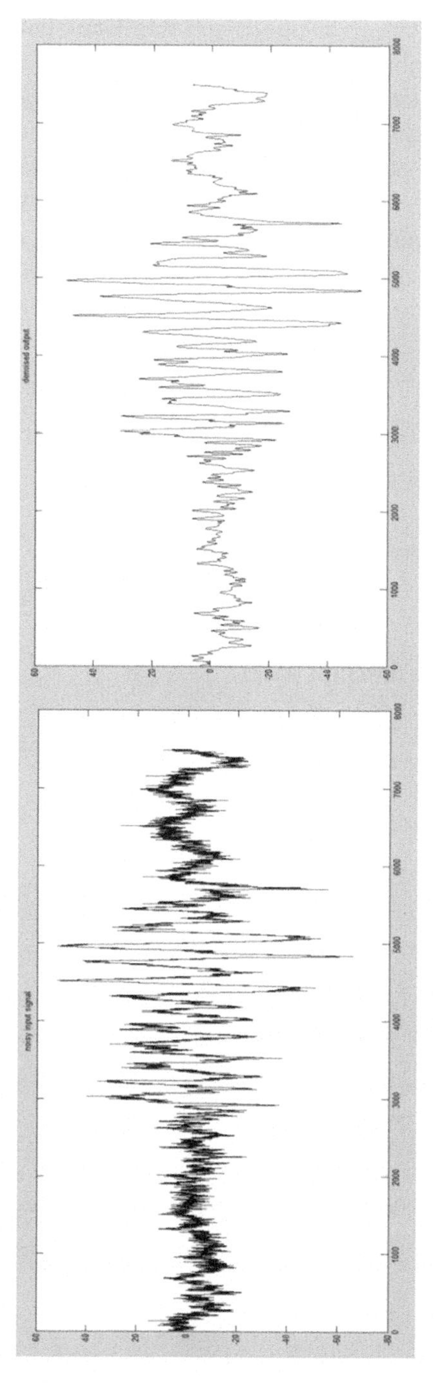

FIGURE 12.3 Noisy signal and filtered signal using EMD method.

analyzing nonstationary, non-linear, and non-Gaussian signals [12] and base function is not obligatory [13].

Though, the data taken from physio Bank is not of good quality because of bad quality of sensing elements, quality results are not achieved. On real data testing, fetal age performs a vital role. Throughout the process of pregnancy, the heart rate gradually expands and fixes in the 18th week. However, it is advised to examine the heart sound in the 36th week of gestation at that time heart sound is perfectly audible and sound is stable too.

KEYWORDS

- **coronary diseases**
- **denoising**
- **empirical mode decomposition (EMD)**
- **fetal PCG**
- **FIR filter**
- **intrinsic mode function (IMF)**

REFERENCES

1. Rene, J., Radek, M., Radana, K., Vanus, Marcel, F., & Jan, N., (2018). Comparison of fetal phonocardiography denoising by wavelet transform and the FIR filter. In: *2018 IEEE 20th International Conference on e-Health Networking, Applications and Services (Healthcom)*.
2. Moghavvemi, M., Tan, B. H., & Tan, S. Y., (2003). A non-invasive pc-based measurement of fetal phonocardiography. *Sensors and Actuators A: Physical, 107*(1), 96–103.
3. Adithya, P. C., Sankar, R., Moreno, W. A., & Hart, S., (2017). Trends in fetal monitoring through phonocardiography: Challenges and future directions. *Biomedical Signal Processing and Control, 33*, 289–305.
4. Samieinasab, M., & Sameni, R., (2015). Fetal phonocardiogram extraction using single channel blind source separation. In: *Electrical Engineering (ICEE), 2015 23rd Iranian Conference*. IEEE.
5. Ana, G., Marijeta, S., Irini, R., & Branimir, R., (2013). *Application of Wavelet and EMD-Based Denoising to Phonocardiograms*. 978-1-4673-6143-9/13/$31.00 ©2013 IEEE.

6. Beya, O., Jalil, B. B., Fauvet, E., & Laligant, O., (2010). Empirical modal decomposition applied to cardiac signals analysis. *Proc. of SPIE-IS&T Electronic Imaging, SPIE, 7535*. 75350B -1-11.
7. Huang, N. E., Shen, Z., Long, S. R., Wu, M. L., Shig, H. H., Quanan, Z., Yen, N. C., et al. (1998). The empirical mode decomposition and the Hilbert spectrum for nonlinear and non-stationary time series analysis. *Proc. Roy. Soc. A, 454*, 903–955.
8. Kais, K., & Abdel-Ouahab, B., (2013). Audio watermarking via EMD. *IEEE Transactions on Audio, Speech and Language Processing, 21*(3).
9. Cesarelli, M., Ruffo, M., Romano, M., & Bifulco, P., (2012). Simulation of foetal phonocardiographic recordings for testing of FHR extraction algorithms. *Computer Methods and Programs in Biomedicine, 107*, 513–523.
10. Abdelkader, R., Kaddour, A., & Derouiche, Z., (2018). Enhancement of rolling bearing fault diagnosis based on improvement of empirical mode decomposition denoising method. The *International Journal of Advanced Manufacturing Technology, 97*(5–8), 3099–3117.
11. Sengottuvel, S., Khan, P. F., Mariyappa, N., Patel, R., Saipriya, S., & Gireesan, K., (2018). A combined methodology to eliminate artifacts in multichannel electrogastrogram based on independent component analysis and ensemble empirical mode decomposition. *SLAS Technology: Translating Life Sciences Innovation, 23*(3), 269–280.
12. Sunan, Z., Jianyan, T., Amit, B., & Jiangli, L. (2019). An efficient porcine acoustic signal denoising technique based on EEMD-ICA-WTD. *Hindawi Mathematical Problems in Engineering, 2019*, 12. Article ID 2858740.
13. Goldberger, A., Amaral, L., Glass, L., Hausdorff, J., Ivanov, C., Mark, P. R., Mietus, J., et al., (2000). Components of a new research resource for complex physiologic signals. *PhysioBank PhysioToolkit Physio Net, 101*(23), e215–e220.
14. Gabriel, R., Patrick, F., & Paulo, G., (2013). On empirical mode decomposition and its algorithms. *ISSR Journals.*
15. Dharmendra, P., Uma, M. T., Prabakaran, K., & Suguna, A., (2013). Detection of heart diseases by mathematical artificial intelligence algorithm using phonocardiogram signals. *ISSR Journals.*
16. Ryan, M., (2019). *Decomposing Signal Using Empirical Mode Decomposition Algorithm Explanation for Dummy*. Article from towards data science.
17. Fateh, B. A., Abdelouhed, S., & Mohamed, B. M., (2014). Choosing interpolation RBF function in image filtering with the bidimensional empirical modal decomposition. In: *1st International Conference on Advanced Technologies for Signal and Image Processing-ATSIP'2014*. Sousse, Tunisia.

CHAPTER 13

SYSTEMATIC SURVEY, PERFORMANCE EVALUATION, AND TRUTH FLOW ANALYSIS OF TWO SUBSONIC WIND TUNNELS WITH TWO-HOLE SPHERICAL FLOW ANALYZER

AKHILA RUPESH[1] and J. V. MURUGA LAL JEYAN[2]

[1]*PhD Scholar, Department of Aerospace Engineering, School of Mechanical Engineering, Lovely Professional University, Punjab–144111, India, E-mail: akhilarupesh56@gmail.com*

[2]*Professor, Department of Aerospace Engineering, School of Mechanical Engineering, Lovely Professional University, Punjab–144111, India, E-mail: jvmlal@ymail.com*

ABSTRACT

As a part of the design process, it is essential to predict the aerodynamics of the flow as it moves over a streamlined body. This is executed with the use of wind tunnels of appropriate Mach numbers, with the influence of computational analysis. But it is mandatory to verify the theoretical results to the experimental results carried out in a wind tunnel. Before testing, it is necessary that the flow parameters, namely the velocity distribution, angularity of the flow, turbulence, are evaluated and are within the margins. This is important when we are evaluating the flow in three dimensions. Thus, ensuring the reliability of the experimental

Optimization Methods for Engineering Problems. Dilbagh Panchal, Prasenjit Chatterjee, Mohit Tyagi, Ravi Pratap Singh (Eds.)

results. In order to evaluate the truthiness of the flow, a two-hole spherical flow analyzer is being used. The truthiness of two wind tunnels across India is being evaluated by a two-hole spherical flow analyzer. From the experiment, the data map will be plotted for the flow parameters using the yaw head constant and the respective orientations of the two-hole spherical flow analyzer. Their nature is evaluated and the findings are discussed.

13.1 INTRODUCTION

Scrutiny of flow patterns around scale models and the measure of aerodynamic forces or pressure upon or the velocity around these structures is the primary purpose of a wind tunnel [1, 2]. Determination of mean values and also the uniformity of various flow parameters say, stagnation pressure, temperature, velocity, Mach number and flow angularity in the region where the model is tested is the major step involved in wind tunnel calibration. However, determination of these flow parameters by currently available methods, demands the flow to be maintained isentropic and laminar, which has been one of the major problems faced during calibration. Determination of flow angle is a crucial requirement for maintaining a laminar flow inside the test section and the operation involves wind tunnel calibration [3, 4]. Though there are wide ranges of instruments available for the calibration of wind tunnel, accurate results are not guaranteed. The major part of calibration is achieved through the measurement of pressures and temperatures and in general, the most basic and useful tool in the calibration of wind tunnels is the measurement of pressures which are sensitive to hole size of the probes and hence it is necessary to keep them as small as possible [5, 6]. Basically, pressure probes are instruments used to determine the associated pressures and also the flow angle within a fluid stream. Design of pressure probes for fluid measurements involves the consideration of the factors, say blockage effects, hole geometry and size, frequency response, local Mach, and Reynolds numbers, etc. Better accuracy is obtained when smaller probes are used. However, they require longer time to respond and also face greater problems regarding contamination [7]. Pressure probes must be designed to be susceptible to the flow direction, and in consequence, the probes are essentially calibrated to determine the effects when they are aligned to the flow field which when rotated about their axis can also give the effects of pitch and yaw [8, 9].

13.2 METHODOLOGY

13.2.1 DESIGN OF TWO-HOLE SPHERICAL FLOW ANALYZER

A two-hole spherical flow analyzer is an instrument used for the measurement of flow parameters in order to justify the flow angularity in a wind tunnel. The holes on the instrument are connected to manometer with gives us the pressure measurements. In a two-hole spherical flow analyzer if the reading on the two ports is the same, this indicates that the nature of the flow is the same in the vicinity of the ports. In a case where it is different, it points towards an angularity in the flow. The two-hole spherical flow analyzer used here has two ports at an angle of 45° from the axis of the instrument and both holes are at 90° w.r.t each other. The specification of the instrument is given in Table 13.1.

TABLE 13.1 Specifications of the Instrument

Features	Dimensions
The internal span of the sphere	25
The external span of the sphere	29
The internal span of the pipe 2	2
The external span of the pipe 3	3
Extent of cylinder pipe 75	75
Extent of pipe from the center of the sphere	90
Depth of sphere	2
The angle between curved pipes	135°

The designed and modeled instrument is shown in Figure 13.1.

13.2.2 MOUNTING OF TWO-HOLE SPHERICAL FLOW ANALYZER IN TEST FACILITY

The procedure involves positioning the instrument in a wind tunnel at a measured distance in the test section. It is made sure the longitudinal axis and the axis of the instrument are parallel to each other. This is again clarified by the readings on the manometer being the same. The angle between the axis of the two-hole spherical flow analyzer and the axis of the wind tunnel gives us the inclination Ψ measured in degrees. The pressure measurements are recorded at each successive inclination of the instrument. Simultaneously a pitot static tube connected to a sensor is used to find the dynamic pressure of the flow. A curve of Yaw head constant v/s Angle is plotted and the curve is evaluated for truthiness. Figure 13.2 shows two-hole spherical flow analyzer mounted in the specified test facility.

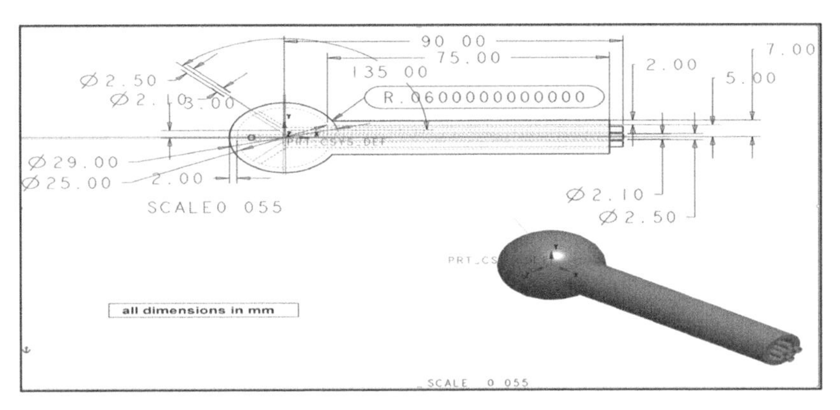

FIGURE 13.1 Two-hole spherical flow analyzer (design and model).

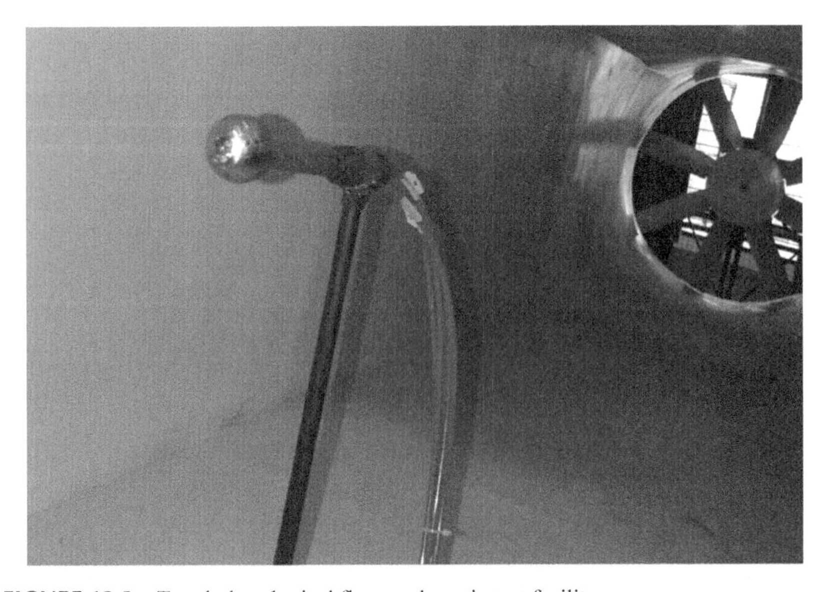

FIGURE 13.2 Two-hole spherical flow analyzer in test facility.

13.3 RESULTS AND DISCUSSION

13.3.1 SURVEY RESULT-1

Table 13.2 gives the wind tunnel specification for survey 1 and Table 13.3 gives test results for survey 1 at 30 m/s.

TABLE 13.2 Wind Tunnel Specification for Survey 1

Organization	MVJ College of Engineering
Test section size	600 × 600 mm
Velocity max	70 m/s
Contraction ratio	9:1
Power supply	3 phase AC – 440 W–64 Amps

TABLE 13.3 Test Results for Survey 1 at 30 m/s Test Section Velocity

SL. No.	L (cm)	Velocity (V) (m/s)	q (P) Pascals	Ψ	P_A (Pa)	P_B (Pa)	ΔP = (P_A ~ P_B) (Pa)	ΔK = (ΔP/q)
1.	10	30	551	0.0	397.05	397.05	0	0
	10	30	551	0.2	397.12	397.01	0.11	0.0002
	10	30	551	0.4	397.17	396.97	0.2	0.000363
	10	30	551	0.8	397.21	396.92	0.29	0.000526
	10	30	551	1.0	397.28	396.88	0.4	0.000726
2.	20	30	551	0.0	406.31	406.31	0	0
	20	30	551	0.2	406.38	406.27	0.11	0.0002
	20	30	551	0.4	406.43	406.21	0.22	0.000399
	20	30	551	0.8	406.48	406.16	0.32	0.000581
	20	30	551	1.0	406.53	406.09	0.44	0.000799
3.	30	30	551	0.0	369.29	369.29	0	0
	30	30	551	0.2	369.35	369.22	0.13	0.000236
	30	30	551	0.4	369.39	369.17	0.22	0.000399
	30	30	551	0.8	369.44	369.14	0.3	0.000544
	30	30	551	1.0	369.49	369.08	0.41	0.000744
4.	40	30	551	0.0	386.24	386.24	0	0
	40	30	551	0.2	386.29	386.2	0.09	0.000163
	40	30	551	0.4	386.33	386.16	0.17	0.000309
	40	30	551	0.8	386.39	386.11	0.28	0.000508
	40	30	551	1.0	386.42	386.7	0.28	0.000508

From the result shown in Figures 13.3 and 13.4, it indicates a steady level increase in turbulence level for speed of 30 m/s at 0.1 m from the entry section of the test section. The curve at 0.2 m is showing a steep increase in turbulence level as the yaw angle is increased. A similar pattern curve is obtained when the instrument is placed at 0.3 m from the inlet of the test section. But as the instrument is moved further to a distance 0.4 m away from the test section inlet, the curve for yaw head constant is showing a sudden decrease which depicts clearly about the flow angularities existing in the tunnel. The results are clearly showing that the yaw head varies for even a small change in the degree of yaw. Table 13.4 gives test results for survey 1 at 45 m/s velocity inside the test section.

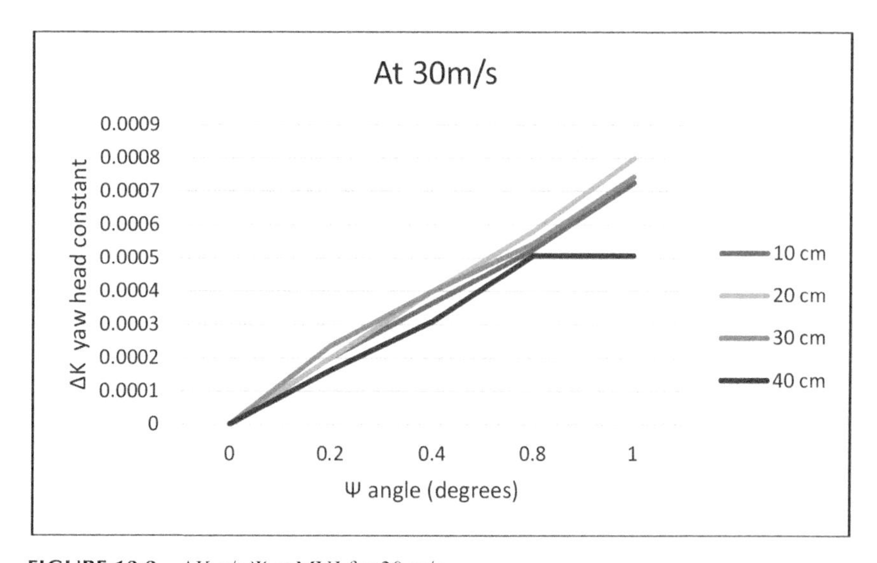

FIGURE 13.3 ΔK v/s Ψ at MVJ for 30 m/s.

From Figures 13.5 and 13.6, the nature of the curve indicates a linear progression for a velocity of 45 m/s at 0.1 m. As the position is changed to 0.2 m there is similar nature with a reduction in the yaw head. At 0.3 m distance, the curve is relatively same with a slight dip as the angle is increased. Further as the two-hole spherical flow analyzer is placed at 0.4 m the nature remains the same with a rise in yaw head confident. The result does not indicate any abrupt variations but there are noticeable deviations.

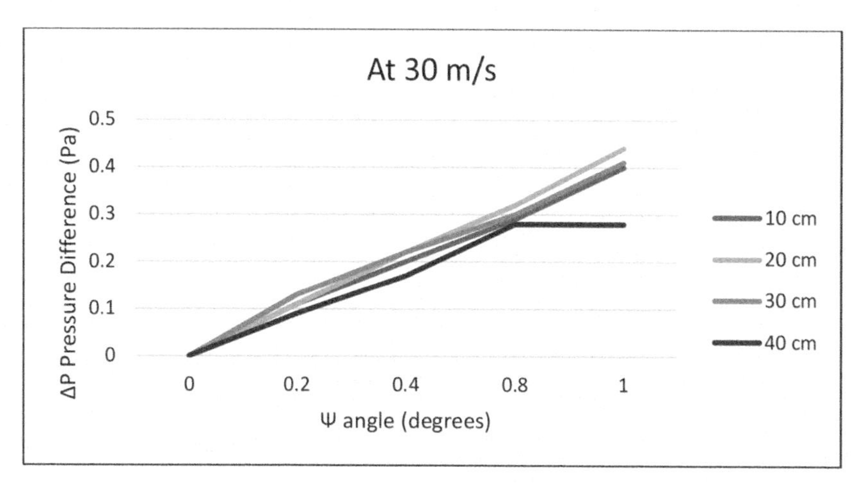

FIGURE 13.4 ΔP v/s Ψ at MVJ for 30 m/s.

TABLE 13.4 Test Results for Survey 1 at 45 m/s Test Section Velocity

SL. No.	L (cm)	Velocity (V) (m/s)	q (P) Pascals	Ψ	P_A (Pa)	P_B (Pa)	ΔP = (P_A ~ P_B)(Pa)	ΔK = (ΔP/q)
1.	10	45	1,240	0.0	664.79	664.79	0	0
	10	45	1,240	0.2	664.74	664.82	0.08	0.0000645
	10	45	1,240	0.4	664.69	664.87	0.18	0.000145
	10	45	1,240	0.8	664.62	664.92	0.3	0.000242
	10	45	1,240	1.0	664.58	664.96	0.38	0.000306
2.	20	45	1,240	0.0	590.89	590.89	0	0
	20	45	1,240	0.2	590.93	590.86	0.07	0.0000565
	20	45	1,240	0.4	590.97	590.81	0.16	0.000129
	20	45	1,240	0.8	591.02	590.76	0.26	0.00021
	20	45	1,240	1.0	591.08	590.72	0.36	0.00029
3.	30	45	1,240	0.0	649.84	649.84	0	0
	30	45	1,240	0.2	649.87	649.79	0.08	0.0000645
	30	45	1,240	0.4	649.91	649.72	0.19	0.000153
	30	45	1,240	0.8	649.95	649.68	0.27	0.000218
	30	45	1,240	1.0	649.99	649.63	0.36	0.00029
4.	40	45	1,240	0.0	720.19	720.19	0	0
	40	45	1,240	0.2	720.22	720.12	0.1	0.0000806
	40	45	1,240	0.4	720.28	720.08	0.2	0.000161
	40	45	1,240	0.8	720.33	720.02	0.31	0.00025
	40	45	1,240	1.0	720.39	719.99	0.4	0.000323

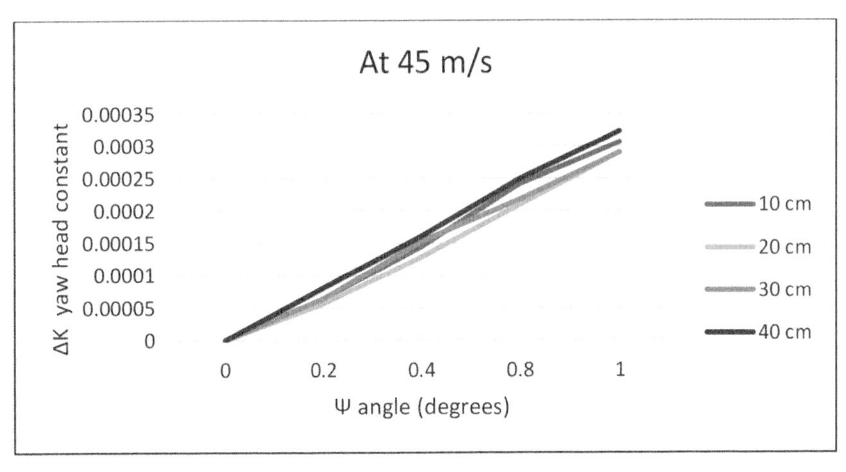

FIGURE 13.5 ΔK v/s Ψ at MVJ for 45 m/s.

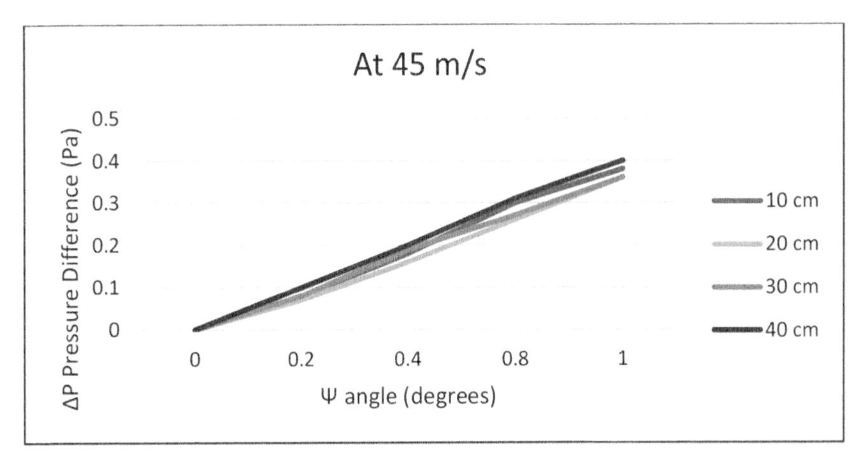

FIGURE 13.6 ΔP v/s Ψ at MVJ for 45 m/s.

13.3.2 SURVEY RESULT-2

Table 13.5 gives the wind tunnel specification for survey 2 and Table 13.6 gives test results for survey 2 at 15 m/s velocity inside the test section.

Figures 13.7 and 13.8 shows a curve for the experiment carried out at 15 m/s. For 0.1 m distance it is seeming that the curve increases steadily till an inclination of 0.2, beyond that there is a sharp decline in the slope. At a location of 0.2 m, the nature of the curve changes very abruptly, with a sharp increase in the slope till the inclination of 0.4 after which there is a sudden decline. The curve at 0.3 m and 0.4 m shows a relatively smaller slope which remains constant with the increase in inclination. It can be concluded that there is an angularity in the flow which is causing this random fluctuation of the curve. Table 13.7 gives test results for survey 2 at 25 m/s velocity inside the test section.

TABLE 13.5 Wind Tunnel Specification for Survey 2

Organization	**Satyam college of engineering and technology–Tamil Nadu**
Test section size	300 × 300 mm
Velocity max	65 m/s
Contraction ratio	6:1
Power supply	3 phase AC – 440 W–64 Amps

TABLE 13.6 Test Results for Survey 2 at 15 m/s Test Section Velocity

Sl. No.	**L (cm)**	**Velocity (V) (m/s)**	**q (P) Pascals**	**Ψ**	**P_A (Pa)**	**P_B (Pa)**	**ΔP = (P_A ~ P_B) (Pa)**	**ΔK = (ΔP/q)**
1.	10	15	138	0.0	37.02	37.02	0	0
	10	15	138	0.2	37.07	36.19	0.88	0.00638
	10	15	138	0.4	37.13	36.15	0.98	0.0071
	10	15	138	0.8	37.18	36.1	1.08	0.00783
	10	15	138	1.0	37.22	36.05	1.17	0.00848
2.	20	15	138	0.0	48.05	48.05	0	0
	20	15	138	0.2	48.1	48.01	0.09	0.000652
	20	15	138	0.4	48.6	47.95	0.65	0.00471
	20	15	138	0.8	48.21	47.9	0.31	0.00225
	20	15	138	1.0	48.27	47.88	0.39	0.00283
3.	30	15	138	0.0	55.39	55.39	0	0
	30	15	138	0.2	55.42	55.32	0.1	0.000725
	30	15	138	0.4	55.47	55.28	0.19	0.00138
	30	15	138	0.8	55.52	55.22	0.3	0.00217
	30	15	138	1.0	55.58	55.18	0.4	0.0029
4.	40	15	138	0.0	62.18	62.18	0	0
	40	15	138	0.2	62.22	62.13	0.09	0.000652
	40	15	138	0.4	62.27	62.08	0.19	0.00138
	40	15	138	0.8	62.32	62.01	0.31	0.00225
	40	15	138	1.0	62.36	61.98	0.38	0.00275

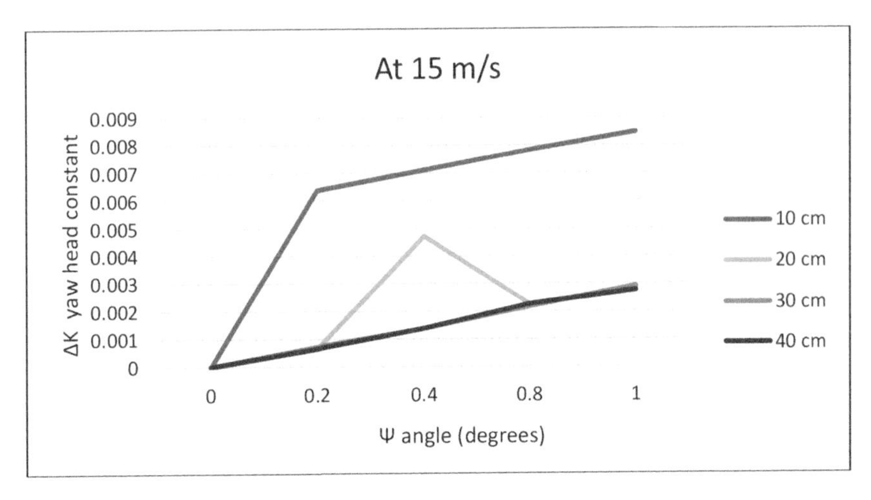

FIGURE 13.7 ΔK v/s Ψ at SCET for 15 m/s.

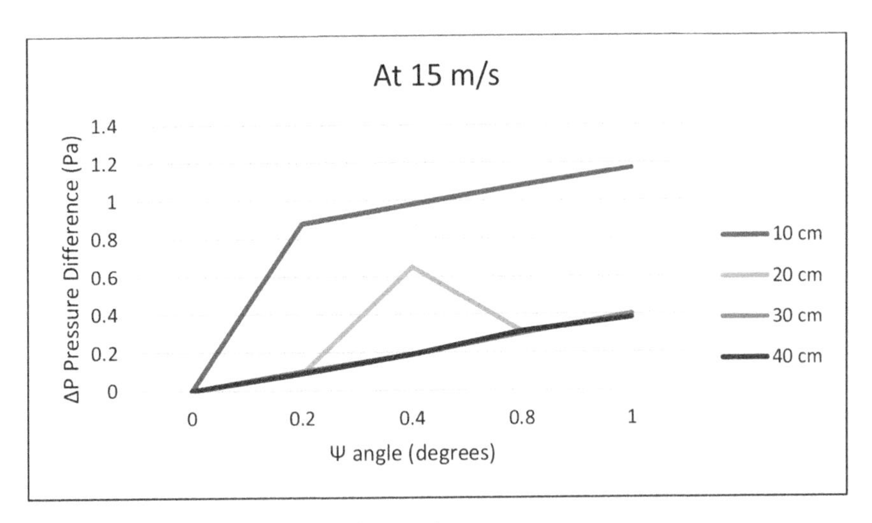

FIGURE 13.8 ΔP v/s Ψ at SCET for 15 m/s.

Figures 13.8 and 13.9 shows a curve for the experiment carried out at 25 m/s. The curve remains almost same for all the distance till the inclination of 0.4. After this point the curve for 0.1 m and 0.2 m remains nearly the same with small fluctuations. But for 0.4 there is a slight increase till an inclination of 0.8 and then a decline. Thus, there is a noticeable shift in the flow properties after an inclination of 0.4.

TABLE 13.7 Test Results for Survey 2 at 25 m/s Test Section Velocity

SL. No.	L (cm)	Velocity (V) (m/s)	q (P) pascals	Ψ	P_A (Pa)	P_B (Pa)	ΔP = (P_A ~ P_B) (Pa)	ΔK = (ΔP/q)
1.	10	25	383	0.0	35.69	35.69	0	0
	10	25	383	0.2	35.72	35.63	0.09	0.000235
	10	25	383	0.4	35.78	35.59	0.19	0.000496
	10	25	383	0.8	35.82	35.55	0.27	0.000705
	10	25	383	1.0	35.88	35.5	0.38	0.000992
2.	20	25	383	0.0	48.54	48.54	0	0
	20	25	383	0.2	48.59	48.49	0.1	0.000261
	20	25	383	0.4	48.63	48.44	0.19	0.000496
	20	25	383	0.8	48.67	48.39	0.28	0.000731
	20	25	383	1.0	48.72	48.35	0.37	0.000966
3.	30	25	383	0.0	54.64	54.64	0	0
	30	25	383	0.2	54.69	54.59	0.1	0.000261
	30	25	383	0.4	54.73	54.54	0.19	0.000496
	30	25	383	0.8	54.79	54.5	0.29	0.000757
	30	25	383	1.0	54.82	54.46	0.36	0.00094
4.	40	25	383	0.0	66.18	66.18	0	0
	40	25	383	0.2	66.22	66.13	0.09	0.000235
	40	25	383	0.4	66.27	66.08	0.19	0.000496
	40	25	383	0.8	66.32	66.01	0.31	0.000809
	40	25	383	1.0	66.36	65.97	0.39	0.00102

13.4 CONCLUSION

A systematic flow parameter prediction has been conducted successfully for the two different wind tunnel testing facility. From the above survey and report, it can be seen that a two-hole spherical flow analyzer with two pressure ports is capable of determining the flow angularity and we could get the exact flow variation for all the two testing facilities. The systematic validation of the instrument has thus been done successfully. It is found that using this instrument, the wind tunnel turbulence level has been predicted quite normal, but few useful observations has obtained in the two tunnels which depicts the yaw head constant variation for various ranges of Mach numbers with slight variations in yaw angles. This data will assist in measuring and explaining the findings obtained during actual model testing to be conducted in these three wind tunnels in the future. Moreover, from the results obtained, clear idea about the truth flow of the two tunnel and systematic understanding about the flow characteristic throughout the test section axis and flow angularity at different distance from the entry of the test section has been obtained.

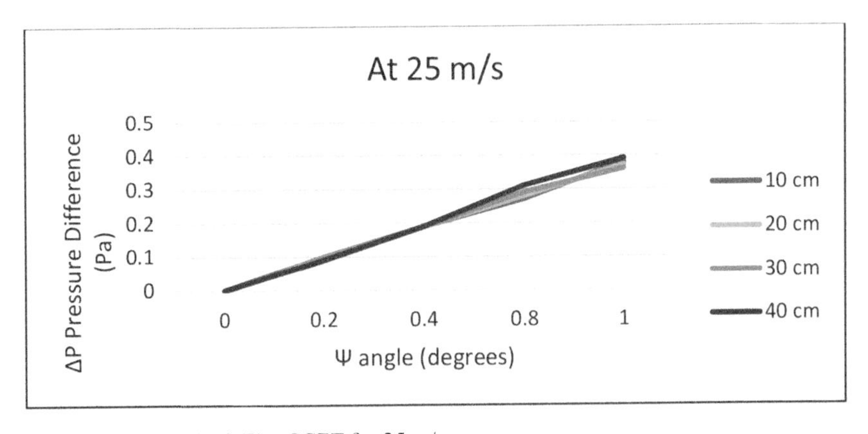

FIGURE 13.9 ΔP v/s Ψ at SCET for 25 m/s.

KEYWORDS

- aerodynamic forces
- air data
- calibration
- computational analysis
- truth flow
- two-hole spherical flow analyzer
- wind tunnel testing

REFERENCES

1. Yasar, M., & Melda, O. C., (2011). A multi-tube pressure probe calibration method for measurements of mean flow parameters in swirling flows. *Journal of Flow Measurement and Instrumentation, 9*, 243–248.
2. Muruga, L. J. J. V., & Senthil, K. M., (2014). Performance evaluation of yaw meter with the aid of computational fluid dynamic. *International Review of Mechanical Engineering, 8*(2).
3. Main, J., Day, C. R. B., Lock, G. D., & Oldfield, M. L. G., (2016). Calibration of a four-hole pyramid probe and area traverse measurements in a short-duration transonic turbine cascade tunnel. *International Journal of Innovative Research in Science, Engineering and Technology, 21*, 302–311.
4. Akhila, R., Muruga, L. J. J. V., & Sanas, U., (2020). Design and analysis of five probe flow analyzer for subsonic and supersonic wind tunnel calibration. *IOP Conference Series: Materials Science and Engineering, 01*(715), 1–7. ISSN 1757-899X.
5. Akhila, R., Muruga, L. J. J. V., Ram, M. V. M., Praveen, K. K., Abhishek, T., Ashish, T., Reddy, K. V. V. M., & Greeshma, M. R., (2020). Comparative study on wind tunnel calibrating instruments. *Advances in Metrology and Measurement of Engineering Surfaces: Lecture Notes in Mechanical Engineering Book Series* (LNME) (Vol. 2, pp. 139–147). Springer, Singapore. ISSN 2195-4356.
6. Akhila, R., Clavin, W. S., Nithyashree, U., & Sanjay, M. V., (2020). Aerodynamic design, analysis, fabrication and testing of a claw yaw sphere for subsonic flow. *AIP Conference Proceedings* (Vol. 2311, No. 01, pp. 030012:1–030012:12). ISSN 1551-7616.
7. Akhila, R., & Muruga, L. J. J. V., (2020). Aerodynamic design and computational analysis of yaw sphere for subsonic wind tunnel. *AIP Conference Proceedings* (Vol. 2311, No. 01, pp. 030016:1–030012:9). ISSN 1551-7616.
8. Akhila, R., & Muruga, L. J. J. V., (2020). Experimental and computational evaluation of five-hole five probe flow analyzer for subsonic wind calibration. *International Journal of Aviation, Aeronautics, and Aerospace, 7*(4), 1–42. ISSN 2374-6793.
9. Akhila, R., & Muruga, L. J. J. V., (2020). Performance evaluation of a two hole and five hole flow analyzer for subsonic flow. *International Journal of Advanced Science and Technology, 29*(5), 7512–7525. ISSN 2207-6360.

CHAPTER 14

THERMAL ANALYSIS OF BALL-END MAGNETORHEOLOGICAL FINISHING TOOL

MOHAMMAD OWAIS QIDWAI,[1] FAIZ IQBAL,[2] and ZAFAR ALAM[3]

[1]*Department of Mechanical Engineering, Delhi Skill and Entrepreneurship University, Okhla-III campus, New Delhi–110020, India*

[2]*School of Engineering, University of Lincoln, Lincoln, LN6 7TS, United Kingdom*

[3]*Department of Mechanical Engineering, Indian Institute of Technology (Indian School of Mines), Dhanbad, Jharkhand–826004, India, E-mail: zafar@iitism.ac.in*

ABSTRACT

Ball end magnetorheological finishing (BEMRF) process requires an electromagnet to precisely control the magnetic field and consequently the forces during the finishing action. The heat generated in electromagnet in the BEMRF tool, pose thermal management issues, which restricts the maximum current supply in the tool. For better finishing process, higher magnetic field strength is desirable, which is directly proportional to the current supplied in the electromagnet. The BEMRF tool uses the direct liquid system, for which the maximum current supplied in the coil is limited up to 8 A, after which the coil damages due to heat. The objective

Optimization Methods for Engineering Problems. Dilbagh Panchal, Prasenjit Chatterjee, Mohit Tyagi, Ravi Pratap Singh (Eds.)

of this chapter is to understand the thermal path, hotspot, and effectiveness of the cooling system, based on which further design modifications can be suggested. Entry of fluid in the direction normal to the cylindrical coil causes fluid structure near inlet region to achieve non-uniform heat dissipation, therefore need arises to modify the inlet port. A comparison case without gaps, is simulated to check the claim, that even, negligible flow rate may improve overall working condition of electromagnet. Based on the analysis of results, some improvements are suggested for the design of the cooling system.

14.1 INTRODUCTION

Ball end magnetorheological finishing (BEMRF) is a nanofinishing technology developed by Singh et al. [19] at the beginning of this decade. It is a variant of magnetic field assisted finishing processes that make use of smart magnetorheological polishing fluid to smoothen the roughness peaks in a controlled manner [1]. BEMRF process has the capability to finish plane, nonplanar, and freeform surfaces up to the order of nanometer (Iqbal and Jha, 2016). Since its inception, this technology has been employed by various researchers to finish a variety of materials ranging from ferromagnetic materials like mild steel [10, 11] to nonmagnetic materials like additive manufactured polylactic acid (PLA) workpiece [14], copper mirrors [2], polycarbonates [12], etc. The capability of BEMRF process to finish freeform complex surfaces of a variety of materials can be attributed to its unique electromagnet-based tool design mounted on a five axis CNC machine tool and having a customized controller developed by Alam et al. in 2019 [2] to provide in-process control of finishing forces by varying the input current supplied to the electromagnet-based tool. This tool requires higher magnetic field for surface finishing operation. Since, magnetic field strength is directly proportional to the current density in the coil, which is limited due to thermal issues. The magnets with air cooling, have limited magnetic field, due to thermal issues arising from Joule's heating effect.

In literature, cooling issues in electromagnet and electrical machines are tackled by several methods such as using spray cooling [15] (El-Rafaie, 2014), flooded stator or direct liquid cooled stator [4, 6], embedded cooling tubes [16]. Spray nozzle cooling provides uniform cooling and high heat flux suffers some serious drawbacks such as nozzle blocking, nozzle erosion, which may cause failure in electrical machines reducing the reliability of the system [21]. Khan et al. in 2020 [13] developed an improved BEMRF tool with direct liquid cooling system. Zhang et al. in 2014 [23] performed experimental analysis on totally enclosed water-cooled permanent magnet machine by experimental method and proposed a combined electromagnetic Finite Element Analysis (FEA) with thermal resistance network thermal model. Ponomarev et al. in 2012 [17] compared the effectiveness of oil cooling in permanent magnet synchronous machine using three thermal models which include Lumped Parameter Thermal Network (LPTN) model, thermostatic FEA model, and Computational Fluid Dynamics (CFD) model. They found that for quick estimation of hotspot LPTN model is best suited. In case detailed insight of thermal problem is needed with realistic temperature profile, CFD with conjugate heat transfer analysis is the best option even when it is a time-consuming process. Yang et al. in 2016 [22] summarized that for effective thermal management, different aspects in electrical machines must be accounted which includes various heat generation sources, paths traverse and sinks. Copper loss can be lowered with lower winding temperature. The rule of thumb suggests that liquid cooling is a highly efficient cooling system with high complexity involved in the process. In the improved BEMRF tool [13], 3-D conjugate heat transfer analysis is conducted to understand the complexity and effectiveness of the cooling system and to identify thermal path and hotspot, based on which design modifications can be suggested so that higher current can be achieved in the coil.

14.2 NUMERICAL ANALYSIS

The numerical model comprises of three domains bobbin, copper coil and cooling fluid surrounding the magnetic winding. The material used to manufacture bobbin is aluminum with constant thermophysical properties, while transformer oil is used for cooling four sets of layers of magnetic winding as shown in Figure 14.1(a). The dimensions are taken as in the experimental model as in Khan et al. [13] shown in schematic diagram in Figure 14.2. The connecting tube attached with jacket have inlet and outlet diameter of 10 mm. Height of the coil and nylon jacket are taken as 140 mm and 150 mm, respectively. The thickness of bobbin is 1.75 mm. Each set consists of 8 layers of coil with a gap of thickness equivalent to the diameter of coil wire between two sets, to flush transformer oil between them. The outer diameter of the coil turns out to be 100 mm approximately. The detailed specification of the coil used in the experiment is given in Table 14.1. The energy dissipated from the coil is equal to the equivalent heat generation in the coil, obtained by using Joule's equation of electrical heating, according to the following relation:

$$Q = I^2R \tag{1}$$

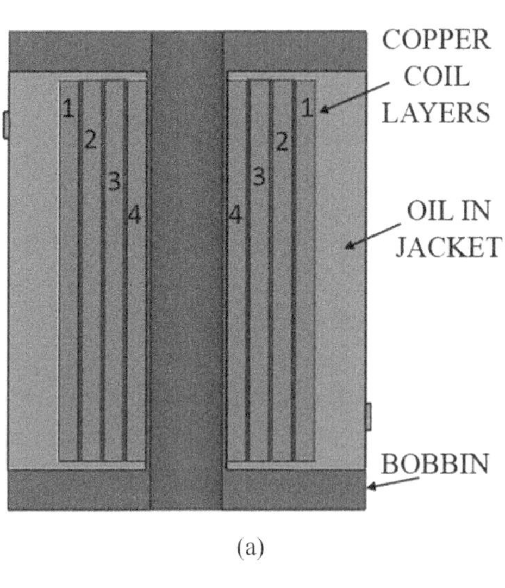

(a)

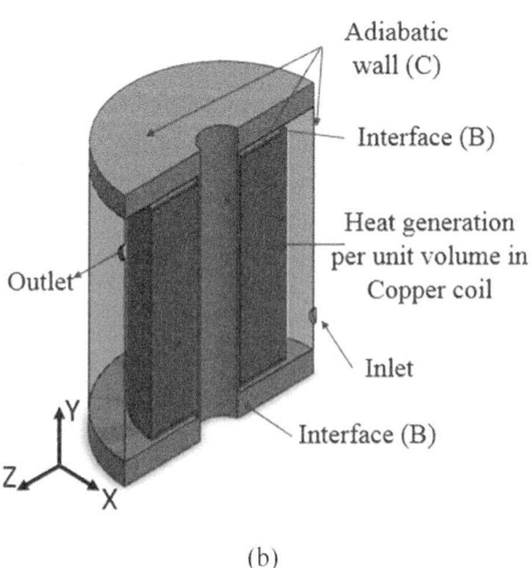

(b)

FIGURE 14.1 (a) Copper layer of winding surrounded with transformer oil in the jacket; (b) boundary conditions applied on the mode.

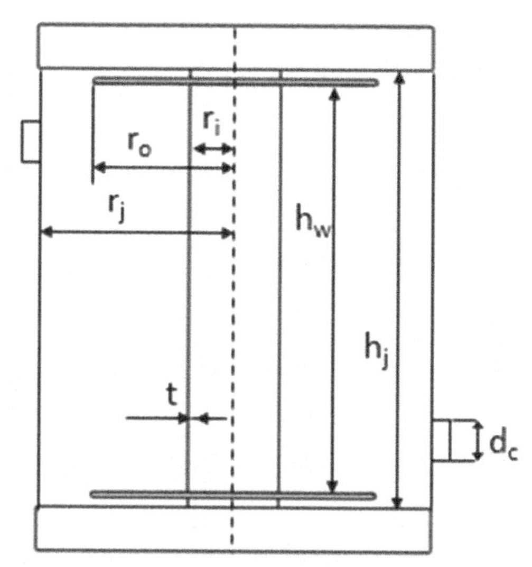

FIGURE 14.2 Dimensions of model in schematic diagram.

TABLE 14.1 Specification of Copper Windings

Standard Wire Gauge	Wire Nominal Resistance (Ω/m)	Number of Turns in Each Layer	Number of Layers in Coil	Length of Wire (m)
18	0.0146	100	24	475

14.3 ASSUMPTIONS AND GOVERNING EQUATIONS

Following are the assumptions considered for the numerical model:

- Incompressible, laminar flow under steady-state conditions;
- No slip boundary condition;
- Radiation heat transfer is negligible;
- Body forces are neglected for cooling fluid;
- Temperature dependent thermophysical properties of fluid.

Based on the assumptions, governing equations required for fluid and heat transfer analysis are as follows [5].

➢ **Continuity Equation:**

$$\nabla . V = 0 \tag{2}$$

➢ **Momentum Equation:**

$$\rho_f \left(\mathrm{V}.\nabla\right)\mathrm{V} = -\nabla P + \mu_f \nabla^2 V \tag{3}$$

➢ **Energy Equation:**

- **For Fluid:**

$$\rho_f c_p \left(\mathrm{V}.\nabla T_f\right) = k_f \nabla^2 T_f \tag{4}$$

- **For Solid:**

$$k_s \nabla^2 T_s = 0 \tag{5}$$

The boundary conditions in conjugate heat transfer are shown in Figure 14.1(b). At inlet, uniform fluid temperature and mass flow rate are applied as constant as in Eqns. (6) and (7), respectively.

$$T = T_{f,in} \tag{6}$$

where; $T = T_{f,in} = 281$ K.

$$\dot{m} = \rho_f \dot{V}_f \tag{7}$$

At the interface boundary, heat flux Eqn. (8) and temperature Eqn. (9) are used to obtain equilibrium between solid and fluid.

$$k_s \nabla T_{s,\Gamma} = k_f \nabla T_{f,\Gamma} \tag{8}$$

$$\mathrm{T}_{s,\Gamma} = T_{f,\Gamma} \tag{9}$$

The heat generation per unit volume is applied in the winding domain as:

$$q" = \frac{Q}{V_w} \tag{10}$$

For flow at the outlet is given by:

$$P = P_{atm} \tag{11}$$

$$\frac{\partial T}{\partial \eta} = 0 \tag{12}$$

The flow rate of transformer oil into the nylon jacket is 5 L/min with the inlet temperature of 8°C. Thermophysical properties of transformer oil are computed based on correlations summarized in Table 14.2.

TABLE 14.2 List of Correlations for Temperature-Dependent Thermophysical Properties of Transformer Oil [8]

Properties	Empirical Correlations
ρ_f	$1098.72 - 0.712T$
c_p	$807.163 + 3.58T$
μ_f	$0.08467 - 0.0004T + 5 \times 10^{-7}T^2$
k_f	$0.1509 - 7.101 \times 10^{-5}T$

For a 90°C change in temperature of oil [13], change in density, specific heat, dynamic viscosity, and thermal conductivity are found to be 7.13%, 17.77%, 56.68% and 4.88%, respectively. Therefore, all the properties are considered as temperature dependent, except thermal conductivity.

The governing equations are solved using coupled CFD solver Ansys CFX, with prescribed boundary conditions. The convergence criteria between two consecutive iterations are set to be less than 1×10^{-6} for the root mean square residuals of continuity, velocities, and energy.

The case under consideration can be classified as the fundamental thermal problem of the second kind in doubly connected ducts, for which the boundary conditions are specified as the constant heat flux in the inner wall and adiabatic condition on the outer wall [18]. As the combined percentage cross-sectional area between three gaps is only 1.2% of the total cross-sectional area of flow, therefore, Reynolds number is defined based on the hydraulic diameter as defined by Tosun in 2007 [20] for annular space between the nylon cover and the outer layer of cylindrical copper coil (layer 1).

$$d_h = d_o - d_i \tag{13}$$

$$Re = \frac{\rho_f v d_h}{\mu_f} \tag{14}$$

Heat transfer coefficient is calculated as:

$$h = \frac{q'}{\left(T_{o,w} - T_{m,f}\right)} \tag{15}$$

where;

$$q' = -k_s \frac{\partial T_s}{\partial \eta} \tag{16}$$

where; $T_{o,w}$ is the temperature of the outer surface wall of the coil; and $T_{m,f}$ is the bulk mean fluid temperatures of the oil.

Average Nusselt number is given as:

$$Nu = \frac{hd_h}{k_f} \tag{17}$$

14.4 GRID INDEPENDENCE TEST

The model is discretized using unstructured and non-uniform mesh. In Figure 14.3, cut section view of the fluid domain with intermediate oil layers is shown.

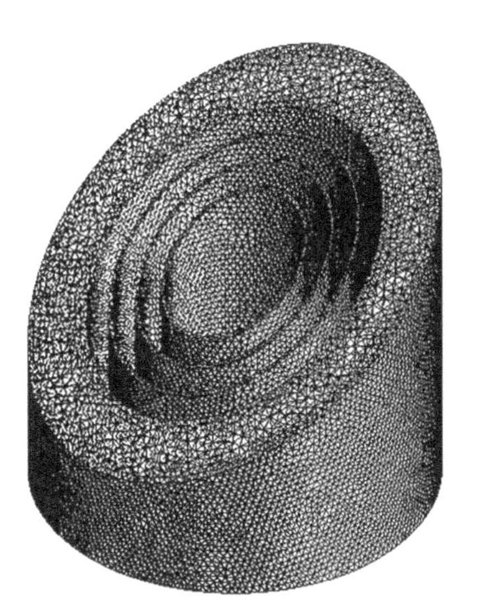

FIGURE 14.3 Unstructured mesh in the fluid domain with cut-section view.

Table 14.3 consists of details of mesh with average Nusselt number as parameter, for case with equivalent heat flux of 2 A current in the coil. The relative error is calculated based on the following:

$$e\% = \left|\frac{j_2 - j_1}{j_1}\right| \times 100 \tag{18}$$

where; j_1 and j_2 represents the parameter value for finest mesh and parameter values acquired from any other mesh, respectively. It is clearly visible relative error for average Nusselt number value of mesh 3 is obtained with reasonable accuracy, therefore it is selected for the study. The adopted mesh 3 is finer near walls to allow for an accurate representation of steep velocity and temperature gradients. Therefore, the elements are used with the edge of minimum size of 0.0001249 m and maximum size of 0.005 m.

TABLE 14.3 Grid Independence Test

Mesh	Number of Elements	Nu	%error	Number of Iteration	Time Taken (hh:mm:ss)
Mesh 1	194,591	106.43	718.36	1,088	00:21:08
Mesh 2	300,244	24.07	85.15	2,045	01:06:03
Mesh 3	574,876	13.15	1.16	4,732	05:24:41
Mesh 4	981,052	13.00	–	5,731	11:35:30

14.5 VALIDATION

To validate the numerical scheme in current study temperature of winding in layer 3 as shown in Figure 14.1(a), are compared with temperature of inner layer of electromagnet, measured by Shah et al. [18]. The experimental values of temperature are obtained using Pt100 resistance temperature detector placed in the central region inside the coil. The maximum deviation in the experimental and numerical values in Figure 14.4 is 3.62%.

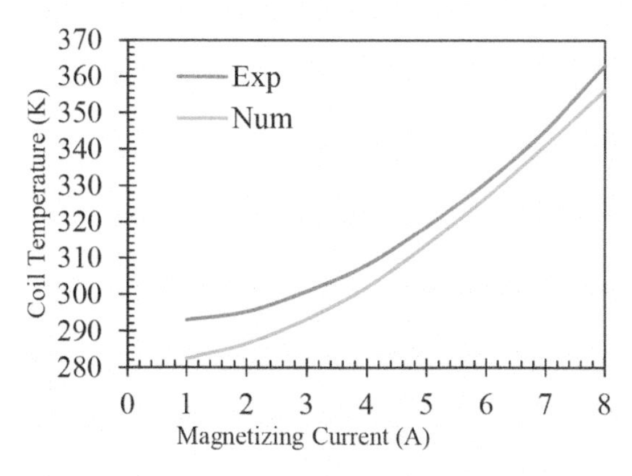

FIGURE 14.4 Validation of temperature variation of electromagnet coil.

14.6 RESULTS AND DISCUSSION

14.6.1 FLOW STRUCTURE

To obtain flow structure of cooling oil in and around the electromagnet coil, streamlines are plotted on plane x = 0 m and z = 0 m shown in Figure 14.5. At plane x = 0 m, oil flows into the jacket from the bottom right side and the outlet is in the top left side. Two vortical structures are formed when the fluid enters through the inlet boundary and hits the outer surface of layer 1 of copper coil. Vortical structures appear below and above the inlet region, which causes mixing of oil from the interface into the bulk fluid, so that larger heat is convected from the outer layer. In the region near the inlet boundary, the temperature of fluid rises above the vortical structure, which may be attributed to the less fluid passing through that region owing to recirculating structure of fluid. This feature, however, causes non-uniform temperature distribution and reduction in average heat transfer coefficient in the outer surface of the coil. Therefore, the direction of the inlet, which is normal to the cylindrical wall, must be modified, so that temperature non-uniformity arises due to fluid interaction, may be avoided.

It is evident from the temperature profile that very less quantity of fluid flows through the gaps between different layers of the coil which causes temperature non-uniformity in different layers. In Figure 14.5, plane z = 0 m, lies normal to the fluid inlet and outlet position. No fluid flow structure is obtained on this plane. Though the Reynolds number is quite low (Re $\cong$ 33), the random fluidic structure suggests the flow separation region around the outer cylindrical coil, helps in dissipating heat through the cooling fluid.

14.6.2 TEMPERATURE DISTRIBUTION

In Figure 14.5, the temperature of oil and bobbin interface suggests, that bobbin participates quite prominently in heat dissipation of coil, wounded over it. In Figure 14.6, temperature distribution in coils with 8 A current flowing through it, is shown. As the interface area is quite large between bobbin and oil, heat is dissipated from the bobbin which helps lowering the temperature of innermost coil, which suggests that lower temperature of innermost coil facilitates heat transfer through itself by conduction mode. The outermost coil is in direct contact with cooling fluid, and therefore it also has lower temperature. However, the heat dissipation from the intermediate layers does not appear to be prominent which is highlighted as one of the drawbacks of this design. In real working apparatus, for a maximum current of 8 A, the temperature reaches 90°C, after which failure occurs. Mass flow rate of transformer oil through the gaps is quantitatively negligible, less than 2.0% of total mass flow rate. In the design phase, it is expected that oil flows through these layers, may cause effective heat dissipation. To check this claim, a comparison case with gaps filled with solid aluminum metal as heat conductor filler material, is simulated with similar boundary conditions. The purpose of adding aluminum filler material is to increase the heat conduction rate between the two layers of coil. On comparing temperature contour in Figure 14.6(a) and (b), not much difference in maximum temperature is obtained, but oil in gaps cool slightly effectively compared to the case without gaps. In the case with oil flowing through intermediate gaps, the maximum temperature reaches 356 K, while in the case with aluminum as the filler material, the maximum temperature is 360 K. It is suggested that, some modifications are necessary in the cooling system with gaps filled with oil, to increase its effectiveness, so that uniform temperature distribution may be obtained and temperature in intermediate layers of coil may decrease further.

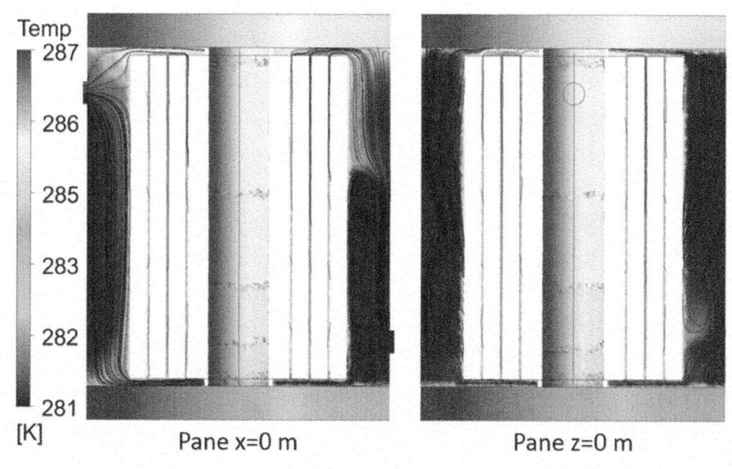

FIGURE 14.5 Temperature contour and streamline of oil flow inside the nylon cover.

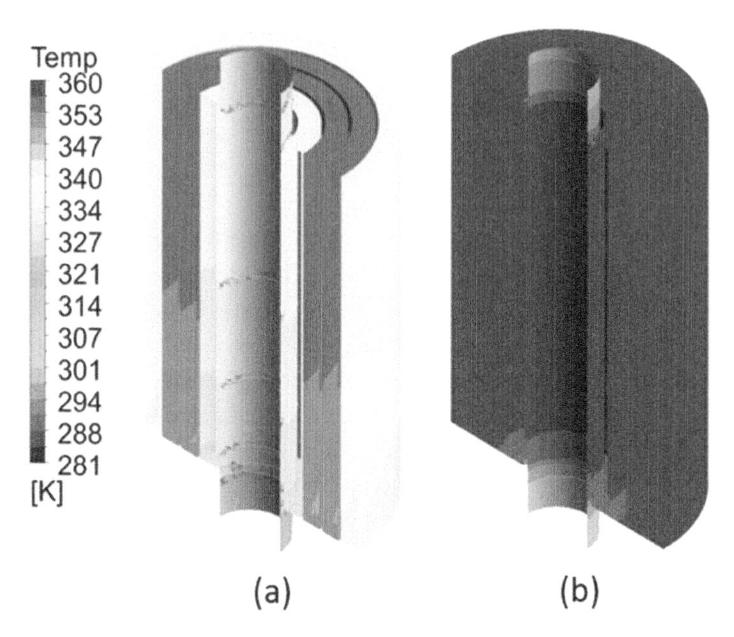

FIGURE 14.6 Temperature in different layers of coil: (a) with oil flowing in gaps; (b) without gaps filled with aluminum.

To obtain heat flow path through the magnetic coil, wall heat flux is obtained for each layer. Wall heat flux in layer 4 is quite different for up to 6 A. Heat generated in coil must be dissipated to the cooling fluid. The positive values of wall heat flux on the layer 4 shown in Figure 14.7, suggests, at lower current values, heat flows from transformer oil to the innermost layer of the coil, which is conducted to the bobbin and ultimately dissipated through the cooling fluid. This behavior suggests that bobbin can be used effectively to dissipate heat from the inner most region of coil, but a larger path means higher resistance and lower efficiency of the cooling system. For higher, values of current greater than 6 A, negative values of wall heat flux are obtained, which signifies that layer 4 also contributes heat directly to the cooling coil, but with higher temperatures in the intermediate layers. The cooling must remain efficient for all values of current, since coil is vulnerable to peak temperatures and may get damaged, if the current reaches beyond a critical temperature in any region of the coil.

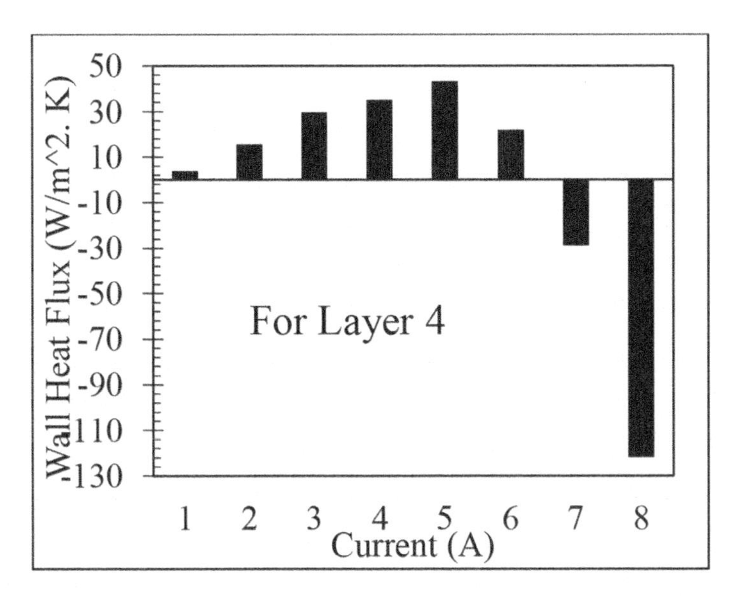

FIGURE 14.7 Wall heat flux values from surface of layer 4.

14.6.3 NUSSELT NUMBER

The Nusselt number is obtained for inner layers 2, 3, and 4 combined, and separately for layer 1 (outer coil). As shown in Figure 14.8, Nusselt number for intermediated layers are very high compared to that of the outer layer, which suggests that even though the mass flow rate of transformer oil inside the gaps is very low, its effectiveness in heat dissipation is quite large. For inner layers, Nusselt number value decreases with an increase in heat generation in the coil, however, it remains invariant for the outer layer. Lower values of Nusselt number on the outer layer indicates that ineffective heat transfer, which means that, cooling fluid is not utilized properly.

14.6.4 FUTURE ASPECT

To improve Nusselt number, for outer layer of coil, cooling fluid should effectively mix into the bulk fluid, to convect larger amount of heat. For which various design modifications, can test in future, such as:

- Entry and exit of cooling fluid may be provided tangentially, to the outer coil;
- number of inlets and outlets can be varied to obtain uniform temperature distribution;
- Axial flow arrangement can be preferred over the radial inlet or outlet.

The following modifications must consider that, gaps in the coil cannot be increased further, as it will decrease the magnetic field strength.

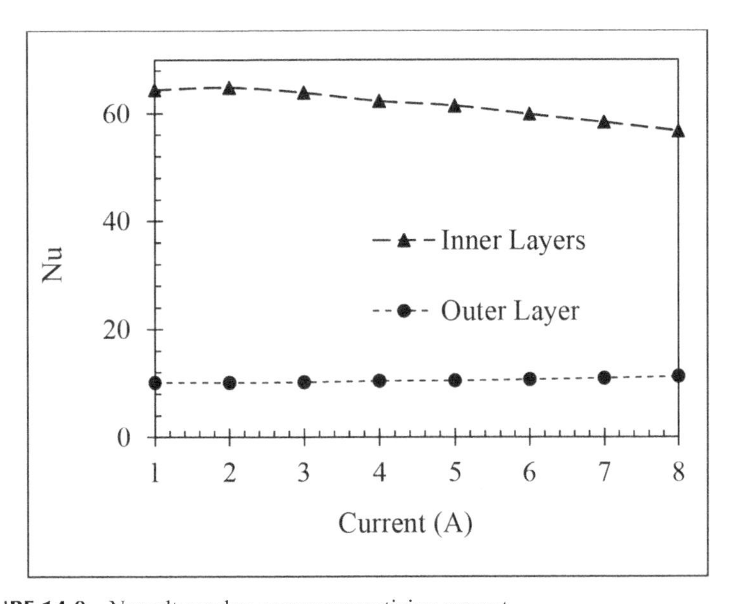

FIGURE 14.8 Nusselt number verses magnetizing current.

14.7 CONCLUSION

The case study is performed to analyze the cooling effect of transformer oil for maintaining the temperature of magnetic copper coil with 4 layers, wounded over a bobbin. This analysis highlights vulnerable points in the cooling system. As per the expectation, the innermost coil must have the highest temperature, but it is found that intermediate layers 2 and 3 have higher peak temperature. This case study compares heat transfer with the coil in which gaps are filled with Aluminum as filler material. The addition of filler material worsens the case and overall temperature rise in the coil by 4 K. The heat flow adopts larger route before dissipating into the cooling fluid, at lower values of current. This makes the cooling system less reliable, as the electromagnet coils are susceptible to peak temperatures anywhere inside. The modified design should ensure that shortest route must offer least thermal resistance, to ensure reliability of machine.

KEYWORDS

- **computational fluid dynamics**
- **electromagnet**
- **finishing tool**
- **nanofinishing**
- **numerical simulation**
- **thermal analysis flow structure**

REFERENCES

1. Alam, Z., & Jha, S., (2017). Modeling of surface roughness in ball end magnetorheological finishing (BEMRF) process. *Wear, 374–375C*, 54–62.
2. Alam, Z., Iqbal, F., Ganesan, S., & Jha, S., (2019). Nanofinishing of 3D surfaces by automated five-axis CNC ball end magnetorheological finishing machine using customized controller. *The International Journal of Advanced Manufacturing Technology, 100*(5–8), 1031–1042.
3. Alam, Z., Khan, D. A., & Jha, S., (2019). MR fluid-based novel finishing process for nonplanar copper mirrors. *The International Journal of Advanced Manufacturing Technology, 101*(1–4), 995–1006.
4. Camilleri, R., Howey, D. A., & McCulloch, M. D., (2015). Predicting the temperature and flow distribution in a direct oil-cooled electrical machine with segmented stator. *IEEE Transactions on Industrial Electronics, 63*(1), 82–91.
5. Chen, W., Ju, Y., Yan, D., Guo, L., Geng, Q., & Shi, T., (2019). Design and optimization of dual-cycled cooling structure for fully-enclosed permanent magnet motor. *Applied Thermal Engineering, 152*, 338–349.
6. Degano, M., Arumugam, P., Fernando, W., Yang, T., Zhang, H., Bartolo, J. B., ... & Gerada, C. (2014, April). An optimized bi-directional, wide speed range electric starter-generator for aerospace application. In *7th IET International Conference on Power Electronics, Machines and Drives* (PEMD 2014) (pp. 1–6). IET.

7. El-Refaie, A. M., Alexander, J. P., Galioto, S., Reddy, P. B., Huh, K. K., De Bock, P., & Shen, X., (2014). Advanced high-power-density interior permanent magnet motor for traction applications. *IEEE Transactions on Industry Applications, 50*(5), 3235–3248.
8. Hannun, R. M., Hammadi, S. H., & Khalaf, M. H., (2018). Heat transfer enhancement from power transformer immersed in oil by earth air heat exchanger. *Thermal Science, 23*(6 Part-A), 3591–3602.
9. Iqbal, F., & Jha, S., (2016). Nanofinishing of freeform surfaces using BEMRF. In: *Nanofinishing Science and Technology* (pp. 255–284). CRC Press.
10. Iqbal, F., & Jha, S., (2018). Closed loop ball end magnetorheological finishing using in-situ roughness metrology. *Experimental Techniques, 42*(6), 659–669.
11. Iqbal, F., & Jha, S., (2019). Experimental investigations into transient roughness reduction in ball-end magneto-rheological finishing process. *Materials and Manufacturing Processes, 34*(2), 224–231.
12. Khan, D. A., Kumar, J., & Jha, S., (2016). Magneto-rheological nano-finishing of polycarbonate. *International Journal of Precision Technology, 6*(2), 89–100.
13. Khan, D. A., Alam, Z., Iqbal, F., & Jha, S., (2020). Design and development of improved ball end magnetorheological finishing tool with efficacious cooling system. In: *Advances in Simulation, Product Design and Development* (pp. 557–569). Springer, Singapore.
14. Kumar, A., Alam, Z., Khan, D. A., & Jha, S., (2019). Nanofinishing of FDM-fabricated components using ball end magnetorheological finishing process. *Materials and Manufacturing Processes, 34*(2), 232–242.
15. Li, Z., Guo, J., Fu, D., Gu, G., & Xiong, B., (2009). Research on heat transfer of spraying evaporative cooling technique for large electrical machine. In: *2009 International Conference on Electrical Machines and Systems* (pp. 1–4). IEEE.
16. Madonna, V., Walker, A., Giangrande, P., Serra, G., Gerada, C., & Galea, M., (2018). Improved thermal management and analysis for stator end-windings of electrical machines. *IEEE Transactions on Industrial Electronics, 66*(7), 5057–5069.
17. Ponomarev, P., Polikarpova, M., & Pyrhönen, J., (2012). Thermal modeling of directly-oil-cooled permanent magnet synchronous machine. In: *2012 XXth International Conference on Electrical Machines* (pp. 1882–1887). IEEE.
18. Shah, R. K., & London, A. L., (2014). *Laminar Flow Forced Convection in Ducts: A Source Book for Compact Heat Exchanger Analytical Data*. Academic Press.
19. Singh, A. K., Jha, S., & Pandey, P. M., (2012). Magnetorheological ball end finishing process. *Materials and Manufacturing Processes, 27*(4), 389–394.
20. Tosun, I., (2007). *Modeling in Transport Phenomena: A Conceptual Approach*. Elsevier.
21. Visaria, M., & Mudawar, I., (2009). Application of two-phase spray cooling for thermal management of electronic devices. *IEEE Transactions on Components and Packaging Technologies, 32*(4), 784–793.
22. Yang, Y., Bilgin, B., Kasprzak, M., Nalakath, S., Sadek, H., Preindl, M., Cotton, J., et al., (2016). Thermal management of electric machines. *IET Electrical Systems in Transportation, 7*(2), 104–116.
23. Zhang, B., Qu, R., Xu, W., Wang, J., & Chen, Y., (2014). Thermal model of totally enclosed water-cooled permanent magnet synchronous machines for electric vehicle applications. In: *2014 International Conference on Electrical Machines (ICEM)* (pp. 2205–2211). IEEE.

CHAPTER 15

A REVIEW ON DEVELOPMENT AND TECHNOLOGY OF VARIOUS TYPES OF SOLAR PV CELL

AMAN SHARMA and VIJAY KUMAR BAJPAI

Department of Mechanical Engineering,
National Institute of Technology, Kurukshetra, Haryana, India,
E-mail: aman_6180081@nitkkr.ac.in (A. Sharma)

ABSTRACT

Solar PV technology is demonstrating its application all over the globe, as a green and renewable energy source. With the research efforts, the power generation efficiency has increased a lot from a mere 6% in the beginning. The technology has enormous scope for increasing its efficiency with integration and applications of various methods. The introduction of new generation solar cells has widened this scope. The second, third, and fourth generation PV cells have properties which make them viable for commercial use. The objective of this review is to show the current state of art of PV technology, material, manufacturing techniques and their efficiency and also highlights different studies done globally in this area of research.

15.1 INTRODUCTION

Chapin et al. [40] at Bell Telephone laboratories USA developed PV cell technology with 6% power generation from solar radiation. It was

Optimization Methods for Engineering Problems. Dilbagh Panchal, Prasenjit Chatterjee, Mohit Tyagi, Ravi Pratap Singh (Eds.)

based on P-N type generation with Silicon as material. The generation efficiency was low, and the onward efforts have been made to enhance the efficiency with the use of low-cost and easily available cell material. With the continuous research effort worldwide, the researchers arrived at first-generation solar PV cell based on (a) mono crystalline Silicon (m-Si), (b) polycrystalline Silicon (p-Si), and (c) non-crystalline or amorphous silicon (a-Si). The m-Si is most costly with highest efficiency of 20%, the p-Si is relatively cheaper with 15% efficiency and a-Si is cheapest with lowest efficiency of 6%. For commercial installation, polycrystalline silicon is favored because of efficiency in relatively low cost. Further, the research was concentrated on low-cost material and high efficiency. The 3rd generation technology is based on nanocrystalline films, active quantum dots, tandem or stacked multi layers of inorganic material such as GaAs/GalnP; organic polymer based solar cell, dye-sensitized solar cells, etc. Most recently the 4th generation cells are made from organic based nanomaterials like carbon graphene and its derivatives.

This chapter reviews the recent developments in solar cell material science and technology based on material, manufacturing techniques and power generation efficiency.

15.2 SOLAR PV CELL

Solar cells are mainly named based on semiconducting material used in their manufacture. The PV cell absorbs incident photons and knocks out electrons which create current. The efficiency of solar cell is dependent on 'diffusion length' of its material. Diffusion length is the average distance carrier moves between generation and recombination. PV cells are either made of one layer of light absorbing material (single junction) or multi-layers (multi-junction) of absorbing material. More the number of P-N junction more the efficiency of power generation.

Semiconductors are crystalline materials which have their electrical properties between conductors and insulators. Therefore, the best semi-conductors have four valance electrons. Moreover, the properties are determined at low temperature by impurities or dopants [26]. Almost 80% of solar cells in the world use silicone-based material [28]. In order to reduce the cost of solar cell, low-cost semiconductor materials such as titanium dioxide (TiO_2) [30], zinc oxide [11, 36, 37] and Tin dioxide SnO_2 [25, 33, 34] are used.

15.3 VARIOUS TYPES OF SOLAR CELL

The solar cells are classified based on the material used in their fabrication and the period of development of such solar cells. Precisely we classify this in four generations:

1. **First Generation:** This is most prominent commercial technology using crystalline Silicon as base material:
    i. **Monocrystalline Silicone:** These are the most expensive and most efficient. These are made from high grade Silicon and have the highest efficiency in the range of 15 to 24%. These are long life, space-efficient, non-hazardous, heat resistant and easy to install [10, 13, 14].
    ii. **Polycrystalline Silicon:** These are manufactured on Poly Silicon wafers. The production process is simple and cost effective. The process requires less energy and have efficiencies in the range of 15%. These are mainly used in commercial installations of the solar power generation.
    iii. **Non-Crystalline or Amorphous Silicon:** This is made from non-crystalline Silicon, which is cheap and available naturally. The efficiency is around 6% and lowest in Silicon cells. It is thin wafer and used in building and multi-story installations.
    iv. **Gallium Arsenide:** This semiconductor material is used for single-crystalline thin film solar cells. These are costly, but also holds the highest efficiency of 28.8% in single junction solar cells. These are especially used in multijunction solar PV for concentrated photovoltaic (CPV). GaAs have highest conversion efficiency due to its ideal bandgap of 1.43 electron volt, it is insensitive to heat and keep its efficiency even at quite high temperature [4]. In multijunction solar cells of gallium arsenide absorb different wavelength lights and give higher energy conversion efficiencies [8].
2. **Second Generation PV Solar Cell:** These are thin film solar cells made from non-crystalline Silicon, Cadmium Telluride and CIGS. These are commercially used in grid connected PV power stations, BIPV, and in small standalone power station. These PV cells are cheaper small amount of material required, high absorption coefficient. However, these have lower efficiency, 20.3% for CIGS [38] light-induced degradation in outdoor uses.

i. **Cadmium Telluride:** CdTe with a band gap of 1.45 electron volt is good to make a simple junction cell for converting incident light into electric current. CdTe-based PV system can achieve lab efficiencies of around 21% [10]. CdTe PV cells can withstand elevated temperatures better than crystalline Silicon cells and can trap radiation efficiently even in humid environment. The elements which are required to make this semiconductor are in short supply compared to Silicon. Moreover, CdTe is potentially toxic material.

ii. **Copper-Indium Selenide (CIS) and Copper Indium Gallium Diselenide (CIGS):** CIS solar cell production has been successfully commercialized by many firms. Current module efficiencies are in the range of 7% to 16%, although efficiencies of 20.3% have been achieved in the laboratory [9]. CIGS modules are light in weight and has low static weight. These are suitable to absorb direct and indirect sunlight hence useful for the roofs and in winter.

3. **Third Generation PV Solar Cells:** This generation of PV cells are aimed to increase device efficiency. It focuses on manufacturing cells with high efficiency using thin layer deposition technique. Third gen PV cell technologies include Dye-sensitized solar cells (DSSC), organic, and polymeric solar cells, perovskite cells, quantum dots solar cells and multijunction cells. The benefits of these cells are better manufacturing technologies which suits large scale manufacturing and are quite sturdy with good efficiency at elevated temperature.

i. **Dye-Sensitized Solar Cells (DSSC):** These are easy to manufacture, have low cost and has higher power conversion efficiency. The average efficiency range ranges from 7–8% [1, 3, 6, 7]. It consists of a combination of semiconductor film such as titanium dioxide TiO_2, Zinc oxide ZnO, Tin oxide SnO_2, niobium peroxide Nb_2O_5, a dye-sensitized transparent conducting substrate, an electrolyte and counter electrode (CE) [1, 3, 6, 7].

These are thin film-based semiconductor formed between a photo sensitized anode and electrolyte. These types of cells are used for rooftop solar collectors where mechanical robustness is needed. These cells work at low light conditions, say in cloudy skies or indirect radiation. These cells are still being investigated for commercialization. The research issues include stabilization of liquid electrolytes, sensitizer made of quantum dots and perovskites [31].

DSSCs with quantum dots replace dye with organic nanoparticles of quantum dots. Most research is concentrated on cadmium compounds with efficiencies nearing 6.76% [5, 16, 22, 23, 27].

DSSCs based on peroxide sensitizer started in 2009 with the efficiencies of 3.7 to 3.8% [15] however the current efficiencies are close to 20% [2, 31]. This is because of their excellent capacity to absorb light.

ii. **Perovskite Solar Cell:** Research efforts are targeted to improve the semitransparent nature of organic solar cell with the help of absorbing material (with lower bandgap than photons), thereby allowing absorption of near-infrared light whereas letting the visible light pass through [29]. The material like methyl ammonium lead halide perovskite improves transparency and the efficiency of solar cell [20]. Most pair of perovskite solar cells are made of a sandwich of metal oxides (TiO_2 or Al_2O_3) and organic transport material. Perovskite material are easily available which have good electrical properties and is apt for solar cell. These have a high absorption coefficient, higher carrier mobility, direct bandgap, and high stability [20, 29]. These cells have a power conversion efficiency of 13% [17, 19, 21].

4. **Fourth Generation Solar Cell:** These are also known as 'inorganic-in-organic.' It is a combination of low cost and flexible thin film polymer and stable in-organic nanostructures like metal nanoparticles and oxides of metal or organic based nanomaterials like graphene and its derivatives.

   i. **Graphene and its Derivatives:** Graphene is the fundamental unit of graphite structures. Graphene is a 2-dimensional material composed of carbon atoms. Graphene is one of the toughest materials with an elastic modulus of 1 TPa, tensile strength of 130 GPa and breaking strength of ~40 N/m [12, 35]. The chemical configuration of graphene gives the material peculiar chemical, electrical, and mechanical properties. Graphene has extraordinary electrical and thermal conductivity; superior to that of Copper or Silicon [32]. Further graphene sheets are flexible and chemically unreactive, these properties make it a good candidate for solar cells. Table 15.1 summarizes practical and theoretical efficiencies of different types of solar cells and their life span.

TABLE 15.1 Efficiency and Life of Different Types of Solar Cells

Solar Cell Type	Theoretical Efficiency	Practical Efficiency	Life (in Years)
m-Si	19–24%	12–18%	25
CIGS	20%	15%	12
GaAs	28.8%	22%	18
a-Si	10–12%	5–9%	15
CdTe	15–16%	5–10%	20
Perovskite	~ 20%	13%	20

15.4 CONCLUSION

The first-generation Si PV cell are commercialized, and their utility remains before all other alternatives available. These are still preferred owing to their efficiencies. But their cost of production is a bottleneck. The researchers all over the world target the research on reducing material cost and manufacturing technology with matching efficiencies. Presently we have reached 4th generation solar cells. Another issue for researchers is the absorption coefficient of new material so that these solar cells absorb all spectrum of light, e.g., infrared, UV, visible light, to increase the absorption and conversion efficiency of these solar cells. To achieve matched conversion efficiencies, the most promising is a multijunction solar cell with DSSC (hybrid, organic or conventional) technology. These offer low efficiency 6 to 7% but also low cost and weight. However, DSSC solar cell has problems with electrolyte stability and lower shelf life. These problems are being solved with research efforts targeted towards solid phase electrolyte and nanoparticle tube of carbon or graphene. Other concentrated PV cell with these materials has a score for high efficiency power generation as they can operate at high temperature and absorb complete spectrum of light. The technology of nanocrystal/quantum dot of semiconductors-based solar cell can theoretically convert more than 60% of the whole solar spectrum into electric power. All above type solar cells remain at laboratory scale and need to be commercialized in the near future to give a challenge to 1st generation, costly Silicon solar cells. Most research is concentrated on low cost, high efficiency, long life, and ease of manufacturing. The 2nd to 4th generation solar cell can match efficiencies or Silicon cell only with concentrated PV cell. Before an alternative to first generation PV cell technology is reached, an effective research effort may be done to improve the efficiency of first-generation silicon cell by using concentrated PV with the help of mirrors and lenses. The concentration of solar irradiation with mirrors and lens would be a practical solution for increasing power generation efficiency. The problems of high temperature in such efforts can be countered with cooling mechanisms as in PVT technology.

KEYWORDS

- **concentrated PV**
- **copper-indium selenide**
- **dye-sensitized solar cell**
- **graphene**
- **monocrystalline silicon**
- **photovoltaics**

REFERENCES

1. Bahramian, A., & Vashaee, D., (2015). *In-situ* fabricated transparent conducting nanofiber shape polyaniline/coral-like TiO_2 thin film: Application in bifacial dye-sensitized solar cells. *Sol. Energy Mater. Sol Cells, 143*, 284–295. http://dx.doi.org/10.1016/j.solmat.2015.07.011.
2. Bose, S., Soni, V., & Genwa, K. R., (2015). Recent advances and future prospects for dye-sensitized solar cells: A review. *Int. J. Sci. Res. Public, 5*, 1–9.
3. Chen, J. G., Chen, C. Y., Wu, S. J., Li, J. Y., Wu, C. G., & Ho, K. C., (2008). On the photophysical and electrochemical studies of dye-sensitized solar cells with the new dye CYC-B1. *Sol. Energy Mater. Sol. Cells, 92*, 1723–1727.
4. Deyo, J. N., Brandhorst, H. W. Jr., & Forestieri, A. F., (1976). Status of the ERDA/NASA photovoltaic tests and applications project. In: *12th IEEE Photovoltaic Specialists Conf.*

5. Giménez, S., Mora-Seró, I., Macor, L., Guijarro, N., Lana-Villarreal, T., Gómez, R., et al., (2009). Improving the performance of colloidal quantum-dot-sensitized solar cells. *Nanotechnology, 20,* 295204. http://dx.doi.org/10.1088/0957-4484/20/29/ 295204.
6. Gong, J., Liang, J., & Sumathy, K., (2012). Review on dye-sensitized solar cells (DSSCs): Fundamental concepts and novel materials. *Renew. Sustain. Energy Rev., 16,* 5848–5860. http://dx.doi.org/10.1016/j.rser.2012.04.044.
7. Gong, J., Sumathy, K., Qiao, Q., & Zhou, Z., (2017). Review on dye-sensitized solar cells (DSSCs): Advanced techniques and research trends. *Renew Sustain. Energy Rev., 68,* 234–246. http://dx.doi.org/10.1016/j.rser.2016.09.097.
8. Grandidier, J., Callahan, D. M., Munday, J. N., & Atwater, H. A., (2012). Gallium arsenide solar cell absorption enhancement using whispering gallery modes of dielectric nanospheres. *IEEE J. Photovolt., 2,* 123–128.
9. Green, M. A., et al., (2011). *Solar Cell Efficiency Tables progress in Photovoltaics: Research and Applications, 19,* 84–92.
10. Green, M. A., Hishikawa, Y., Warta, W., Dunlop, E. D., Levi, D. H., Hohl-Ebinger, J., & Ho-Baillie, A. W. H., (2017). Solar cell efficiency tables (version 50). *Prog. Photovolt., 25,* 668–676.
11. Janotti, A., & Van De, W. C. G., (2009). Fundamentals of zinc oxide as a semiconductor. *Rep. Prog. Phys., 72,* 126501. http://dx.doi.org/10.1088/0034-4885/72/12/ 126501.
12. Jiang, J., Wang, J., & Li, B., (2009). Young's modulus of graphene: A molecular dynamics study. *Phys. Rev., 80,* 113405.
13. Kirk-Othmer, (2007). Silicon. In: *Kirk-Othmer Encyclopedia of Chemical Technology* (5th edn.). Wiley: Hoboken, NJ, USA. ISBN 978-0-471-48494-3.
14. Kivambe, M., Aissa, B., & Tabet, N., (2017). Emerging technologies in crystal growth of photovoltaic silicon: Progress and challenges. *Energy Procedia, 130,* 7–13.
15. Kojima, A., Teshima, K., Shirai, Y., & Miyasaka, T., (2009). Organometal halide perovskites as visible-light sensitizers for photovoltaic cells. *J. Am. Chem. Soc., 131,* 6050, 6051. [PubMed].
16. Lee, H., Wang, M., Chen, P., Gamelin, D. R., Zakeeruddin, S. M., Grätzel, M., et al., (2009). Efficient CdSe quantum dot sensitized solar cells prepared by an improved successive ionic layer adsorption and reaction process. *Nano Lett., 9,* 4221–4227. http://dx.doi. org/10.1021/nl902438d.
17. Liu, M., Johnston, M. B., & Snaith, H. J., (2013). Efficient planar heterojunction perovskite solar cells by vapor deposition. *Nature, 501,* 395–398. http://dx.doi. org/10.1038/nature12509.
18. Mathew, S., Yella, A., Gao, P., Humphry-Baker, R., Curchod, B. F. E., Ashari-Astani, N., et al., (2014). Dye-sensitized solar cells with 13% efficiency achieved through the molecular engineering of porphyrin sensitizers. *Nat. Chem., 6,* 242–247. http:// dx.doi.org/ 10.1038/nchem.1861.
19. McGehee, M. D., (2014). Perovskite solar cells: Continuing to soar. *Nat. Mater., 13,* 845–846. http://dx.doi.org/10.1038/nmat4050.
20. Mei, A., Li, X., Liu, L., Ku, Z., Liu, T., Rong, Y., et al., (2014). *A Hole-Conductor–Free, Fully Printable Mesoscopic Perovskite Solar Cell with High Stability,* 295. doi: http://dx.doi. org/10.1126/science.1254763.

21. Nicholson, B., Verma, S., & Med, S. S. P., (2014). Interface engineering of highly efficient perovskite solar cells. *Science, 27*(80), 238–242. http://dx.doi.org/10.1126/science.1254050.
22. Nozik, A. J., (2002). Quantum dot solar cells. *Phys. E Low-Dimens Syst Nanostruct., 14*, 115–120. http://dx.doi.org/10.1016/S1386-9477(02)00374-0.
23. Pan, Z., Zhao, K., Wang, J., Zhang, H., Feng, Y., & Zhong, X., (2013). Near-infrared absorption of CdSexTe1–x alloyed quantum dot sensitized solar cells with more than 6% efficiency and high stability. *ACS Nano, 7*, 5215–5222. [PubMed].
24. Parisi, M. L., Maranghi, S., & Basosi, R., (2014). The evolution of the dye-sensitized solar cells from Grätzel prototype to up-scaled solar applications: A life cycle assessment approach. *Renew Sustain. Energy Rev,. 39*, 124–138. http://dx.doi.org/10.1016/j.rser.2014.07.079.
25. Park, M. S., Kang, Y. M., Wang, G. X., Dou, S. X., & Liu, H. K., (2008). The effect of morphological modification on the electrochemical properties of SnO_2 nanomaterials. *Adv. Funct. Mater., 18*, 455–461. http://dx.doi.org/10.1002/adfm.200700407.
26. Phillips, J., (2012). *Bonds and Bands in Semiconductors*. 1st Edition – August 28, 1973, pp. 1–288, eBook ISBN: 9780323156974.
27. Quantum, M. H., Sensitized, D., Cells, S., Nano, A. C. S., & Bisquert, J., (2010). *Artic Model High-Effic. Quantum Dot, 4*, 5783–5790. http://dx.doi.org/10.1021/nn101534y.the.
28. Rahman, M. Z., (2014). Advances in surface passivation and emitter optimization techniques of c-Si solar cells. *Renew. Sustain. Energy Rev., 30*, 734–742. http://dx.doi.org/10.1016/j.rser.2013.11.025.
29. Roldán-Carmona, C., Malinkiewicz, O., Betancur, R., Longo, G., Momblona, C., Jaramillo, F., et al., (2014). High efficiency single-junction semitransparent perovskite solar cells. *Energy Environ. Sci., 7*, 2968. http://dx.doi.org/10.1039/C4EE01389A.
30. Serpone, N., (2006). Is the band gap of pristine TiO_2 narrowed by anion-and cation-doping of titanium dioxide in second-generation photocatalysts? *J. Phys. Chem. B, 110*, 24287–24293. http://dx.doi.org/10.1021/jp065659r.
31. Shalini, S., Balasundaraprabhu, R., Kumar, T. S., Prabavathy, N., Senthilarasu, S., & Prasanna, S., (2016). Status and outlook of sensitizers/dyes used in dye-sensitized solar cells (DSSC): A review. *Int. J. Energy Res., 40*, 1303–1320.
32. Sharma, S., Kumar, P., & Chandra, R., (2016). Mechanical and thermal properties of graphene–carbon nanotube-reinforced metal matrix composites: A molecular dynamics study. *J. Compos. Mater., 51*, 3299–3313.
33. Tennakone, K., Kumara, G. R. R. A., Kottegoda, I. R. M., & Perera, V. P. S., (1999). An efficient dye-sensitized photoelectrochemical solar cell made from oxides of tin and zinc. *Chem. Commun.*, 15, 16. http://dx.doi.org/10.1039/a806801a.
34. Wang, H., & Rogach, A. L., (2014). Hierarchical SnO_2 nanostructures: Recent advances in design, synthesis, and applications. *Chem. Mater, 26*, 123–133. http://dx.doi.org/10.1021/cm4018248.
35. Weiss, N. O., Zhou, H., Liao, L., Liu, Y., Jiang, S., Huang, Y., & Duan, X., (2012). Graphene: An emerging electronic material. *Adv. Mater., 24*, 5782–5825. [PubMed].
36. Zhang, Q., Cao, G., & Gratzel, M., (2011). Hierarchically structured photoelectrodes for dye-sensitized solar cells. *J. Mater. Chem., 21*, 6769. http://dx.doi.org/10.1039/c0jm04345a.

37. Zhang, Q., Chou, T. P., Russo, B., Jenekhe, S. A., & Cao, G., (2008). Aggregation of ZnO nanocrystallites for high conversion efficiency in dye-sensitized solar cells. In: *Inteditor. Angew Chemie* (Vol. 47, pp. 2402–2406). http://dx.doi.org/10.1002/anie.200704919.
38. Zin, N. S., McIntosh, K., Fong, K., & Blakers, A., (2013). High efficiency silicon solar cells. *Energy Procedia, 33*, 1–10.
39. Husain, A. A. F., et al., (2018). *Renewable and Sustainable Energy Reviews, 94*, 779–791.
40. Chapin, D. M., Fuller, C. S., & Pearson, G. L. (1954). A new silicon p-n junction photocell for converting solar radiation into electrical power. *Journal of Applied Physics*, *25*(5), 676–677.

CHAPTER 16

CROP PREDICTION TECHNIQUES WITH K-MEANS ALGORITHMS

LAXMAN THAKRE[1] and MAYUR NIKHAR[2]

[1]*Department of Electronics Engineering, G. H. Raisoni College of Engineering, Nagpur, Maharashtra, India, E-mail: laxman.thakare@raisoni.net*

[2]*PG Scholar, MTech VLSI, G. H. Raisoni College of Engineering, Nagpur, Maharashtra, India*

ABSTRACT

This chapter focuses on strategies and steps taken to improve farming by focusing on technical knowledge and development to make the agricultural sector more reliable and easier for the farmers by predicting the suitable crop by using Machine learning techniques by sensing parameters like soil, weather, and market trends. Gathering accurate data from these sensors is a relatively easy task, but the prediction of crop type from the gathered data requires knowledge and the implementation of high-level algorithms. In the method of data mining technique, farming is a relatively new technique for predicting of horticulture crops, forecasting or animal management, etc. The method is given by the article applications of data-mining methods in the area of horticulture and associated sciences.

Optimization Methods for Engineering Problems. Dilbagh Panchal, Prasenjit Chatterjee, Mohit Tyagi, Ravi Pratap Singh (Eds.)

16.1 INTRODUCTION

In this chapter, the main point is data mining and relativeness is the removal of unseen divining data from huge datasets, which is an effective new technique with an incredible capability to help industries focus on the very essential information in their data cloud. In the present, data mining is used to determine future scopes as well as behaviors, allowing businesses development, knowledge-driven decisions. Instead of automation, the expected analysis provided by data mining is established as a major step beyond studying previous events provided by standard retrieval tools for decision support systems. Current crop data and forecasts based on this data are used for the allocation of resources such as operating time such as snow extraction, the selection of bulk materials such as high-quality soil fertilizers and the efficient integration of personnel and equipment. In better plant production the artificial neural network has proven to be an effective modeling and predictive tool. These technical aspects emphasize improving the use of technology in the agricultural sector to achieve better yields. Obtaining information or details in the field of agriculture, such as soil conditions, climate, plant physiology, and more processes that take place in the field, can be achieved by using different sensors, satellites, etc. These datasets are extremely helpful when it comes to agricultural production.

Application of data mining method in farming for the resulting the more methods for studying adjoined with standards and correlations automatically creates various data groups were formed, to clarify the often endless and error-prone process of obtaining knowledge from experimental data. In fronted of mainly, these methods are logically retarded justifiable and accomplish well on big or small artificial analysis of information sets, and they count on their capability to sense of real-time information. The chapter shows that explain a system that is applying a territory of machine learning methods to issues in farming as well as floriculture. They quickly studied techniques that are rising out of machine-learning research, portray a product workbench for trying different things with an assortment of strategies on real-time data collection and explain an analyst investigation of dairy production management in which extract rules were implicit from an average-sized information set of very large data.

16.1.1 ARTIFICIAL NEURAL NETWORK

Artificial neural networks, as the name suggests "neural" is a word inspired by the brain. It works in the same way as the human brain works. In Neural networks, it contains inputs, outputs, and layers, in which neurons are inputs assigned to ANN and are made up of hidden layers by other units and used by the output layers to produce output. The accuracy of neural networks increases as data increases. They adapt to the difficulty without knowing the principles of the lower layers. In artificial intelligence (AI) and machine learning techniques such as ID3 and other marketing algorithms used to determine tomato yield. Tomatoes are a widely used crop worldwide, grown in almost every part of the world. To design a tomato specialist program, invest the help of computer engineers to design and organize an agricultural scientist and a professional tomato expert. In maize farming, machine learning techniques are used. Maize is also a popular crop with a major source of grain and genotypes of rice that are adapted and adapted to drought conditions which should be grown under controlled conditions and marginal law should be applied.

16.1.2 ANALYTICAL EXPLORATION

A very large number of farming information can be supported by a suitable application. For given data set and Online Analytical Processing Methods for capable use of farming information. Better data provides a flexible yet efficient and reliable storage structure for a huge amount of data while techniques provide mechanisms for improvised and in-depth observation of data. Old tools and database methods will not use to succeed here because of their inflexible nature. But methods utilized in the task are evenly useful for integrated systems at any location supplies related information is accessible such as GPS.

16.1.3 PROJECT SCOPE

The aim of the study is that to get a program to learn neural system exploitation and a building program that predicts seed classifications supported machine learning method. The system has experimented exploitation seed dataset and so seed categories are foretold exploitation of the developed system. The separate parts like weather, type of soil and its arrangement, crop yield, district topography, and market price affect the selection of crops. We think about, the weather and soil elements. Machine-learning provides various successful counts that rely on various factors. It is a troublesome undertaking that recognizes the finest suitable when there is an alternative accessible. Thus, with the help of AI algorithm, an exact crop can be anticipated. We considered the weather factors and soil element attributes of the area to detect the appropriate yields.

16.2 LITERATURE REVIEW

The chapter written by Ranjeet et al. suggested that the main components of Artificial Neural Network models are ready with varied variant neurons in hid-layer, back-propagation learning prediction. Changing these characteristics inspired the model to create a perfect capability to predict crop yield. Forming rate, educational rate as well as the variety of unknown nodes which gives the most part affect show conduct. For the foremost half, fewer unknown nodes were needed because of quantitative knowledge diminished. The most effective system has fewer unknown nodes than the start variety of nodes. Artificial Neural Network models along with a lot of nodes could have appeared in overfitting as hostile examine connections off them. Root mean square error was utilized to evaluate the execution of the created system.

The researcher Meenaeta said in paperback propagation, Artificial Neural Network is used for harvesting crop gauging and set migrated systems available in the Artificial Neural Network. The area is considered 18 factors area unit thought of because of the data elements and therefore the harvest field because the yield variable and result are not found. The 24 factors area unit traditional month-to-month rainfall, monthly higher temperature, monthly less temperature, month-to-month traditional humidity, average wind speed (kilometers per hour), and extreme speed (kilometers per hour).

So as to anticipate the amount of the crops, the field is examined based on observation they are classified. This classification is dependent on data-mining algorithms. This chapter gives knowledge of different grouping standards like K-Nearest Neighbor and Naive Bayes. Utilizing this chapter, we studied categorizing standards and recognized which helps to fit for data set that we will use in our task.

The result of N. Hemageetha's paper how's different data-mining methods like Decision Trees Market-based Analysis, Association Rule Mining, Clustering, and Classification so forth. It completely covers the Data-Mining idea. Different data mining algorithms, for example, K-Mean, Naive Bayes classifier, J48 are clarified in this chapter. It shows that grouping of soil-based Genetic Algorithm, Naive Bayes, Association Rule Mining and Components of ANN. In the end, it contains Clustering in the soil database. This chapter helped to understand and analysis of various data-mining algorithms and categorization systems. This will end up being incredibly recipient while building our system and will help in mining the data set acquired from various sensors utilized remotely.

Awanit Kumar, Shiv Kumar shows a model for the forecast of production of yields in the present year and it minted. To decide the harvested crop production, it utilizes a data-mining algorithm-Means. This system likewise utilizes and coming to the resulting valve of yields expectation mechanism in the form of fuzzy logic. Crops Fuzzy logic is a standard based expectation logic wherein a lot of rules are applied on the land for cultivating, rainfall, and production of yields results are obtained is more. Utilizing this chapter, a clear insight into how K-Means can be utilized to examine data sets is acquired. It randomly utilized one and more to set of standards and they are applied in form of fuzzy logic, we apply the set of standards to determine which crop will grow maximum more cost got at a dependent on earlier years cost of yields and current soil and climate information.

The best harvest for farmers to do this work is collecting data from different technologies such as IoT and web services and can handle a large amount of data. GPS is used to take pictures of the agricultural field to detect the emerging number of plants and plants for testing and is stored in archives and location. And it is talked about in the cloud.

16.3 METHODOLOGY

The strategy is based on a modern decision-making tree algorithm is given in IBLE that it, for the most part, utilizes in the IT (Information theory). Coming about the capacity of the channel, the concept to take the important characteristic to the entity in the measure and intimated. For giving the standard with various characteristics the point to identifying the example can effectively the correct distinction. The application is that this algorithm is applied in the oral cavity disease determination, the exploratory outcome showed, the algorithm has a very strong recognition capability to agriculture case finding to very good assistance analysis function (Figure 16.1).

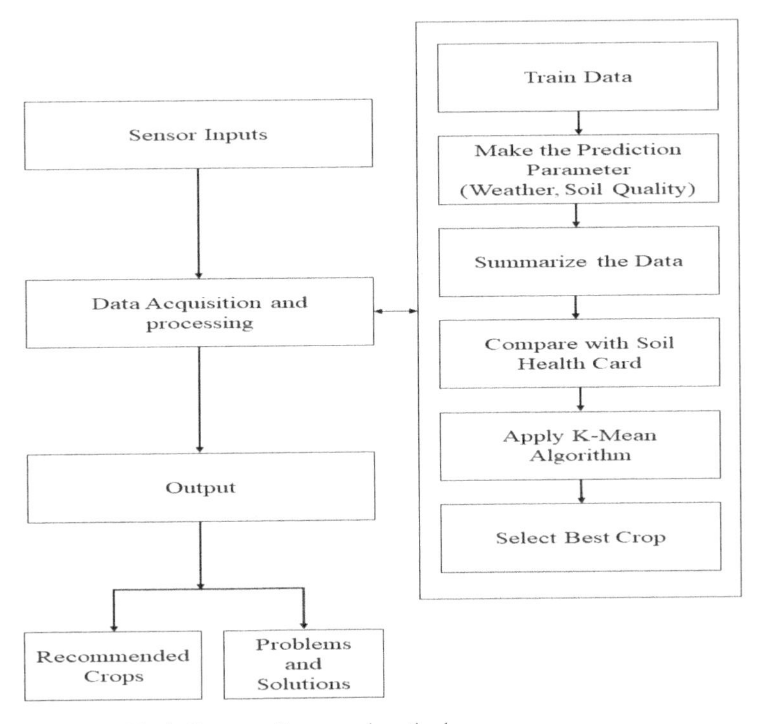

FIGURE 16.1 Block diagram of proposed method.

Data mining has two main methods, classification, and prediction. Future data is classified and predicted using classification and prediction which are two methods of data analysis. The aim is to set the accuracy of the Hugh test rather than the training accuracy of the class algorithms.

The three methods of data-mining are supervised learning, Semi-supervised learning, and unsupervised learning. In this chapter several methods that are used to find intelligence are described below.

This analysis has incontestable the determination of rice crop yield by applying one of the machine learning techniques, SVM (support vector machine). In terms of test accuracy and quality conjointly Multilayer Perceptron and Bayes internet showed the very best accuracy and very best quality and social media optimization showed the bottom accuracy and worst quality.

16.3.1 IoT SYSTEM DESIGN

IoT sensors has the primary job of extricating real-time information within territory which is anticipated. As realized that IoT is also a sector in current innovative improvement in that important data can drive. Thus, cost-effective sensors, for example, pH meter, temperature sensor, humidity sensor and soil moisture sensor are utilized for estimating parameters, for example, soil dampness level, humidity, surrounding weather conditions, and pH level of soil. Those sensors were enormously useful in detecting climate changes also supplement nutrient into land. The detected data by assistance of communication system arrives at the database by assistance of a micro-controller. The information in the database utilized for creating further weather conditions studies. These sensors are likewise utilized for recognizing qualities of the farming field at a prior level. Checking those attributes appears like simple one. Be that as it may, this data can be utilized to roll out incredible improvements in the farming work. The suggested plan for the IoT system is appeared in Figure 16.2.

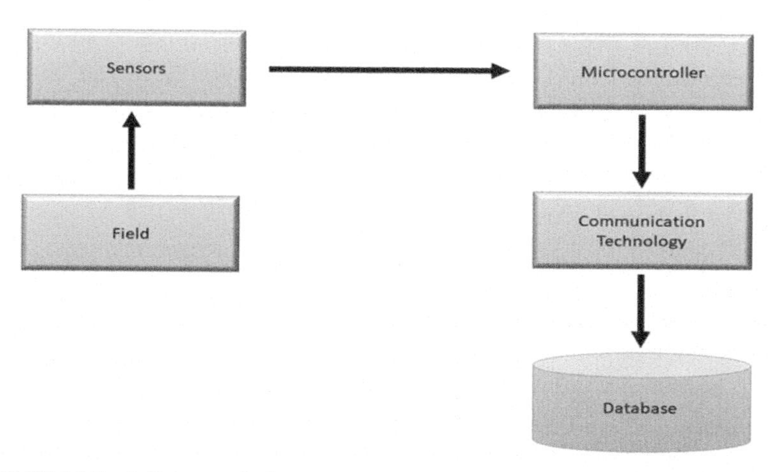

FIGURE 16.2 IoT system design.

In this case, the study of how agricultural integration analyzes the seed database using K-Means, Fuzzy C Means is done by passing different parameters such as perimeter, area, length, and width of the kernel, unequal coefficient, compactness, and length of kernel groove using the system of R programming.

The advantage in recent technology develops with the utilization of the Deep Learning and IoT (Internet of Things) has created all things possible. The Internet of Things technique is considerably needful in managing ad real data with real-time information by utilizing sensors. Most important information can be used in a way and emergent are nursery which is given to the trained Deep Learning algorithm like ANN for forecast decisions. Result and outcome considerably needful in recommending an appropriate crop to be planted in the specific area. This part explains the preprocessing phase, features, dataset, Deep Neural Network and IoT design.

- **Step 1: Data Acquisition:** Wheat seeds are vital in cultivating and different varieties of wheat seeds are giving us distinctive yields which must made expanded every year for take care of demand for the general population. The seed dataset gives the names of three distinct seeds as Kama, Rosa, and Canadian. These following wheat seeds give diverse yields in cultivating. The chose data from UCI machine learning repository contains both categorical and continuous characteristic esteems. The dataset contains following properties like Area, Perimeter, compacting radius, Length, Width, asymmetric Coefficient, lk Groove, type wheat seed.
- **Step 2: Preprocessing:** The selected dataset contains no missing values in the table. Identifying the missing values can be done either by eliminating the record or by replacing the missing values by calculating the mean. In this chapter utilizing the R programming aids in finding the missing values with the help of neon function (Figure 16.3).

16.4 STATISTICAL ANALYSIS

In this section, we define a set of criteria for analyzing the algorithms. The set of algorithms are categorized as per the length of the input, the type of features extracted, the accuracy, and the probable application of the system under various real-time scenarios.

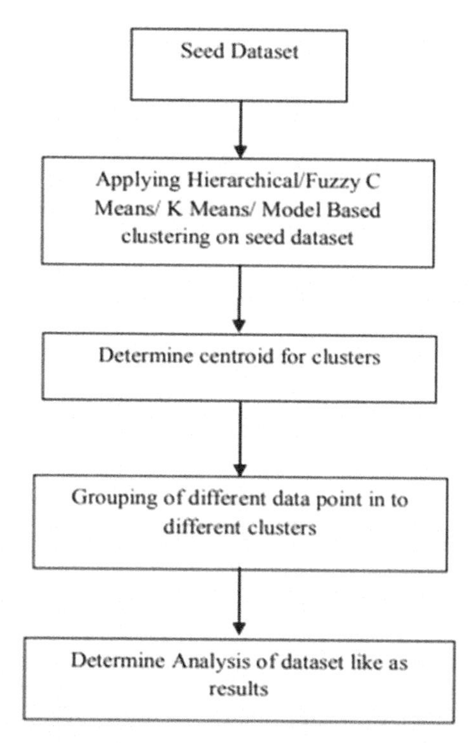

FIGURE 16.3 Analyzes clustering algorithms that are different from the seed.

16.4.1 PARALLEL K-MEANS ALGORITHMS

Is the result of K-Means algorithms technique has shown to be effectively in producing good cluster for intimate conclusion for numerous practical applications are development applications in agro-field? After that the K-Means technique is very familiar for its satisfactorily decent and simple results. The graph direct algorithm of K-Means technique needs consume time proportionate to the product of documents counts of the vectors and cluster counts per iteration for the integration. The K mean method is more consumptive and readable method in database.

16.4.2 INITIALIZATION OF PHASE

Select more set of K beginning end m j k j = 1 in Rd.

- This selection may be done in a random manner or making use of some heuristic;
- Phase Distance Calculation n, compute its ≤ I ≤ For each data point Xi 1 k and then find ≤ j ≤ Euclidean distance to each cluster m j 1 the closest cluster centroid;
- Resultant Phase Centric Recalculation k recomputed cluster centroid m j as ≤ j ≤ For each 1 the average of data points assigned to it;
- This step is Convergence Condition. Repeat steps as possible.

In this, a set of criteria is defined for analyzing the algorithms. The set of algorithms are categorized as per the length of the input, the type of features extracted, the accuracy, and the probable application of the system under various real-time scenarios (Figures 16.4 and 16.5). Table 16.1 showcases this comparison in detail.

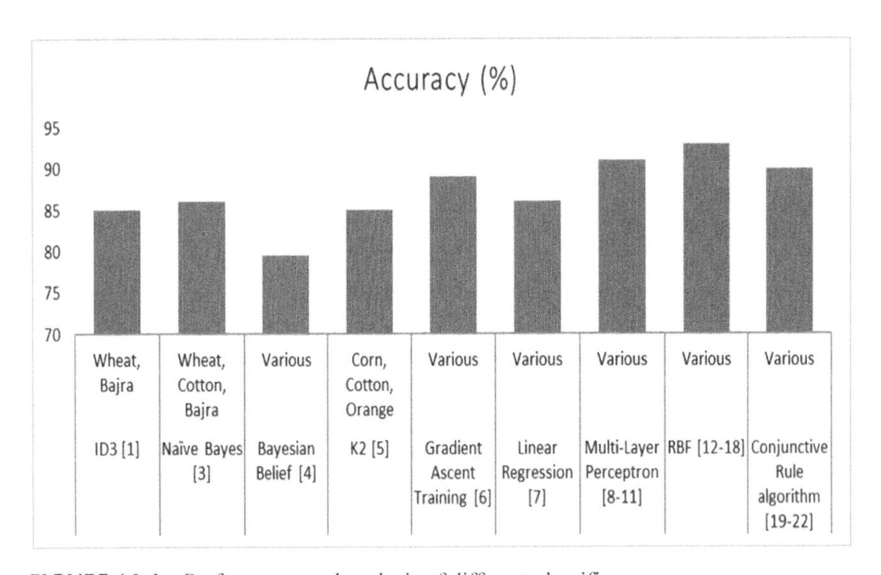

FIGURE 16.4 Performance and analysis of different classifiers.

16.5 CONCLUSIONS

For the methods and increase in the number of utilization of data-mining techniques in farming and an increasing number of data that are presently available from many assets. This consideration is a novel and research area and it is expected to develop further. Maximum no of work to be done on this emerging and interesting research area. For the normal technique of combining GPS and computer science technology with farming will help in managing crops and forecasting effectively. Number of methods via the multi-linear regression, artificial neural network, and SVM (support vector machine) are studied. So, it is concluded that an artificial neural network is an appropriate technique for the project, as the inputs are less in number.

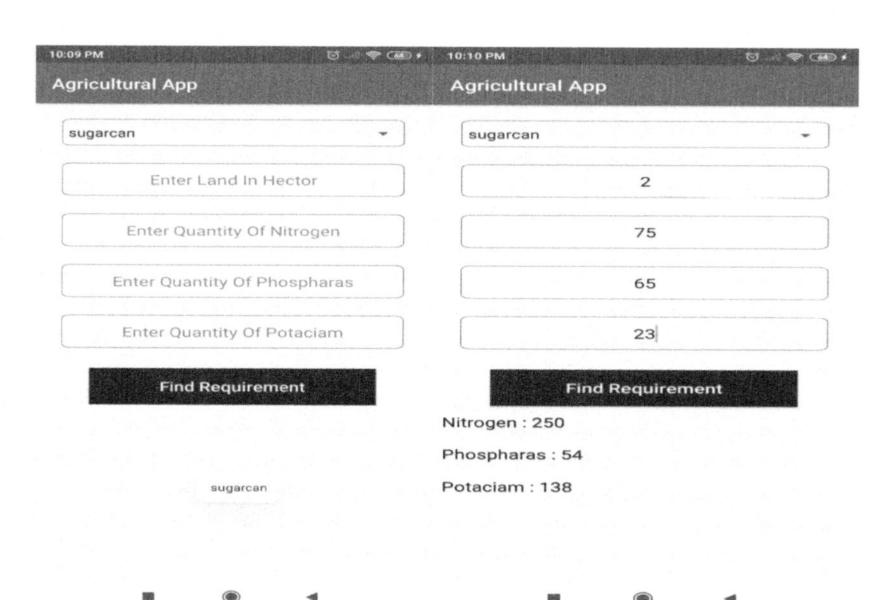

FIGURE 16.5 Mobile agriculture app showing result for sugarcane.

TABLE 16.1 Performance and Analysis of Different Classifiers

Algorithm	Crop Types	Accuracy (%)
ID3 [1]	Wheat, Bajra	85
Naïve Bayes [3]	Wheat, Cotton, Bajra	86
Bayesian belief [4]	Various	79.5
K2 [5]	Corn, Cotton, Orange	85
Gradient ascent training [6]	Various	89
Linear regression [7]	Various	86
Multi-layer perceptron [8–11]	Various	91
RBF [12–18]	Various	93
Conjunctive rule algorithm [19–22]	Various	90

KEYWORDS

- **artificial neural networks**
- **crop productivity classification**
- **data mining**
- **information theory**
- **internet of things**
- **recommend soil factors**

REFERENCES

1. Bendre, M. R., Thool, R. C., & Thool, V. R., (2015). *Big Data in Precision Agriculture*. NGCT.
2. Monali, P., Santosh, K. V., & Ashok, V., (2015). Analysis of soil behavior and prediction of crop yield using *Data Mining approach. International Conference on Computational Intelligence and Communication Networks.*
3. Abdullah Na, William Isaac, Shashank Varshney, & Ekram Khan, (2016). "An IoT Based System for Remote Monitoring of Soil Characteristics," 2016 International Conference on Information Technology (InCITe) – The Next Generation IT Summit on the Theme – Internet of Things: Connect your Worlds DOI: 10.1109/InCITe39245.2016 6-7 Oct. 2016.
4. Suma, N., Sandra, R. S., Saranya, S., Shanmugapriya, G., & Subhashri, R., (2017). IoT based smart agriculture monitoring system. *IJRITCC*.
5. Heemageetha, N., (2016). *A survey on Application of Data Mining Techniques to Analyze the Soil for Agricultural Purpose*. IEEE.
6. Basak, J., Sudharshan, A., Trivedi, D., & Santhanam, M. S., (2004). Weather data mining using independent component analysis. *J. of Machine Learning Research, 5*, 239–253.
7. Dhivya, B., Manjula, Siva, B., & Madhumathi, (2017). A survey on crop yield prediction based on agricultural data. *International Conference in Modern Science and Engineering.*
8. Giritharan, R., & Koteeshwari, R. S., (2016). *Agricultural Crop Predictor and Advisor Using ANN for Smartphones*. IEEE.
9. Nagini, R., Rajnikanth, T. V., & Kiranmayee, B. V., (2016). Agriculture yield prediction using predictive analytic techniques. In: 2nd *International Conference on Contemporary Computing and Informatics (ic3i).*
10. Suresh, S. B., Sudhir, B. L., Umesh, A., & Laxman, T., (2009). Modeling of cantilever based power harvester as an innovative source of power for RFID tag. In: 2nd *IEEE ICETET International Conference.* G. H. Raisoni College Of Engineering, Nagpur.

11. Awan, A., & Md Sap, M., (2012). A framework for predicting oil-palm yield from climate data. *International Journal of Information and Mathematical Sciences.*
12. Kim, S., & Wilbur, (2011). An EM clustering algorithm which produces dual representation. In: *Machine Learning and Applications and Workshops (ICMLA), IEEE 10th International Conference* (Vol. 2, pp. 90–95).
13. Duan, W., Qiu, Duan, L., Liu, Q., & Huan, H., (2012). An improved rough k-means algorithm with weighted distance measure. In: *Granular Computing (GrC), IEEE International Conference* (DOI 10.1109/GrC.2012.6468643 pp. 97–101).

CHAPTER 17

INVESTIGATION OF TRIBOLOGICAL PERFORMANCE OF ELECTROLESS NI-B-W COATINGS AT HIGH TEMPERATURE AND ITS OPTIMIZATION

ARKADEB MUKHOPADHYAY,[1] TAPAN KUMAR BARMAN,[2] and PRASANTA SAHOO[2]

[1]*Department of Mechanical Engineering, Birla Institute of Technology, Mesra, Ranchi–835215, Jharkhand, India, E-mail: arkadebjume@gmail.com*

[2]*Department of Mechanical Engineering, Jadavpur University, Kolkata–700032, West Bengal, India*

ABSTRACT

The present work aims to investigate the high-temperature tribological performance of Ni-B-W coating. Ni-B-based ternary coating with co-deposition of W has been considered in an attempt to achieve higher, hardness, wear resistance, and enhanced friction performance. The coatings were deposited autocatalytically on steel substrate from a sodium borohydride-based alkaline bath. Ni-B-W coatings in heat-treated form (350°C for 1 hour) were subjected to tribo-tests on a pin-on-disc tribometer. Applied normal load, sliding speed and operating temperature

Optimization Methods for Engineering Problems. Dilbagh Panchal, Prasenjit Chatterjee, Mohit Tyagi, Ravi Pratap Singh (Eds.)

was varied at three equally spaced levels within 10–30 N, 0.25–0.42 m/s and 100–500°C, respectively. To improve process capability, perform statistical analysis and draw significant conclusions, Taguchi's L_{27} orthogonal array was adopted to perform experiments. Tribological responses considered were mass loss (signifying wear) and coefficient of friction (COF). Grey relational analysis was implemented to optimize the tribo-test parameters. The objective was to minimize mass loss and COF simultaneously. Optimum mass loss and COF was achieved at 10 N load, 0.25 m/s speed and 500°C operating temperature. Operating temperature was concluded to be of the highest significance in controlling the tribo-behavior of the coatings from analysis of variance (ANOVA). Also, proper coating deposition was ensured by characterizing them using various techniques such as energy dispersive spectroscopy, X-ray diffraction and field emission scanning electron microscope (FESEM).

17.1 INTRODUCTION

Nickel boron alloys play an important role in wear reduction of automobile and aerospace components, slurry pump components, gun barrel bores and allow greaseless operation with anti-friction properties [1]. Consequently, their investigation has received immense importance to enhance the lifetime of mating components, especially at demanding conditions [2–4]. Essentially, nickel boron alloys and composites may be obtained by the electroless method where sodium borohydride or dimethylamine borane (DMAB) is used as reducing agent. DMAB-based baths offer higher bath stability and lower temperature of deposition. On the other hand, sodium borohydride is considered to be a strong reducing agent and is preferred [5, 6].

A crucial role is played by the B content in controlling structural aspects, mechanical, and tribological characteristics. Nickel boron alloys can be further sub-classified depending on the B content, such as high-B, mid-B, and low-B [7]. In general, mid-B coatings having B within 5–7 wt.% has been focused in most of the research works [8, 9]. The mid-B coatings show amorphous characteristics when there is around 6 wt.% B in as-plated state [10]. When the B content is around 5%, they show a mixture of amorphous and nano-crystalline phases [11]. Ni-B coatings with B as high as 8 wt.% was deposited by Vitry and Bonin [12] and they were amorphous in as-plated condition. The $NaBH_4$ concentration in the coating bath modifies the amount of B present in the alloy, its morphology, structure, and mechanical properties. With an increase in $NaBH_4$ concentration, the B increased and promoted the amorphous nature of the coatings [13]. In fact, Bulbul [13] compared the morphology of Ni-B coatings with 'pea grains' in dispersed form, grains of 'maize,' cluster of 'grapes' or surface of a 'broccoli'/'cauliflower' for various bath formulations. Barman et al. [14] also concluded an increase in B content with an increase in $NaBH_4$. It was also reported that with B content increasing, the amorphous nature became more profound and the elastic modulus and hardness increased [14].

The coating structure and properties can be further modified and enhanced by annealing, thermo-chemical treatment, vacuum heat treatment, etc. [15–17]. Phase transformation occurs when the coatings are heat treated. Hard phases are formed, such as Ni_3B and Ni_2B. Also, the reflection from Ni (111) becomes sharp with a decrease in peak broadening becoming evident. Ni – 6.4% B coatings heat treated at 20°C per hour within temperature range of 50–550°C revealed formation of 2 exothermic peaks at 306°C and 427°C [18]. Thus, on suitable heat treatment, the coatings crystallized to form the intermetallic phases of Ni_3B and Ni_2B [15]. Consequently, an increase in hardness was observed and attributed to precipitation hardening [8–12]. Thereby, several research works were carried out where heat-treated Ni-B coatings in the mid-B range were investigated and it was concluded that the mechanical and tribological performance of the coatings were enhanced post-heat treatment [8–10, 19–21]. The wear mechanism and phase transformation post sliding wear was also investigated in detail under dry sliding condition [22, 23]. Finally, in a recent study, the ambiguity in the chosen heat treatment temperature range was investigated by Pal and Jayaram [24], and it was concluded that the phase transformations were controlled by local inhomogeneity in composition of the coatings. It was also claimed that complete crystallization of the coatings occurs on heat treating at 385°C for 4 hours. High-B coatings on the other hand showed similar tribological performance compared to mid-B ones in as-plated condition and after heat treating [12]. But the coating morphology and structure was modified. The mid-B coatings fully crystallized to Ni_3B post heat treatment at 400°C for 1 hour. But the high-B coatings were multi-phased [12].

Research is still ongoing to enhance the tribo-performance of the deposits. In this regard, the coatings have been modified either by alloying with a third and fourth element to form ternary/quaternary alloy or micro/nanoparticles co-deposition. The hardness, tribological behavior at ambient conditions as well as a high temperature improves by co-deposition of W and Mo to form Ni-B-W [25, 26] and Ni-B-Mo [27, 28] coatings. In fact, excellent thermal stability was exhibited when both W and Mo was co-deposited [29]. Promising properties was also exhibited by Ni-B-Sn coatings [30, 31]. Apart from poly-alloy coatings, co-deposition of Al_2O_3, TiO_2, SiC, etc., was considered in a quest to improve the mechanical properties and corrosion resistance [32–34].

The Ni-B alloys also had great potential at high temperature. This was revealed by Pal et al. [35]. Since the precipitated Ni_3B and Ni_2B phases have high melting temperature (~1,150°C), they may be considered for applications within 500–600°C. The hardness and modulus of completely crystallized Ni-B coatings were quite consistent at high temperatures when compared to room temperature [35]. It was also established through pin-on-disc tribo-tests that severe wear of Ni-B coatings occurred at 100°C compared to 300 or 500°C [36]. Ni-B-W coating was found to possess high wear resistance at 100–500°C whereas Ni-B-Mo imparts self-lubricity due to formation of lubricious molybdates and oxides [37]. It was also established through X-ray diffraction (XRD) that phase transformation occurred due to in-situ heat treatment effect at 300 or 500°C [38].

Due to the excellent wear reduction capabilities, the Ni-B coatings are being widely studied and newer variants are being explored. Several studies have been carried out to correlate the structure and properties as well. Even, high temperature tribo-performance has been also reported. But optimization of coating performance at elevated temperatures has been scantily reported. The present work therefore reports the wear and friction performance multi-criteria optimization of Ni-B-W coatings. The ternary variant has been considered because of its outstanding thermal stability, wear resistance and hardness. A pin-on-disc tribometer is used to investigate the high temperature performance. Taguchi's design philosophy is employed to carry out experiments by varying load, speed, and operating temperature. Optimization and statistical analysis are carried out using gray relational analysis (GRA) and ANOVA. The present work would prove to be beneficial in improving the cost-effectivity and achieve optimal performance from the coatings.

17.2 EXPERIMENTAL METHODS AND MATERIALS

17.2.1 PREPARATION OF SUBSTRATE

The substrate needs to be carefully prepared prior to coating deposition and freed of any impurities. Autocatalytic Ni-B-W deposition was carried out on steel substrates (AISI 1040). The steel substrates were ground to N5 roughness grade and procured locally. They were cleaned thoroughly in detergent water and running water. Post initial cleaning, the substrates were rinsed in deionized water. Subsequently, degreasing was carried out in acetone. Again, deionized water was used to rinse the specimens. Finally, removal of an oxide layer and activation was carried out in 50% HCl for a few seconds. The samples were finally dipped into the electroless bath. Steel blocks having dimension $20 \times 20 \times 2$ mm^3 was used for characterization purposes. For tribo-tests, pin specimens were coated having 30 mm length and 6 mm diameter.

17.2.2 COATING DEPOSITION

Coating was deposited from an alkaline bath having pH 12.5. The reducing agent used was sodium borohydride. The basic Ni-B-W bath formulation was adopted from the work carried out by Mukhopadhyay et al. [26]. Bath temperature was accurately controlled at 90 ± 2°C. Coating was deposited on the substrates for 4 hours. After initial 2 hours of deposition, the coating bath was replenished by a fresh one. This was done to ensure higher deposition, prevent bath instability and replenishment of exhausted ions required to carry out the autocatalytic process. The coated specimens were withdrawn from the bath after four hours, rinsed in deionized water and dried in warm air. To improve the mechanical properties, they were further heat treated at 350°C. This was carried out in a muffle furnace for a period of 1 hour.

17.2.3 ANALYSIS OF COMPOSITION, MORPHOLOGY, AND STRUCTURE

It is necessary to determine the coating composition. It has been well established in literature that coating characteristics are dependent on plating condition. For composition determination, energy dispersive spectroscopy (EDS) was used (EDAX Corporation). The composition was determined at 10 keV, i.e., a low accelerating voltage for detection of light elements such as B. The structural aspects of the coating were ascertained from X-ray diffraction (XRD) spectrums (RIGAKU, ULTIMA III). The results were analyzed for 2θ range of 20–80° and the scan rate was set at 1°/min. The surface features of the coatings were observed by using a field emission scanning electron microscope (FESEM). It works along with EDS and gives information on the morphological characteristics (FEI QUANTA, FEG 250).

17.2.4 MICROHARDNESS MEASUREMENTS

Microhardness indicates the resistance of the coatings to deformation caused by indentation. Since electroless nickel alloys with B have high hardness, the mechanical properties of the Ni-B-W coating is also identified by its resistance to indentation, i.e., microhardness. There are several methods to measure microhardness but the Vicker's indentation technique has received widespread acceptance for hardness characterization of the electroless coatings. In the present work, Vicker's hardness was measured at 100 gf keeping in mind the coating thickness. The substrate effect needs to be minimized in such a case. The dwell time was set at 15 s and the speed of indentation was 25 μm/s. Average of 6 indentation results are reported.

17.2.5 TRIBOLOGY TESTS AND EXPERIMENTAL DESIGN

Pin-on-disc setup was used for tribology tests (DUCOM, TR-20-M56). In this configuration, either the pin or a disc act as the material whereas the other acts as counterface. They press against each other and the disc rotates resulting in a sliding action. From the tests, frictional force and wear rate may be determined. In this case, electroless Ni-B-W coated and heat-treated pin specimens slide against EN 31 hardened steel counterface. The temperature at which tests are carried out, load, and speed was varied as shown in Table 17.1. The rotating disc was heated inductively and the temperature was continuously monitored by a pyrometer. The frictional force was continuously recorded on time via a load cell. This was converted to coefficient of friction (COF). Mass loss (gm) of the coatings was determined using a high precision weighing balance. The mass loss signified wear of the coatings. Thus, COF and mass loss were considered to be the responses. The process parameters or variables were load, speed, and temperature as shown in Table 17.1. They were varied as per the levels enlisted in Table 17.1. Other parameter such as the sliding duration was kept constant at 5 min. The experiments were designed as per Taguchi's L_{27} orthogonal array (OA). This allowed estimation of the interaction effects of the parameters also. The trends of variation in COF and wear rate were also analyzed based on the data collected as per OA. Furthermore, wear mechanism of the coatings was also investigated. This was done in order to obtain an insight into the predominant wear mechanism affecting the coatings at high temperatures. For this, backscattered electron (BSE) images were taken post sliding wear. The BSE images are bright images of the coating whereas dark regions indicate delamination and substrate exposure due to severe wear.

TABLE 17.1 Variable Parameters for Tribological Tests

Test Parameters	Code	Unit	Levels		
			1	2	3
Operating temperature	A	°C	100	300*	500
Applied normal load	B	N	10	30*	50
Sliding speed	C	m/s	0.25	0.33*	0.42

**Initial test run condition.*

17.2.6 OPTIMIZATION AND STATISTICAL SIGNIFICANCE OF TEST PARAMETERS

Grey relational analysis (GRA) is a very simple and efficient tool that helps in multi-criteria decision making very efficiently at less time. Though, optimization results obtained through this method gives results that are exactly at the specific points or combinations being considered. GRA has been very efficiently utilized for tribo-performance optimization of Ni-B coatings in earlier studies [21]. In the present chapter also, this method has been considered due to its cost-effectiveness and simplicity.

Normalization of responses and results is the first step in GRA. All the data is initially normalized between 0 and 1 for maintaining homogeneity.

Mostly, different responses with different units and range of values are encountered in optimization problems. Hence, normalization is necessary. There are several criteria based on which normalization is carried out. They are nominal the best, lower-the-better, and higher-the-better [21]. Here, it is desired to optimize or rather minimize wear and COF of the coatings. Naturally, it is advised to normalize the values based on lower-the-better criteria. Such a data pre-processing may be calculated as [21]:

$$x_i(k) = \frac{\max y_i(k) - y_i(k)}{\max y_i(k) - \min y_i(k)} \tag{1}$$

where; $x_i(k)$ is the normalized value, max; and min $y_i(k)$ are the largest and smallest values of the kth response, i.e., $y_i(k)$. In this work, there are two responses, i.e., wear, and COF and their corresponding k being 1 and 2, respectively.

After normalization, the data is further processed to find out the gray relational coefficient (GRC). This GRC gives a relationship between the ideal best (=1) and how closer an experimental result is to this ideal best. The GRC $\xi_i(k)$ can be calculated as [21]:

$$\xi_i(k) = \frac{\Delta_{min} + r\Delta_{max}}{\Delta_{0i}(k) + r\Delta_{max}} \tag{2}$$

where; the absolute value of difference between $x_0(k)$ and $x_1(k)$ is denoted by Δ_{0i}. The corresponding largest and smallest values of Δ_{0i} are denoted as Δ_{max} and Δ_{min}, respectively. The value of r lies between 0 and 1 and is called the distinguishing coefficient. Due to good stability of outcomes, it is generally taken as 0.5 [21].

After obtaining the GRC, the gray relational grade (GRG) is obtained. Higher value of GRG denotes near-optimal result. The multi-performances are converted into a single performance index which is the GRG and may be used for further analysis. This GRG denoted as γ, may be calculated as [21]:

$$\gamma_i = \frac{1}{n}\sum_{k=1}^{n} \xi_i(k) \tag{3}$$

where; the total number of performance measures or responses are denoted by n. In the present work, it is 2. Basically, the GRG is average of the kth GRC. Higher GRG closer to 1 denotes better system performance and the run is closer to optima. Optimization of GRG results in simultaneous minimization of multiple responses. In this case, it is wear and COF both.

The final step after obtaining the optimal combination of parameters is carrying out confirmation tests. In the confirmation experimental runs, a GRG is predicted as follows [21]:

$$\hat{\gamma} = \gamma_m + \sum_{i=1}^{o} (\overline{\gamma}_i - \gamma_m) \tag{4}$$

where; the total mean of all the GRG is denoted by γ_m, the mean grades at optimal levels are denoted by $\overline{\gamma}_i$, and the number of design variables is denoted by o which is 3 in this case. The GRG obtained experimentally at the optimal setting of process parameters is compared with this predicted grade. Also, the GRG obtained at an initial test run is compared with the optimal results. Generally, mid-level combination of parameters is taken as initial test run [21]. The predicted and experimentally obtained optimal GRG should have better performance index compared to this initial test run.

After optimization, the statistical significance of process parameters in controlling the responses, ANOVA is carried out. ANOVA is carried out on the GRG since this signifies the tribological characteristics. From the ANOVA results, the parameters that significantly influence wear and COF simultaneously may be analyzed. ANOVA is based on F-value. If it is higher for a certain design variable than a given tabulated value at a particular confidence level, it denotes the significance of that parameter. The F-ratio is the variance of a factor divided by the variance of error. A flowchart of the complete experimental optimization scheme is shown in Figure 17.1.

17.3 RESULTS AND DISCUSSION

17.3.1 COMPOSITION OF THE COATINGS

The composition of the coating was ascertained in as-plated state. The result of EDS analysis, i.e., the spectrum is shown in Figure 17.2. The wt.% of B lies in the mid-boron range. Also range of the different elements deposited is laid down since EDS point analysis was carried out at different areas on the surface. This is also in accordance with the result reported by Pal and Jayaram [24] where it was discussed that there is compositional inhomogeneity in the coatings. The wt.% of B and W is given in Figure 17.2 and the rest amount is Ni. The composition of the coating with respect is important because they control the structural aspects. Therefore, in the very preliminary stage it is necessary to ascertain the coating composition especially the co-deposited B and W content.

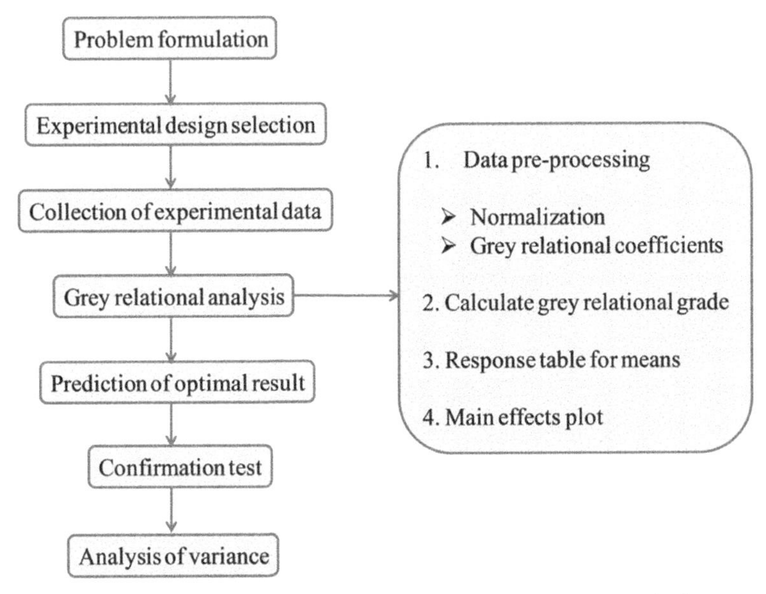

FIGURE 17.1 Experimental design and optimization process – schematic flow.

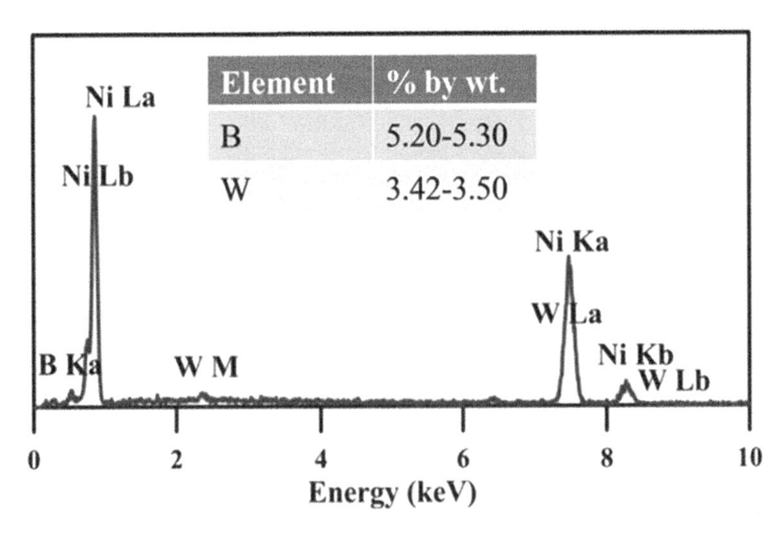

FIGURE 17.2 The EDS result of as-plated Ni-B-W.

17.3.2 COATING STRUCTURE

The structure of coatings pre- and post-heat treatment has been carried out using XRD and shown in Figure 17.3. A broad hump is observed in as-plated state is seen in Figure 17.3. Therefore, the coating is amorphous in un-treated or as-plated condition. Disorderly arrangement in atoms results in a broad hump in XRD spectra [8, 9]. A mixture of amorphous and nano-crystalline behavior has been observed in some research works for mid-B coatings in as-deposited state [11]. In the present work, the amorphous phase is promoted by W [39]. On heat treatment, the coating crystallizes. Also, Ni_3B and Ni_2B phases are precipitated. At 2θ value ~ 44.9°, peak of Ni (111) appears. This was also observed in other research works for mid-B variants of Ni-B coatings [8, 9]. Further, due to the compositional modulation, crystallization of Ni-B coatings take place over a range of temperature and different regions tend to crystallize at different temperatures [15].

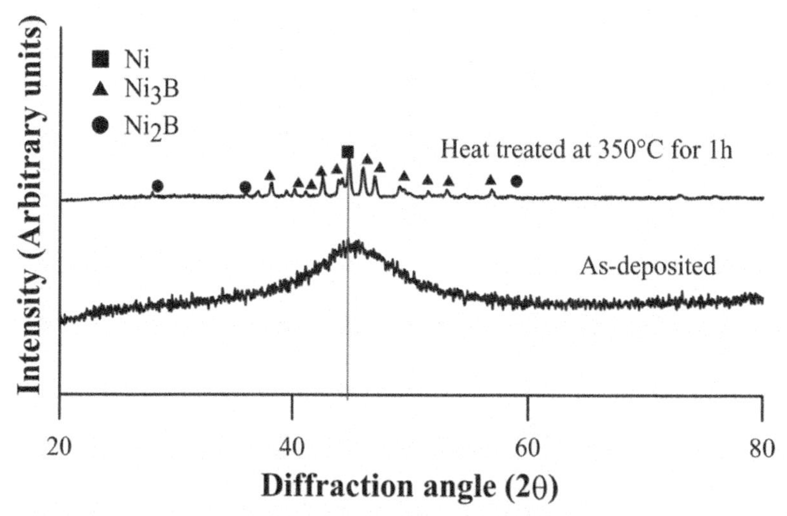

FIGURE 17.3 The XRD results of Ni-B-W coatings.

17.3.3 MORPHOLOGY OF THE COATINGS

The topographical features of Ni-B-W coatings observed through the use of FESEM and they are shown in Figure 17.4. The as-plated coating observed under FESEM (Figure 17.4(a)) present globular features which is quite common in borohydride reduced alloys [8–14]. These globules are densely distributed and the image does not reveal any surface defect. Generally, the coatings tend to appear gray. The compact globular structures tend to provide lubricity to the coatings when sliding under dry condition. On the other hand, a slight inflation in the size of nodules is observed on heat treatment along with the appearance of specific cellular boundaries (Figure 17.4(b)). The inflation of the nodules has been attributed to the fact that crystallization of the coatings occurs on heat treatment [21]. The results are also consistent with XRD where phase transformation has been clearly indicated. Globular structures that resemble the surface of a cauliflower have been observed in several other research works [8–14]. Due to such globular structures, the coatings tend to be inherently self-lubricating and reduce the actual area of contact with the mating surface [19, 20]. The coating cross-section (Figure 17.4(c)) indicates feather like growth in longitudinal direction. The coatings are almost 25 μm thick with a clear demarcation line separating two different plating episodes (double bath deposition scheme).

FIGURE 17.4 Morphological features of Ni-B-W coating: (a) as-deposited; (b) heat treated; and (c) cross-section.

17.3.4 MICROHARDNESS OF THE COATINGS

The as-plated coatings have microhardness ~759 HV_{100}. This microhardness is quite noteworthy compared to the binary Ni-B coatings reported in other works [19, 20]. Krishnaveni et al. [8] observed a microhardness of 570 HV_{100} for Ni-B coatings in as-deposited state. Thus Ni-B-W coatings have a microhardness which is higher than its binary variant. A noteworthy improvement in microhardness takes place when W is co-deposited. In heat-treated condition (350°C for 1 hour), the microhardness of the coatings is ~1,181 HV_{100} which is a remarkable increase compared to as-plated state. This increase in microhardness may be attributed to crystallization of the coatings and precipitation of hard boride phases [19–21]. This is also known as precipitation hardening. Further, W co-deposition also results in solid solution strengthening [38].

17.3.5 TRIBOLOGICAL BEHAVIOR

The tribo-tests and its results, i.e., mass loss and COF was investigated on a pin-on-disc tribo-tester. The combination of process parameters along with the test results has been laid down in Table 17.2. The mass loss was calculated by weighing the coated specimen pre and post sliding wear at different parametric settings and accordingly 27 experiments were carried out based on Taguchi's experimental design, i.e., L_{27} OA.

It may be also noted that since the interaction effect of the parameters are also desired to be investigated, the L_{27} OA was chosen. Else L_9 OA could have been used to carry out the tests. Since there are 3 factors at 3 levels, the L_{27} OA also corresponds to a full factorial design. Consequently, the trends of friction and wear could be evaluated from the result. The plots of COF and mass loss are presented in Figures 17.5 and 17.6, respectively.

As the operating temperature increases, the COF decreases (at all loads) as can be seen in Figure 17.5. The COF at 30 N load and 100°C operating temperature is the highest at all values of sliding speed. At lower temperatures, the oxides of W tend to be brittle and results in an increase of COF due to grinding of the hard coating material between the pin and counterface. The improved COF at high temperature is attributed to the formation of tribo-oxide patches which provide self-lubricating properties. At higher temperatures, the oxide glazes tend to reduce the COF by aggravating chemical reactions. Also, at high temperatures, the WO_x provide lubrication glazes and decreases the COF.

TABLE 17.2 Tribo-Test Results Along with the Parametric Combinations in L_{27} Array

Exp. No.	Operating Temperature (°C)	Normal Load (N)	Sliding Speed (m/s)	Experimental Data	
				COF	Mass Loss (gm)
1.	100	10	0.25	0.575	0.0011
2.	100	10	0.33	0.741	0.0022
3.	100	10	0.42	0.892	0.0026
4.	100	30	0.25	0.95	0.0035
5.	100	30	0.33	0.92	0.00417
6.	100	30	0.42	0.915	0.0053
7.	100	50	0.25	0.561	0.0032
8.	100	50	0.33	0.617	0.0041
9.	100	50	0.42	0.627	0.00480
10.	300	10	0.25	0.518	0.00144
11.	300	10	0.33	0.678	0.00210
12.	300	10	0.42	0.619	0.00170
13.	300	30	0.25	0.451	0.00150
14.	300	30	0.33	0.592	0.00230
15.	300	30	0.42	0.554	0.00180
16.	300	50	0.25	0.421	0.00120
17.	300	50	0.33	0.578	0.00192
18.	300	50	0.42	0.51	0.00130
19.	500	10	0.25	0.439	0.00077
20.	500	10	0.33	0.475	0.00097
21.	500	10	0.42	0.421	0.00120
22.	500	30	0.25	0.475	0.00075
23.	500	30	0.33	0.497	0.00084
24.	500	30	0.42	0.415	0.00100
25.	500	50	0.25	0.392	0.00162
26.	500	50	0.33	0.475	0.00196
27.	500	50	0.42	0.429	0.00219

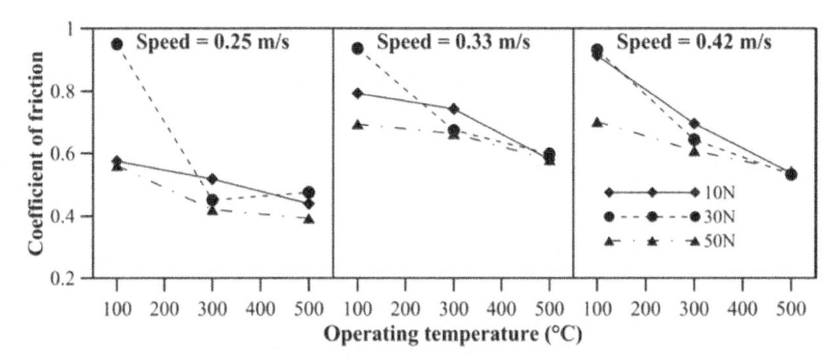

FIGURE 17.5 COF trends of Ni-B-W coatings with tribo-test parameters.

A similar behavior may also be observed for the wear characteristics in Figure 17.6. At all sliding speeds, as the temperature increases, mass loss decreases. While sliding wear, the in-situ thermal treatment causes microstructural changes and phase transformations occur. The mechanical properties of the alloy thus improve further. Consequently, the wear resistance increases. Moreover, at high temperature, the higher sliding speed may replenish the oxide glazes providing more lubricity along with a tough matrix underneath the protective oxide layer. Consequently, less wear and COF of the coating occurs at higher temperature and speed. But higher load tends to increase the contact stresses and a subsequent breakage of the oxide film. Thus, when load is increase, mass loss of the coatings also tends to increase. Hence, synergistic effects of load, speed, and operating temperature control the tribological behavior of the coatings. Further light may be shed on this by inspecting the worn specimen which has been addressed subsequently.

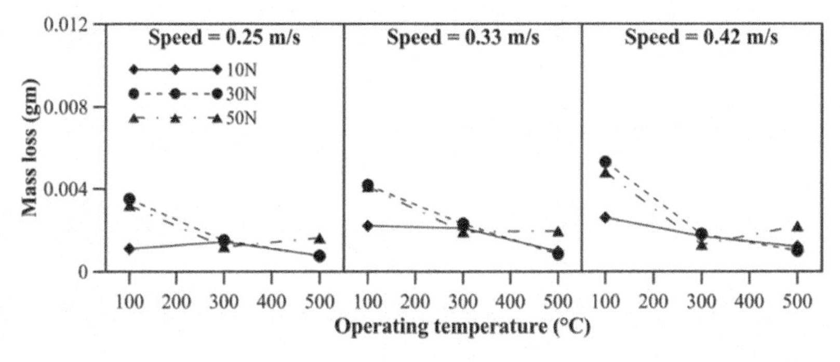

FIGURE 17.6 Mass loss trends indicating wear of Ni-B-W coatings with tribo-test parameters.

17.3.6 TRIBOLOGICAL MECHANISMS

BSE images of the coatings at 50 N and 0.42 m/s, i.e., highest load and sliding speed at different operating temperatures is presented in Figure 17.7. At 100°C, a cracked surface morphology is observed with scattered debris particles (Figure 17.7(a)). The loading action on the coatings during sliding wear is of cyclic type. Hence, due to this, the cracks may propagate, leading to fracture of the nodules. A rolling effect may however be induced by the scattered fine debris particles and this result in a decrease in COF. Therefore, at 50 N, 0.42 m/s and 100°C, the COF seems to decrease. A mixed abrasive as well as adhesive wear mechanism is concluded at 100°C for highest value of speed and load.

At 300°C, the nodules seem to be grinded and the worn surface appears to be smooth (Figure 17.7(b)) in comparison with 100°C. The bright region at the flattened portion indicates tribo-chemical patches and oxide layers. The tough flattened portions act as load bearing areas and improve wear resistance. The surface is comprised of oxide glazes as can be deciphered from the bright BSE image. Further, the wear morphology also exhibits scattered oxidized debris. The wear and friction characteristics were seen to improve at 300°C compared to 100°C because of further crystallization of the coatings leading to a tough matrix [40]. The surface is also characterized by fine scattered debris particles. This enhances the wear performance at high temperatures.

Further, at 500°C (Figure 17.7(c)), the nodules are completely crushed and deformed. They appear flattened. Such flattened portions also act as load bearing sites as discussed previously and results in improvement of tribological characteristics. At high temperatures, tribo-chemical reactions occur and are enhanced further. A stable tribo-oxidative patch form. Such oxide glazes may be clearly inferred from the BSE image of the worn specimen. Since microstructural changes at high temperatures are imminent, they result in toughening of the coating. The tough coating in turn supports the oxide patches formed. Also, higher speeds replenish the oxide layer. Thus, at high temperature, clearly the tribological characteristics are improved than at 100C. High-temperature friction and wear mechanisms are concluded to be characterized by grinding, formation of cracks, formation of oxide glazes and microstructural changes due to the on-time heat treatment. The phase transformations during sliding wear were successfully demonstrated by Madah et al. [23] at room temperature and Mukhopadhyay et al. [26, 40] at high temperatures.

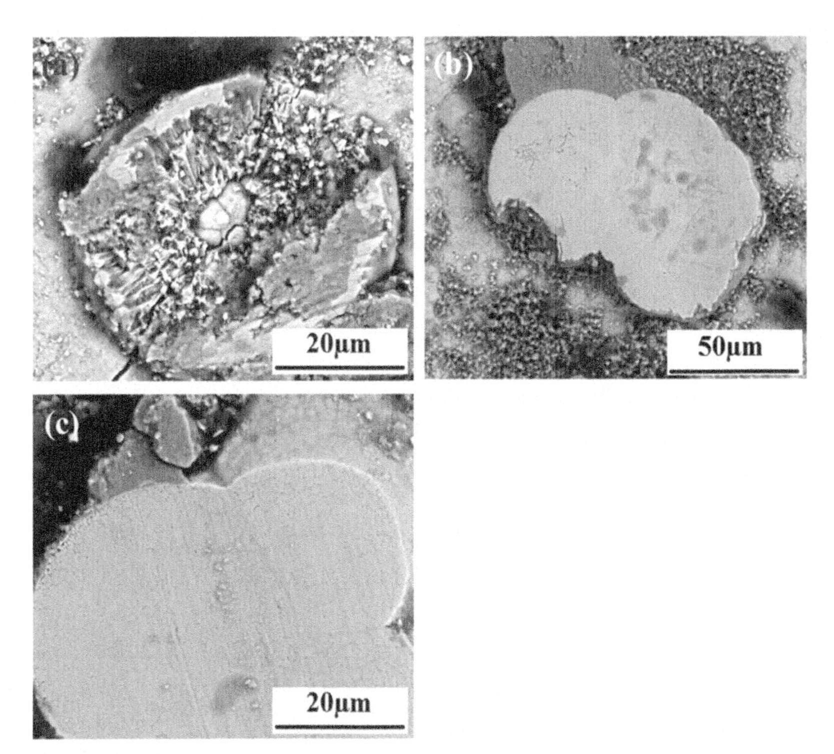

FIGURE 17.7 BSE image of worn Ni-B-W coatings at 50 N load and 0.42 m/s sliding speed at: (a) 100°C; (b) 300°C; and (c) 500°C.

17.3.7 OPTIMIZATION OF TRIBOLOGICAL CHARACTERISTICS

Optimal setting of tribological test parameters for minimum wear and COF has been carried out using GRA. In GRA, initially the data is normalized between 0 and 1. Here minimization of friction and wear is to be carried out. Hence, both are normalized as per lower-the-better quality characteristics. Thus, normalization is carried out as per Eqn. (1). Normalized COF and wear are laid down in Table 17.3. This is also known as gray relational generation. After normalization, GRC is evaluated from Eqn. (2). The GRG is finally evaluated using Eqn. (3) which is the multiple performance index. The calculations of GRC and GRG in the present work are laid down in Table 17.3. In this way, multiple responses are converted to single response. The detailed process of GRA and calculation may be also found in the work done by Das and Sahoo [21].

TABLE 17.3 Calculations of Gray Relational Analysis

SL. No.	Normalized Values		Grey Relational Coefficients		Grade
	COF	Wear	COF	Wear	
1.	0.672	0.923	0.604	0.867	0.735
2.	0.375	0.681	0.444	0.611	0.528
3.	0.104	0.593	0.358	0.552	0.455
4.	0.000	0.396	0.333	0.453	0.393
5.	0.054	0.248	0.346	0.399	0.373
6.	0.063	0.000	0.348	0.333	0.341
7.	0.697	0.462	0.623	0.481	0.552
8.	0.597	0.264	0.554	0.404	0.479
9.	0.579	0.110	0.543	0.360	0.451
10.	0.774	0.848	0.689	0.767	0.728
11.	0.487	0.703	0.494	0.628	0.561
12.	0.593	0.791	0.551	0.705	0.628
13.	0.894	0.835	0.825	0.752	0.789
14.	0.642	0.659	0.582	0.595	0.589
15.	0.710	0.769	0.633	0.684	0.658
16.	0.948	0.901	0.906	0.835	0.870
17.	0.667	0.743	0.600	0.660	0.630
18.	0.789	0.879	0.703	0.805	0.754
19.	0.916	0.996	0.856	0.991	0.924
20.	0.851	0.952	0.771	0.912	0.841
21.	0.948	0.901	0.906	0.835	0.870
22.	0.851	1.000	0.771	1.000	0.885
23.	0.812	0.980	0.727	0.962	0.844
24.	0.959	0.945	0.924	0.901	0.912
25.	1.000	0.809	1.000	0.723	0.862
26.	0.851	0.734	0.771	0.653	0.712
27.	0.934	0.684	0.883	0.612	0.748

One of the important features of OA is that the average effect of each parameter at each level on the responses may be estimated. This has been laid down in Table 17.4 as response table of GRG for the means. Ranks based on delta values have been also provided in Table 17.4. Based on the rank, operating temperature (A) has the highest significance followed by speed (C) and load (B). The corresponding main effects plot for means of GRG is shown in Figure 17.8. From Figure 17.8, the optimal combination of parameters that minimizes both COF and wear is 500°C temperature, 10 N normal load and 0.25 m/s sliding speed, i.e., *A3B1C1*. To a certain extent, the slope of Figure 17.8 denotes the effect of process parameters on the response. Higher slope generally indicates strong correlation [21]. Again, temperature has the highest slope and this is an indication that it has high contribution in controlling the multi-performance index. This means that operating temperature is highly influential in controlling the tribological characteristics.

TABLE 17.4 Means of Gray Relational Grade at Different Levels of Parameters

Level	A	B	C
1.	0.4785	0.6967	0.7487
2.	0.6897	0.6427	0.6173
3.	0.8443	0.6731	0.6464
Delta	0.3658	0.054	0.1314
Rank	1	3	2
Mean grade = 0.671			

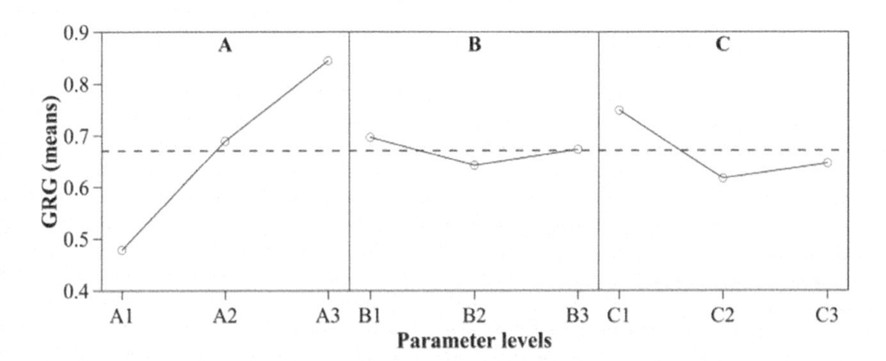

FIGURE 17.8 Main effects plot for determining optimal parametric setting.

The interaction plot showing the influence of interaction effects of tribo-test parameters on the means of GRG is shown in Figure 17.9. In the interaction plot, several lines are indicated. If these lines are almost parallel, then it may be concluded that no noteworthy interaction effects occur. But if the lines intersect, then a strong influence of the interaction effects may be concluded. Though such plots do not give a conclusive decision and statistical analysis is necessary. From the interaction plot, it seems initially that there are no such effects of the interaction between temperature (A), load (B) and speed (C).

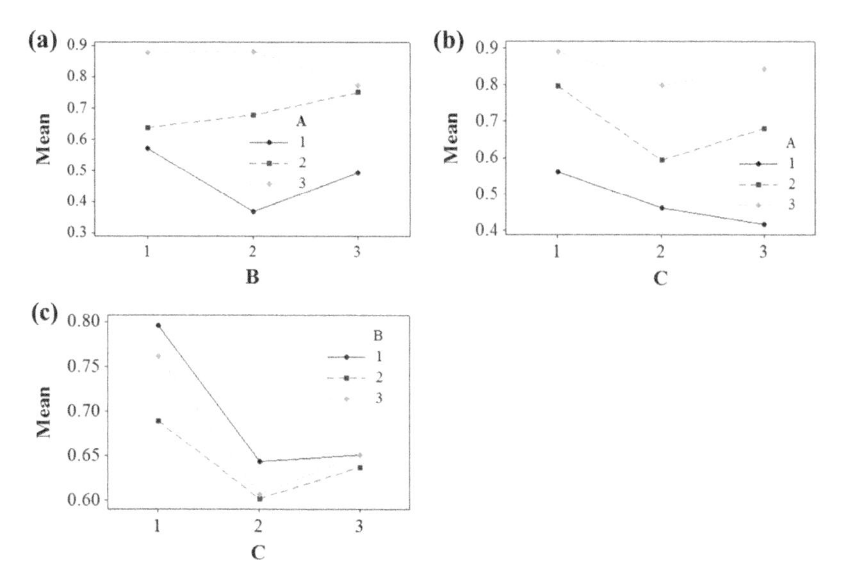

FIGURE 17.9 Interaction plots of the means of gray relational grade.

17.3.8 CONFIRMATION TEST

The final step of GRA is carrying out a validation test run where the results obtained at optimal combination of parameters is checked with a test run chosen arbitrarily. In the present work the mid-level combination of parameters is considered as the initial test run and is based on literature review [21]. The formula for predicting the grade is given in Eqn. (4). This is compared with the grade obtained at optimum condition. The results of the validation test are laid down in Table 17.5. Marked improvement in GRG is achieved through optimization in comparison with the initial test run. Also, the predicted and obtained GRG are in close agreement. Therefore, GRA has been gainfully applied to optimize the tribological characteristics of the deposits.

17.3.9 STATISTICAL SIGNIFICANCE OF PARAMETERS USING ANOVA

The statistical significance of process parameters may be determined from ANOVA. The results are laid down in Table 17.6. Based on F-statistics, the highest contribution is observed for operating temperature (A). At second position, in terms of contribution in controlling tribo-performance of Ni-B-W coatings is sliding speed (C). Comparatively, applied normal load (B) do not seem to have a high significance. Some contribution is also observed from the interaction of operating temperature and load (A × B). Thus, it may be clearly seen that even though no interaction effects were concluded from Figure 17.9, but still ANOVA results show there is some influence of A × B. Further, the results of ANOVA also agree with the trends observed for tribological behavior in Figures 17.5 and 17.6. Since the temperature effect is dominant in deciding the wear mechanisms also, highest significance has been also observed in the ANOVA results of the same.

TABLE 17.5 Results of Validation Test

Response Parameters	Initial Setting	Optimal Setting	
	A2B2C2	A3B1C1	
		Predicted	Experimental
Wear	0.0023	–	0.00077
COF	0.592	–	0.439
Grade	0.589	0.948	0.924
Improvement = 0.335 (56.87%)			

TABLE 17.6 Results of Analysis of Variance

Source	DOF	Seq SS	Adj MS	F-Ratio	% Contribution
A	2	0.607	0.303	155.600	71.977
B	2	0.013	0.007	3.380	1.564
C	2	0.086	0.043	21.970	10.162
A×B	4	0.092	0.023	11.810	10.929
A×C	4	0.022	0.005	2.780	2.569
B×C	4	0.008	0.002	1.030	0.950
Error	8	0.016	0.002	–	1.850
Total	**26**	**0.844**			**100.001**

17.4 CONCLUSION

The tribo-behavior of a ternary Ni-B variant was investigated in the present work. The following conclusions may be inferred from the study:

- Coating characterization revealed that Ni-B-W coatings with 5.2–5.3% by weight B and 3.4% by weight of W has been deposited in as-deposited condition on AISI 1040 steel. Post heat treatment at 350°C for 1 hour in a muffle furnace resulted in formation of a thin oxide film. XRD results revealed that in as-deposited condition, the coatings are amorphous while post heat treatment crystallization occurs. Nickel boride phases, namely Ni_3B and Ni_2B are seen along with Ni (111).
- SEM results show that the coatings present a dense globular morphology in as-deposited condition. Specific cellular boundaries appear post heat treatment which is again a possible indication of occurrence of phase transformation.
- Microhardness of as-deposited Ni-B-W coatings is ~759 HV_{100} which increases to ~1,181 HV_{100} on heat treatment.
- Higher wear and COF of the heat-treated coatings is observed at 100°C in comparison with 300 or 500°C. Several factors are responsible for this, such as the formation of oxide layers and change in microstructure during sliding wear at high temperature.
- GRA predicted optimal COF and wear is obtained for a parametric setting of 10 N load, 0.25 m/s speed and 500°C operating temperatures. ANOVA results indicate that operating temperature is the most significant factor which controls the friction and wear characteristics of Ni-B-W coatings.
- Further, a noteworthy improvement in the multi-performance index, i.e., the GRG in this case is observed. Compared to the initial test run, 56.87% improvement in GRG takes place at optimal condition. Thus, the implementation of GRA and ANOVA is successful for optimizing the tribological performance of the deposits and analyzing the effects of influential parameters.

ACKNOWLEDGMENT

The authors gratefully acknowledge the support of TEQIP II and DST PURSE, Phase II program of Jadavpur University.

KEYWORDS

- **analysis of variance**
- **coefficient of friction**
- **electroless**
- **Ni-B-W**
- **optimization**
- **tribological behavior**

REFERENCES

1. Riddle, Y. W., & Bailerare, T. O., (2005). Friction and wear reduction via an Ni-B electroless bath coating for metal alloys. *JOM, 57*(4), 40–45.
2. Sudagar, J., Lian, J., & Sha, W., (2013). Electroless nickel, alloy, composite and nano coatings–A critical review. *Journal of Alloys and Compounds, 571*, 183–204.
3. Sahoo, P., & Das, S. K., (2011). Tribology of electroless nickel coatings–a review. *Materials & Design, 32*(4), 1760–1775.
4. Loto, C. A., (2016). Electroless nickel plating: A review. *Silicon, 8*(2), 177–186.
5. Hamid, Z. A., Hassan, H. B., & Attyia, A. M., (2010). Influence of deposition temperature and heat treatment on the performance of electroless Ni–B films. *Surface and Coatings Technology, 205*(7), 2348–2354.
6. Anik, M., Körpe, E., & Şen, E., (2008). Effect of coating bath composition on the properties of electroless nickel–boron films. *Surface and Coatings Technology, 202*(9), 1718–1727.
7. Mukhopadhyay, A., Barman, T. K., & Sahoo, P., (2018). Electroless nickel coatings for high temperature applications. In: Kumar, K., & Davim, J., (eds.), *Composites and Advanced Materials for Industrial Applications* (pp. 297–331). Hershey, PA: IGI Global. doi: 10.4018/978-1-5225-5216-1.ch013.
8. Krishnaveni, K., Sankara, N. T. S. N., & Seshadri, S. K., (2005). Electroless Ni–B coatings: Preparation and evaluation of hardness and wear resistance. *Surface and Coatings Technology, 190*(1), 115–121.
9. Balaraju, J. N., Priyadarshi, A., Kumar, V., Manikandanath, N. T., Kumar, P. P., & Ravisankar, B., (2016). Hardness and wear behavior of electroless Ni–B coatings. *Materials Science and Technology, 32*(16), 1654–1665.
10. Delaunois, F., Petitjean, J. P., Lienard, P., & Jacob-Duliere, M., (2000). Autocatalytic electroless nickel-boron plating on light alloys. *Surface and Coatings Technology, 124*(2, 3), 201–209.
11. Vitry, V., Kanta, A. F., Dille, J., & Delaunois, F., (2012). Structural state of electroless nickel–boron deposits (5 wt.% B): Characterization by XRD and TEM. *Surface and Coatings Technology, 206*(16), 3444–3449.

12. Vitry, V., & Bonin, L., (2017). Increase of boron content in electroless nickel-boron coating by modification of plating conditions. *Surface and Coatings Technology, 311*, 164–171.
13. Bulbul, F., (2011). The effects of deposition parameters on surface morphology and crystallographic orientation of electroless Ni-B coatings. *Metals and Materials International, 17*(1), 67–75.
14. Barman, M., Barman, T. K., & Sahoo, P., (2019). Effect of borohydride concentration on tribological and mechanical behavior of electroless Ni-B coatings. *Materials Research Express, 6*(12), 126575.
15. Pal, S., Verma, N., Jayaram, V., Biswas, S. K., & Riddle, Y., (2011). Characterization of phase transformation behavior and microstructural development of electroless Ni–B coating. *Materials Science and Engineering: A, 528*(28), 8269–8276.
16. Kanta, A. F., Vitry, V., & Delaunois, F., (2009). Effect of thermochemical and heat treatments on electroless nickel–boron. *Materials Letters, 63*(30), 2662–2665.
17. Dervos, C. T., Novakovic, J., & Vassiliou, P., (2004). Vacuum heat treatment of electroless Ni–B coatings. *Materials Letters, 58*(5), 619–623.
18. Sankara, N. T. S. N., & Seshadri, S. K., (2004). Formation and characterization of borohydride reduced electroless nickel deposits. *Journal of Alloys and Compounds, 365*(1, 2), 197–205.
19. Çelik, İ., Karakan, M., & Bülbül, F., (2016). Investigation of structural and tribological properties of electroless Ni–B coated pure titanium. *Proceedings of the Institution of Mechanical Engineers, Part J: Journal of Engineering Tribology, 230*(1), 57–63.
20. Bülbül, F., Altun, H., Ezirmik, V., & Küçük, Ö., (2013). Investigation of structural, tribological and corrosion properties of electroless Ni–B coating deposited on 316L stainless steel. *Proceedings of the Institution of Mechanical Engineers, Part J: Journal of Engineering Tribology, 227*(6), 629–639.
21. Das, S. K., & Sahoo, P., (2011). Tribological characteristics of electroless Ni–B coating and optimization of coating parameters using Taguchi based grey relational analysis. *Materials & Design, 32*(4), 2228–2238.
22. Madah, F., Dehghanian, C., & Amadeh, A. A., (2015). Investigations on the wear mechanisms of electroless Ni–B coating during dry sliding and endurance life of the worn surfaces. *Surface and Coatings Technology, 282*, 6–15.
23. Madah, F., Amadeh, A. A., & Dehghanian, C., (2016). Investigation on the phase transformation of electroless Ni–B coating after dry sliding against alumina ball. *Journal of Alloys and Compounds, 658*, 272–279.
24. Pal, S., & Jayaram, V., (2018). Effect of microstructure on the hardness and dry sliding behavior of electroless Ni–B coating. *Materialia, 4*, 47–64.
25. Aydeniz, A. I., Göksenli, A., Dil, G., Muhaffel, F., Calli, C., & Yüksel, B., (2013). Electroless Ni-B-W coatings for improving hardness, wear and corrosion resistance. *Material in Technologie., 47*(6), 803–806.
26. Mukhopadhyay, A., Barman, T. K., & Sahoo, P., (2018). Tribological behavior of electroless Ni–B–W coating at room and elevated temperatures. *Proceedings of the Institution of Mechanical Engineers, Part J: Journal of Engineering Tribology, 232*(11), 1450–1466.

27. Serin, I. G., & Göksenli, A., (2015). Effect of annealing temperature on hardness and wear resistance of electroless Ni–B–Mo coatings. *Surface Review and Letters, 22*(05), 1550058.
28. Mukhopadhyay, A., Barman, T. K., & Sahoo, P., (2018). Effect of heat treatment on tribological behavior of electroless Ni-B-Mo coatings at different operating temperatures. *Silicon, 10*(3), 1203–1215.
29. Mukhopadhyay, A., Barman, T. K., Sahoo, P., & Davim, J. P., (2019). Tribological characteristics of electroless Ni-B-W-Mo coatings under dry sliding condition. *Journal of Manufacturing Technology Research, 11*(1, 2), 13–23.
30. Bonin, L., Vitry, V., & Delaunois, F., (2019). The tin stabilization effect on the microstructure, corrosion and wear resistance of electroless NiB coatings. *Surface and Coatings Technology, 357*, 353–363.
31. Abdel-Gawad, S. A., Sadik, M. A., & Shoeib, M. A., (2019). Preparation and properties of a novel nano Ni-B-Sn by electroless deposition on 7075-T6 aluminum alloy for aerospace application. *Journal of Alloys and Compounds, 785*, 1284–1292.
32. Ghaderi, M., Rezagholizadeh, M., Heidary, A., & Monirvaghefi, S. M., (2016). The effect of Al_2O_3 nanoparticles on tribological and corrosion behavior of electroless Ni–B–Al_2O_3 composite coating. *Protection of Metals and Physical Chemistry of Surfaces, 52*(5), 854–858.
33. Niksefat, V., & Ghorbani, M., (2015). Mechanical and electrochemical properties of ultrasonic-assisted electroless deposition of Ni–B–TiO_2 composite coatings. *Journal of Alloys and Compounds, 633*, 127–136.
34. Georgiza, E., Gouda, V., & Vassiliou, P., (2017). Production and properties of composite electroless Ni-B-SiC coatings. *Surface and Coatings Technology, 325*, 46–51.
35. Pal, S., Sarkar, R., & Jayaram, V., (2018). Characterization of thermal stability and high-temperature tribological behavior of electroless Ni-B coating. *Metallurgical and Materials Transactions A, 49*(8), 3217–3236.
36. Mukhopadhyay, A., Barman, T. K., & Sahoo, P., (2018). Effect of operating temperature on tribological behavior of as-plated Ni-B coating deposited by electroless method. *Tribology Transactions, 61*(1), 41–52.
37. Mukhopadhyay, A., Barman, T. K., Sahoo, P., & Davim, J. P., (2018). Comparative study of tribological behavior of electroless Ni–B, Ni–B–Mo, and Ni–B–W coatings at room and high temperatures. *Lubricants, 6*(3), 67.
38. Mukhopadhyay, A., Barman, T. K., & Sahoo, P., (2017). Tribological behavior of sodium borohydride reduced electroless nickel alloy coatings at room and elevated temperatures. *Surface and Coatings Technology, 321*, 464–476.
39. Drovosekov, A. B., Ivanov, M. V., Krutskikh, V. M., Lubnin, E. N., & Polukarov, Y. M., (2005). Chemically deposited Ni-W-B coatings: Composition, structure, and properties. *Protection of Metals, 41*(1), 55–62.
40. Mukhopadhyay, A., Barman, T. K., & Sahoo, P., (2018). Wear and friction characteristics of electroless Ni-B-W coatings at different operating temperatures. *Materials Research Express, 5*(2), 026526.

CHAPTER 18

EXPERIMENTAL STUDY OF NANOSECOND PULSED FIBER LASER MICRO-DRILLING ON QUARTZ

ABHISHEK SEN,[1] BISWANATH DOLOI,[2] and
BIJOY BHATTACHARYYA[2]

[1]*Department of Mechanical Engineering,*
Calcutta Institute of Technology, Uluberia, West Bengal, India,
E-mail: abhishek.sen1986@gmail.com

[2]*Department of Production Engineering, Jadavpur University,*
Kolkata, West Bengal, India

ABSTRACT

A nanosecond pulsed fiber laser system with an average power of 50 W is utilized in the present research work to generate micro-holes on a transparent quartz material for utilization in optoelectronics. The effect of laser power, pulse frequency, duty cycle, and air pressure on the entry hole circularity of micro-holes on quartz is analyzed with the aid of response surface methodology (RSM). The developed mathematical model for the entry hole circularity is validated through ANOVA analysis for the adequacy of the experimental model. The present research aims to bring about the utilization of nanosecond pulsed fiber laser systems to generate micro-holes on quartz at the fundamental wavelength, i.e., 1,064 nm. The experimental results show that the combination of pulse frequency of 65 kHz, a duty cycle of 50%, laser power of 37 W, and air pressure of 2.50 kgf/

Optimization Methods for Engineering Problems. Dilbagh Panchal, Prasenjit Chatterjee, Mohit Tyagi, Ravi Pratap Singh (Eds.)

cm^2 leads to the maximum circularity of 0.88 at entry side. The conformity tests reveal that the error for circularity is found out to be 3.23%, and the predicted result in a very close agreement with the experimental results.

18.1 INTRODUCTION

The utilization of quartz in microelectronics [1] and biomedical [2] applications has been increasing rapidly in the present-day scenario due to its desirable properties such as moderate thermal expansion coefficient, high biocompatibility along with high wear resistance [3]. In previous research studies, very few research works have been carried out to generate micro-features on quartz for the utilization in different areas mentioned above.

Laser systems are advantageous than other conventional systems due to their flexibility, machining time, and high coverage of a comprehensive range of engineering materials [4]. Compared to Nd-YAG and CO_2 laser systems, fiber lasers have been widely utilized for thin metal cutting to the generation of various micro-features on a wide range of engineering materials [5].

Transferring photon energy of the laser light to glass is challenging, as it is transparent to a wide range of wavelengths [6]. Thus, the generation of micro-hole on quartz using laser systems has not been investigated extensively.

However, Sen et al. [7] carried out preliminary research work on micro-drilling of quartz utilizing the fiber laser system to understand its feasibility. The authors successfully facilitated the utilization of the fiber laser system for the generation of micro-hole on quartz. However, the author suggested that further research works should be conducted to optimize the parametric combinations for achieving desired micro-hole characteristics. Rahman et al. [8] improved laser-induced silicon plasma-assisted ablation of quartz by adding a continuous wave laser. Lin et al. [9] showcased that the quartz's microstructure could be controlled by adjusting the machining parameters by picosecond laser and ultra-precision machining. Picosecond laser machining with a short focal length produced U shaped microfluidic channels on quartz. However, surface quality, along with surface roughness, was found to be inadequate. Swain et al. [10] numerically determined the laser intensity to control the surface property of the fused quartz.

This present research work aims to generate through micro-holes on quartz by fiber laser micro-drilling process and also to conduct successful research analysis to widen the scope of research findings further. The considered process parameters are laser power, pulse frequency, duty cycle, and assist air pressure. The effects of the process parameters mentioned above on entry hole circularity have been analyzed and discussed with the aid of surface and contour plots. A total of 31 experiments have been designed and performed based on response surface methodology (RSM) experimental design. The experimental models are validated through the analysis of variance (ANOVA). The optimized parametric combination for the entry hole circularity has been determined with single-objective optimization. Finally, a set of 5 experiments has been carried out to compare the predicted optimized values with experimental results.

18.2 EXPERIMENTAL METHODS AND MATERIALS

In the present research work, a transparent square quartz material of dimension $25 \times 25 \times 1$ mm^3 is selected for the fabrication of micro-hole, utilizing a fiber laser system of 50 W of average power. The working regime of the fiber laser is at 1,064 nm of wavelength. The laser beam is generated in a fiber laser head. The fiber laser delivery system consists of a collimator, beam bender, beam delivery unit, and focusing lens. A collimator is used to transform the light output from an optical fiber into the fiber end, which is approximately equal to the focal length. A photographic view of a collimator is shown in Figure 18.1. After the collimator, a beam bender with 100% reflectivity is placed at an angle of 45° with the horizontal plane. Because of this, the laser propagates perpendicular to the focus lens. At the top of the beam bender, a charge-coupled device camera (CCD) is placed, further connected to a CCTV. In Figure 18.2, the photographic view of a beam bender is shown.

The laser finally propagates through an F-θ lens of 71 mm diameter of the focusing lens, which is protected by a nozzle for preventing against dust and other contaminations. The spot diameter of the laser beam irradiated on the top surface of the quartz is 21 μm. A CNC controller unit controls the movement of the worktable along the X-Y axis along with the laser nozzle movement along the Z-axis. Servo motors are attached to each of the axes and also connect to the servo interfacing unit. This servo controller is connected to the computer system (interface Software-I mark plus) by which the axis movements of the X–Y worktable are controlled. The jet flow of assisting gas for removing the molten material from the micro-drilled surface is achieved with a compressed air gas supply. A moisture separator, attached with the air compressor, results in the jet flow of dry, pressurized air to the laser micro-drilling zone.

FIGURE 18.1 Photographic view of the collimator.

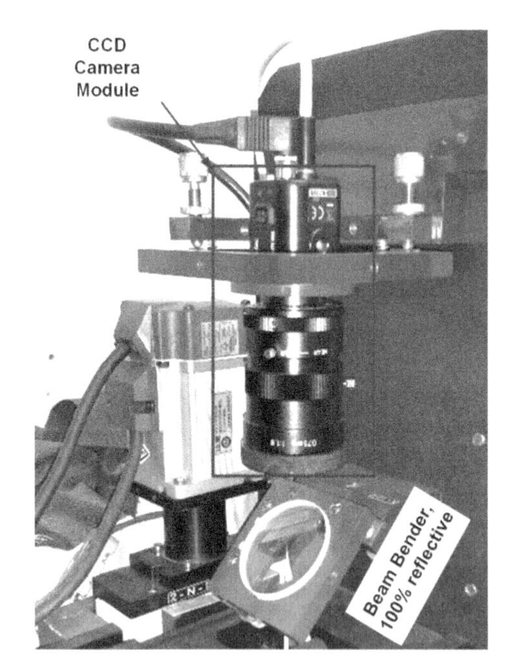

FIGURE 18.2 Photographic view of beam bender.

A focused laser beam is used as a heat source to increase temperature rapidly to the melting and evaporation of the substrate material's temperature during the laser drilling process. In pulsed mode operation, fiber laser's high beam quality combined with high pulse repetition rates and pulse energy can effectively utilize a single pulse for percussion laser drilling. As shown in Figure 18.3, laser percussion drilling is selected as the fabrication method of micro-holes on quartz.

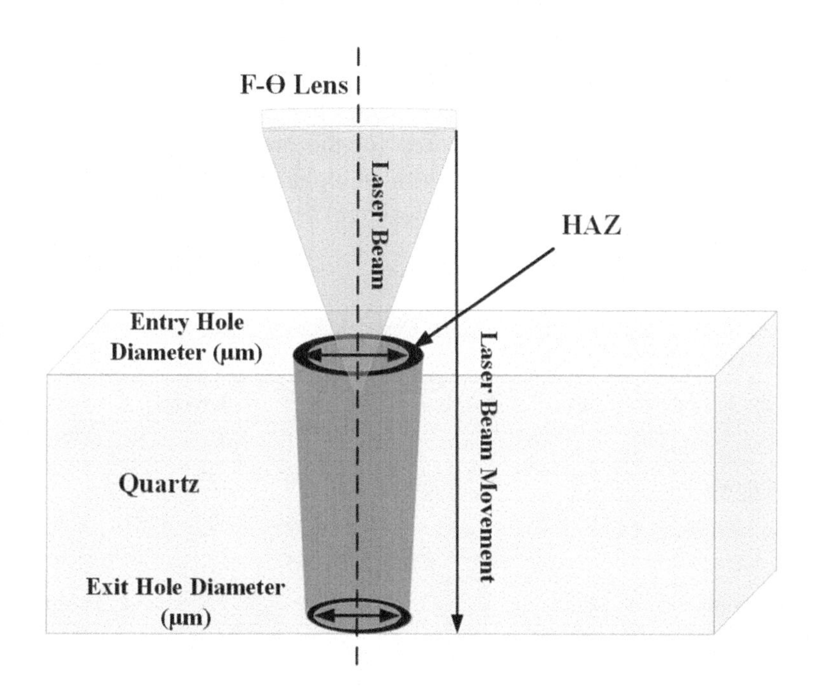

FIGURE 18.3 Laser percussion drilling on quartz.

18.2.1 MEASUREMENTS

The microscopic views of the micro-holes are shown using an Olympus STM6-LM optical measuring microscope using a 50X lens. ImageJ software is utilized to calculate the entry hole circularity.

18.3 RESULTS AND DISCUSSION

18.3.1 MODELING AND ANALYSIS OF FIBER LASER MICRO-DRILLING OF QUARTZ

Response surface methodology (RSM) [11, 12] is an analytical technique utilized for planning experimentation, modeling, and optimizing responses. RSM based modeling and subsequent analyzes of fiber laser micro-drilling process parameters on fiber laser micro-drilling characteristics of micro-hole generation on quartz have been made. In this experimentation, four factors, along with their five levels, have been considered. Table 18.1 represents the actual and corresponding coded values of fiber laser micro-drilling process parameters. A set of 31 experiments has been conducted for the present research study by central composite design (CCD) with an alpha (α) value of 2.000. Table 18.2 shows the design of the experiment's matrix of all the process parameters settings.

TABLE 18.1 Actual and Corresponding Coded Values of Fiber Laser Micro-Drilling Process Parameters

Process Parameters	Unit	Symbol	Levels				
			–2	–1	0	1	2
Laser power	W	X_1	27.5	30	32.5	35	37.5
Pulse frequency	kHz	X_2	50	55	60	65	70
Duty cycle	%	X_3	50	55	60	65	70
Air pressure	Kgf/cm²	X_4	1.5	2	2.5	3	3.5

18.3.2 DEVELOPMENT OF EMPIRICAL MODELING BASED ON RSM

Minitab-17 software has been used for designing the experimental plan. The mathematical model's development has been carried out utilizing the data obtained during experiments, as listed in Table 18.2 for fiber laser percussion micro-drilling on quartz. The corresponding empirical equation for circularity (Y_1) is obtained as follows:

$$\begin{aligned} Y_1 = {} & -0.718+0.1076X_1 -0.0337X_2 -0.0093X_3 -0.350X_4 -0.001501 \\ & X_1\times X_1 +0.000604X_2\times X_2 -0.000743X_3\times X_3 +0.00043X_4\times X_4 \\ & +0.000210X_1\times X_2 +0.001460X_1\times X_3 +0.005750X_1\times X_4 \\ & -0.001110X_2\times X_3 -0.006500X_2\times X_4 +0.01300X_3\times X_4 \end{aligned} \quad (1)$$

TABLE 18.2 Design of Experiments Matrix of Values of Process Parameters and Observed Responses

Exp. No.	Pulse Frequency (kHz)	Duty Cycle (%)	Laser Power (W)	Air Pressure (Kgf/cm2)	Circularity
1.	55	55	30	2	0.801
2.	65	55	30	2	0.755
3.	55	65	30	2	0.825
4.	65	65	30	2	0.816
5.	55	55	35	2	0.748
6.	65	55	35	2	0.780
7.	55	65	35	2	0.702
8.	65	65	35	2	0.758
9.	55	55	30	3	0.795
10.	65	55	30	3	0.782
11.	55	65	30	3	0.790
12.	65	65	30	3	0.834
13.	55	55	35	3	0.790
14.	65	55	35	3	0.861
15.	55	65	35	3	0.719
16.	65	65	35	3	0.832
17.	50	60	32.5	2.5	0.679
18.	70	60	32.5	2.5	0.724
19.	60	50	32.5	2.5	0.854
20.	60	70	32.5	2.5	0.853
21.	60	60	27.5	2.5	0.825
22.	60	60	37.5	2.5	0.772
23.	60	60	32.5	1.5	0.785
24.	60	60	32.5	3.5	0.831
25.	60	60	32.5	2.5	0.820
26.	60	60	32.5	2.5	0.808
27.	60	60	32.5	2.5	0.815
28.	60	60	32.5	2.5	0.805
29.	60	60	32.5	2.5	0.812
30.	60	60	32.5	2.5	0.811
31.	60	60	32.5	2.5	0.825

18.3.3 ANOVA TEST RESULTS OF THE DEVELOPED MODELS

ANOVA test has been performed using MINITAB software to test the adequacy of the developed model. Table 18.3 shows the result of ANOVA for the circularity of fiber laser-generated micro-hole on quartz.

TABLE 18.3 ANOVA Table for Circularity

Source	DOF	Adjusted SS	Adjusted MS	F-Value	p-Value
Model	14	0.058882	0.004206	103.91	0.000
Linear	4	0.012933	0.003233	79.88	0.000
Pulse frequency	1	0.004760	0.004760	117.61	0.000
Duty cycle	1	0.000060	0.000060	1.49	0.240
Laser power	1	0.004108	0.004108	101.50	0.000
Assist air pressure	1	0.004004	0.004004	98.93	0.000
Square	4	0.028115	0.007029	173.66	0.000
2-way interaction	6	0.017834	0.002972	73.44	0.000
Error	16	0.000648	0.000040	–	–
Lack-of-fit	10	0.000360	0.000036	0.75	0.671
Pure error	6	0.000287	0.000048	–	–
Total	**64**	0.059529			
Model Summary			**R^2**	**R^2 Adjusted**	**R^2 Predicted**
			98.91	**97.96**	**95.86**

18.3.4 ANOVA FOR CIRCULARITY

Table 18.3 represents the ANOVA results for the entry hole circularity of the micro-holes generated on quartz. Table 18.3 shows that the associated p-value for linear effect and four process parameters are less than 0.05 for circularity. The lack of fit value for the model is found as which is insignificant. As a result, the fit value for the developed model is significant. Thus, the developed empirical model on circularity represented by Eqn. (1) is highly significant and adequate at a confidence level of 95% to represent the relationship between circularity and the fiber laser process parameters.

18.3.5 ANALYSIS OF FIBER LASER PROCESS PARAMETERS BASED ON RESPONSE SURFACE AND CONTOUR PLOTS

The various response surfaces have been plotted between one process criteria and two process parameters at a time, while the other two remaining process parameters are kept at constant values. The results of parametric analyzes of circularity for process variables such as laser power, pulse frequency, duty cycle, and air pressure have been discussed with the aid of surface and contour plots.

18.3.6 PARAMETRIC INFLUENCES ON HOLE CIRCULARITY OF MICRO-HOLES ON QUARTZ

Figure 18.4 illustrates the influence of the duty cycle and pulse frequency on the entrance hole circularity. It is observed that the circularity increases with an increase in pulse frequency up to 65 kHz, decreases with a further increase in pulse frequency. It is also found that the hole circularity gradually decreases with an increase in duty cycle at a lower value of pulse frequency, but increases with an increase in duty cycle at a higher value of pulse frequency. High Intensity of pulse energy in conjunction with high peak power at the lower value of pulse frequency and duty cycle may lead to a uniform spattering of molten debris on the edge of the periphery of the drilled hole. As a result, an increment in the hole circularity of fiber laser-generated micro-holes is observed.

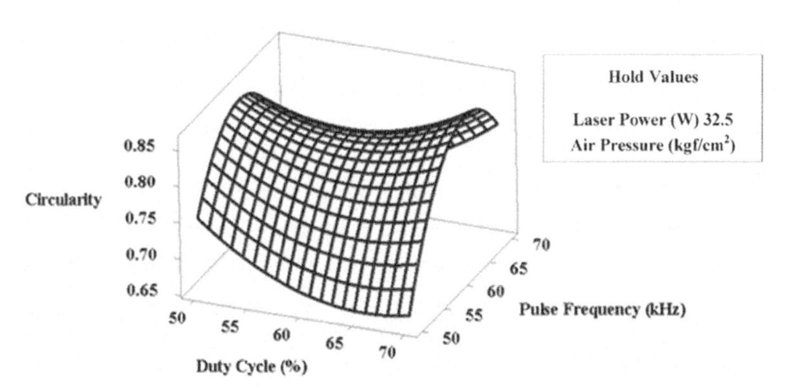

FIGURE 18.4 Surface plot for influences of duty cycle and pulse frequency on hole circularity.

Figure 18.5 shows the contour plot for the hole circularity due to the influences of pulse frequency and duty cycle. The contour plot reveals that a significant interaction between two process variables, i.e., pulse frequency and duty cycle. The largest confined region indicates the most favored outcome from the present experimental range, whereas the second smallest confined zone indicates at the most preferred parametric condition to achieve the highest value of hole circularity.

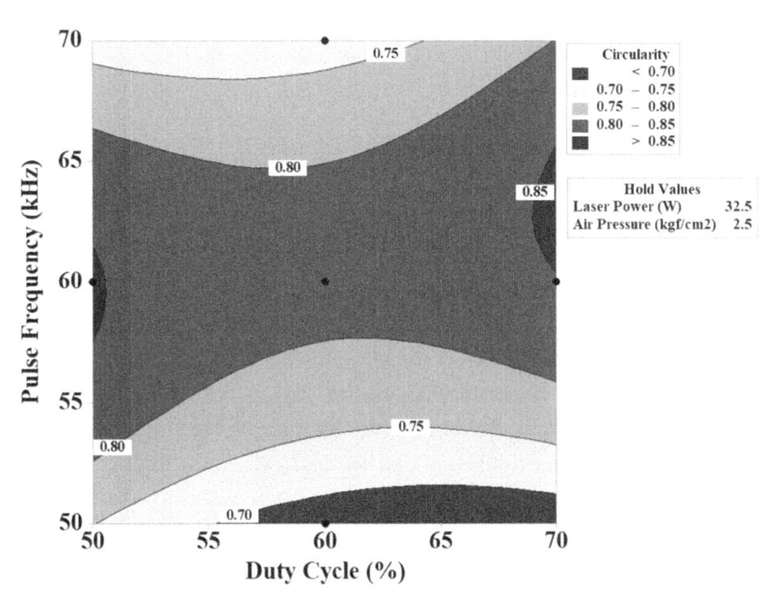

FIGURE 18.5 Contour plot for influences of duty cycle and pulse frequency on hole circularity.

Figure 18.6 reveals a parabolic nature of the surface plot for the hole circularity due to the interaction between pulse frequency and air pressure. The highest value of hole circularity can be achieved at a moderate value pulse frequency, i.e., 55 to 60 kHz. It is also observed from Figure 18.6 that at the lower value of pulse frequency, air pressure has a very small or no effect on the hole circularity. Conversely, the hole circularity linearly increases with the increase in air pressure at a higher value of pulse frequency. A decrease in pulse off time due to an increase in pulse frequency results in a decrease in agitation, forming a melt pool at the top surface of the workpiece. Lesser agitation on the melt pool in terms of surface tension leads to form a smooth edge of the micro-drilled hole.

Pulse frequency is inversely proportional to pulse energy when the laser power and duty cycle are kept as constants at their moderate values. The high pulse energy of a Gaussian laser beam produces an adequate melt pool and increases the spattering effect. The lower value of air pressure is incapable of sweeping away the spatters or properly removes the melt pool from the micro-drilled zone, which leads to the formation of the non-uniform edge of the micro-drilled hole. As a result, low circularity is observed, as shown in Figure 18.7. As observed in Figure 18.7, the heat affected zone around the periphery of the micro-hole has also contributed to the lowering of the circularity of micro-hole.

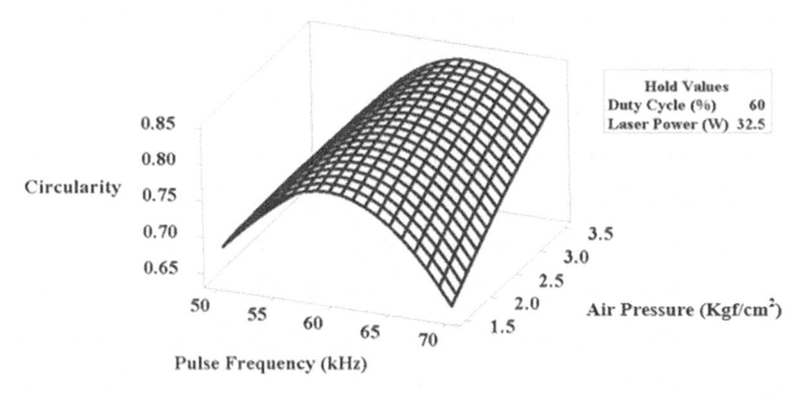

FIGURE 18.6 Surface plot for influences of pulse frequency and air pressure on hole circularity.

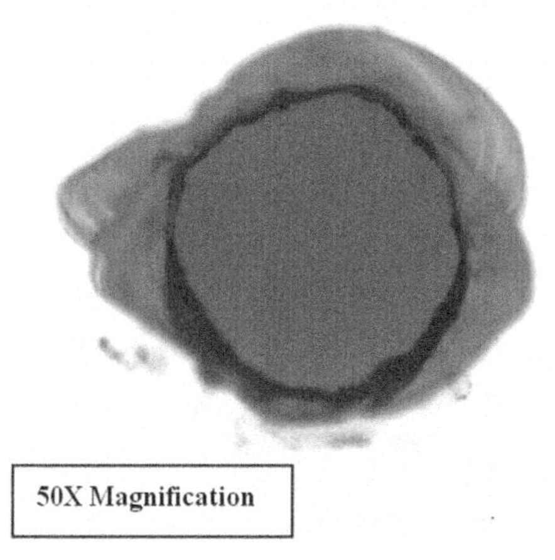

FIGURE 18.7 Microscopic image of quartz micro-hole at 50X magnification at pulse frequency of 60 kHz, 60% duty cycle, laser power of 32.5 W and air pressure 1.5 kgf/cm^2.

Figure 18.8 indicates the interaction between pulse frequency on hole circularity via a contour plot. The elliptic shape of contours shows a functional interaction between air pressure and pulse frequency. The second-largest confined zone indicates the parametric options at which the highest value of hole circularity can be attained.

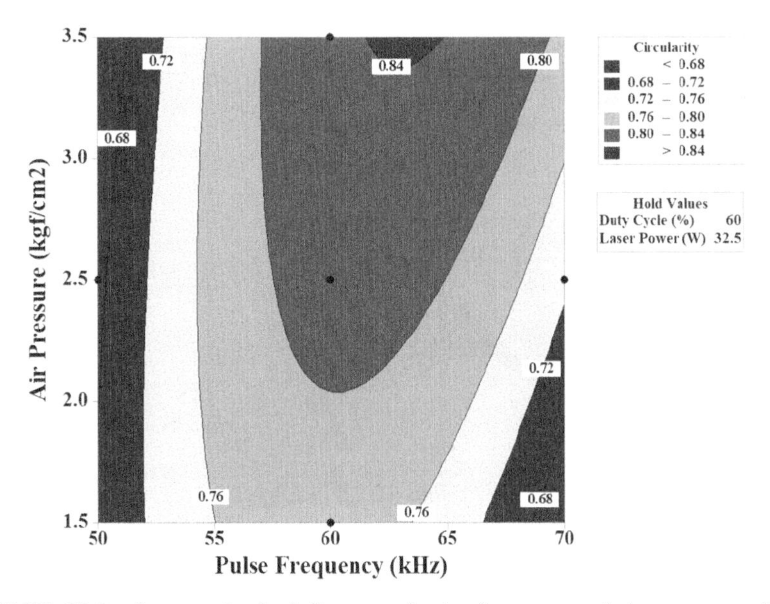

FIGURE 18.8 Contour plot for influences of pulse frequency and air pressure on hole circularity.

18.3.7 PARAMETRIC OPTIMIZATION OF FIBER LASER MICRO-DRILLING OF QUARTZ

For achieving optimal parametric combinations during fiber laser micro-drilling of quartz, single-objective optimization has been performed using MINITAB software. Figure 18.9 illustrates that the parametric settings required to achieve the maximum hole circularity of 0.88 at entry side are pulse frequency of 65 kHz along with duty cycle of 50%, the laser power of 37 W, and air pressure of 2.50 kgf/cm^2. Figure 18.10 showcases the microscopic image of the fiber laser-generated micro-hole on quartz at an optimized parametric combination.

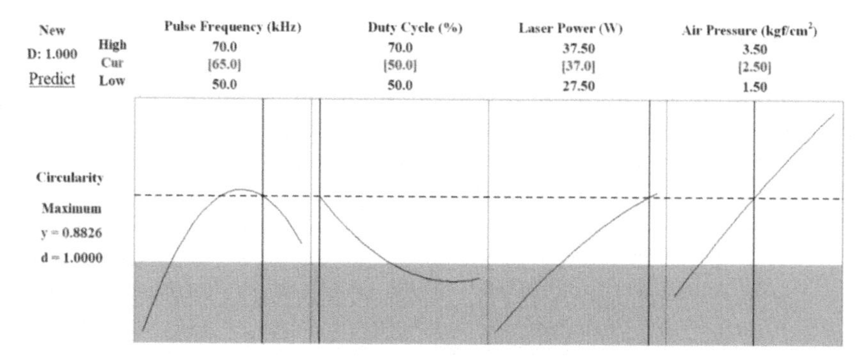

FIGURE 18.9 Single-objective optimization results for hole circularity.

FIGURE 18.10 Microscopic image of quartz micro-hole at the optimal condition.

18.3.8 CONFIRMATION EXPERIMENTS

Confirmation experiments are conducted to validate the predicted results at the optimized experimental parameter settings. The mean prediction error of the experimental values is evaluated to compare the experimental values with their optimized values.

A total of five confirmation experiments are conducted. The experimental values of the entry hole circularity are shown in Table 18.4. The mean prediction error for circularity is found out to be 3.23%. The calculated error values indicate that the produced results are a close agreement with the predicted results.

TABLE 18.4 Comparison between Single-Objective Optimization Results with Actual Results for Circularity

Exp. No.	Circularity		
	Maximum Value	**Actual Value at Optimum Condition**	**Percentage of Errors**
1.	0.8826	0.8556	3.06
2.	0.8826	0.8510	3.58
3.	0.8826	0.8595	2.62
4.	0.8826	0.8546	3.17
5.	0.8826	0.8497	3.73

Note: Mean percentage of Errors (%) = 3.23.

18.4 CONCLUSIONS

In the present research work, a nanosecond pulsed fiber laser system is utilized to generate micro-holes on transparent quartz. An RSM based experimental design is utilized to carry out a total of 31 experiments to analyze the effect of various process parameters on the circularity of micro-hole drilled on quartz. From the ANOVA table, the p-value of the lack of fit for hole circularity is 0.671. The developed mathematical model on entry hole circularity is adequate for prediction during fiber laser micro-drilling on quartz.

Single objective optimization results show that for obtaining the maximum hole circularity (0.88), the combination of the process parameters is pulse frequency of 65 kHz along with duty cycle of 50%, laser power of 37 W, and air pressure of 2.50 kgf/cm^2. The conformity tests reveal that the error for circularity is found out to be 3.23%, and the predicted results in a very close agreement with the experimental results.

The present work facilitates the utilization of a nanosecond pulsed fiber laser system at 1,064 nm wavelength to fabricate micro-holes on quartz. Further, an experimental investigation can be carried out into fiber laser micro-drilling of polymers and glass materials, considering different biomedical applications.

ACKNOWLEDGMENT

The authors would like to express their gratitude towards the assistance aided by CAS Ph-IV program of the Production Engineering Department of Jadavpur University under the University Grants Commission, New Delhi, India, and Calcutta Institute of Technology, Uluberia, Howrah.

KEYWORDS

- **circularity**
- **exit hole diameter**
- **fiber laser**
- **heat affected zone**
- **percussion drilling**
- **quartz**
- **response surface methodology (RSM)**

REFERENCES

1. Hannah, J. J., Uswachoke, C., Baig, A. S., Adeyemo, S. O., Boland, J. L., Damry, D. A., Davies, C. L., et al., (2018). Quantum dots and nanostructures: Growth, characterization, and modeling XV. *International Society for Optics and Photonics, 10543*, 105430I.
2. Reza, E. M., & Alizadeh, N., (2015). A dual usage smart sorbent/recognition element based on nanostructured conducting molecularly imprinted polypyrrole for simultaneous potential-induced nanoextraction/determination of ibuprofen in biomedical samples by quartz crystal microbalance sensor. *Sensors and Actuators B: Chemical, 220*, 880–887.
3. Xie, X. L., Tang, C. Y., Kathy, Y. Y. C., Wu, X. C., Tsui, C. P., & Cheung, C. Y., (2003). Wear performance of ultrahigh molecular weight polyethylene/quartz composites. *Biomaterials, 24*(11), 1889–1896.
4. Takahisa, M., (2000). State of the art of micromachining. *CIRP Annals-Manufacturing Technology, 49*(2), 473–488.
5. Sen, A., Doloi, B., & Bhattacharyya, B., (2015). Fiber laser micro-machining of Ti-6Al-4V. In: *Lasers Based Manufacturing* (pp. 255–281). Springer, New Delhi.
6. Okazaki, K., Torii, S., Makimura, T., Niino, H., Murakami, K., Nakamura, D., & Okada, T., (2011). Sub-wavelength micromachining of silica glass by irradiation of CO_2 laser with Fresnel diffraction. *Applied Physics A, 104*(2), 593–599.

7. Sen, A., Doloi, B., & Bhattacharyya, B., (2018). An experimental investigation into fiber laser micro-drilling of quartz. *International Journal of Mechatronics and Manufacturing Systems, 11*(2, 3), 182–202.
8. Rahman, T. U., Huagang, L., Qayyum, A., & Hong, M., (2020). Enhancement of pulsed laser-induced silicon plasma-assisted quartz ablation by continuous wave laser irradiation. *Journal of Laser Applications, 32*(2), 022064.
9. Lin, Y. C., Lee, C. C., Lin, H. S., Hong, Z. H., Hsu, F. C., Hung, T. P., & Lyu, Y. T., (2017). Fabrication of microfluidic structures in quartz via micro machining technologies. *Microsystem Technologies, 23*(6), 1661–1669.
10. Swain, S., Kar, S. P., Swain, A., Sarangi, R. K., & Sekhar, P. C., (2020). Numerical prediction of solid-liquid front during laser melting of fused quartz. *Materials Today: Proceedings*.
11. Kibria, G., Doloi, B., & Bhattacharyya, B., (2012). Optimization of Nd: YAG laser micro–turning process using response surface methodology. *International Journal of Precision Technology, 3*(1), 14–36.
12. Moradi, M., & MohazabPak, A. R., (2018). Statistical modeling and optimization of laser percussion microdrilling of Inconel 718 sheet using response surface methodology (RSM). *Lasers in Engineering, 39*. Old City Publishing.

CHAPTER 19

GREY RELATIONAL ANALYSIS BASED ON TAGUCHI FOR OPTIMIZATION OF BEAD GEOMETRY AND PROCESS VARIABLES IN TUNGSTEN INERT GAS BEAD-ON-PLATE WELDING

DEBRAJ DAS,[1] SUBASH CHANDRA SAHA,[2] PANKAJ BISWAS,[3] and JHUMA MITRA[4]

[1]*Department of Mechanical Engineering, Tripura Institute of Technology, Narsingarh, Agartala–799009, Tripura, India, E-mail: er.debraj@gmail.com*

[2]*Department of Mechanical Engineering, NIT, Agartala–799046, Tripura, India*

[3]*Department of Mechanical Engineering, IIT Guwahati–781039, Assam, India*

[4]*Department of Civil Engineering, Tripura Institute of Technology, Narsingarh, Agartala–799009, Tripura, India*

ABSTRACT

It is a significant effort to pick up perfect welding process variables for improving the quality of weld bead geometry in any welding process. Using process optimization for welding variables is a complicated matter, reason is that, it depends on multi-factors, which is influence by the

Optimization Methods for Engineering Problems. Dilbagh Panchal, Prasenjit Chatterjee, Mohit Tyagi, Ravi Pratap Singh (Eds.)

circumstances. In this study, a tungsten solid wire having 2.5 mm diameter has been used as an electrode with direct current electrode negative (DCSP) polarity. An argon gas has been supplied for the shielding purpose. A multi-response optimization technique has been employed to optimize the weld bead geometry and process parameters for Tungsten Inert Gas (TIG) bead-on-plate welding of IS 2062B mild steel. Taguchi's experimental design method has been used to define the significant sets of experiments of the process variables with the intention of reducing the number of experimental runs. Three important welding process variables viz. current (I), welding speed (S) and gas flows or passes rate (G_f) with a constant voltage were considered in this work. The bead geometry viz. penetration, bead width and HAZ thickness; and hardness of HAZ were computed. The weightage of the responses on overall outcome characteristic of the weld bead has also been assessed numerically by ANOVA (analysis of variance) method. The optimal process variables, which was defined from the Signal to Noise ratios plot, has been validated from the experimental sets. It shows the workability of the gray-based Taguchi method in manufacturing world for on-going development in product quality. It has been noticed that the process variables have a remarkable impact on weld bead geometry as well as hardness of the HAZ.

19.1 INTRODUCTION

In today's manufacturing world, the continuous requirement of quality outputs has given rise to the quick evolution of today's mechanized manufacturing domains. The satisfactory level of customers has been quantified by the quality of the output or feasibility of product. The output quality mainly depends on the required specifications obtained in the output product that be outfit its useful needs in several fields of practice. Quality of welding primarily rests on metallurgical properties of the weld metal and heat affected zone, which actually are characterized by physical and chemical behavior of metallic elements, their inter-metallic compounds, and mixtures of the weldment. Additionally, the mechanical characteristics and metallurgic characteristics of the weld rest on weld bead geometry and weld bead geometry have straight way allied to welding process standards. Hence, it has been required to find out an optimized variables setting, that is competent to give the suitable weld quality. However, this optimization should be carried out in such a way that all the objectives should complete simultaneously. Such an optimization technique is called multi-variable optimization.

So many researchers had been followed different formulae to investigate the optimization process, such as Taguchi analysis, regression models, response surface methodology and artificial neural networks analysis, etc., like Arvind and Ajay [3]. In general, processes that could be adopted to optimized the process parameters settings are orthogonal array design (reduced number of experimental runs but sound) that are full factorial design, fractional factorial design, response surface methodology (RSM), gray relation analysis (GRA) and Taguchi technique [4]; and Taguchi and Regression modeling are used to actively resolve the optimum welding conditions for getting the highest ultimate strength in range of parameters [3]. In case of multi-response optimization, GRA needs [4]. It is well known that high efficiency and high quality are the developing targets of welding technology. The quality of a weld joint is straight way affected by the welding heat input, that is, related to current, voltage, and traverse speed during welding, and the quality joint can be explained in terms of weld-bead geometry, mechanical properties, and distortion [5, 16].

For the optimization process, the demand of genetic algorithm and GRA based Taguchi method have been increased gradually in the last decade. But optimization based on genetic algorithm asks for a good number of tests [18]. On the other hand, the analytical design of experiments introduced by Taguchi has been used to select the optimum combination of variables for improving productivity as well as quality of products with least experimental tests [2, 6].

Since last three decades the Taguchi method had been efficiently used to optimized the products design or procedures which has single objective variable. Many analyzers have been also chosen Taguchi's design of experiment and gray relation analysis (GRA) for optimizing the welding process variables [7–12]. Anawa et al. selected Taguchi's technique to reducing the weldment fusion zone in laser welding process [7]. The Taguchi's orthogonal array technique had been employed successfully in the SAW process for obtaining desirable weld bead geometry by renowned researcher Datta et al. Subsequently ANOVA technique was used for evaluating the percentage contributions of every factor [8].

In case of tungsten inert gas (TIG) welding operation Taguchi's technique has been applied to finding the optimized weld bead geometry [9]. However, the research work in the field of TIG welding on IS 2062B grade mild steel is rarely found in published literature. The optimization technique for the effect of the process variables viz. welding current, traverse speed, and gas flow rate on weld bead geometry and hardness of the HAZ in TIG welding on mild steel, so far have not been discussed in detail in any previous published work. Therefore, in this present study, the investigations of TIG welding have been executed on IS 2062B steel to analyze the effect of process variables on weld bead geometry and hardness of the HAZ. The Taguchi-based gray relational analysis (GRA) has been followed up to optimize the welding process variables, and finally, ANOVA has been employed to find the individual significance of process variables on degree of weld quality.

19.2 DESIGN OF EXPERIMENT (DOE)

The traditional experimental design methods are too complex and difficult to use. Additionally, a large number of experiments have to be carried out when the number of machining parameters increase. Therefore, the factors causing variations should be determined and checked under laboratory conditions. These studies are considered under the scope of off-line quality improvement.

Taguchi method is a common optimization technique employed to solve the engineering problems. In the present study, an array has been designed based on the Dr. Genichi Taguchi's orthogonal arrays to ensure that all level of all factors is considered equally and moreover to reduce the expensive testing cost. The welding current, welding speed and gas flow have been considered, at four different levels, respectively, are shown in Table 19.1.

Orthogonal array is a "table" (array) whose entries come from a fixed finite set of symbols (typically, {1, 2, …, n}), arranged in such a way that there is an integer t so that for every selection of t columns of the table, all ordered t tuples of the symbols, formed by taking the entries in each row restricted to these columns, appear the same number of times. An orthogonal array is type of experiment where the column represents the independent variables are orthogonal to one another.

Taguchi's L16 orthogonal array has been grouping the samples into 16 numbers of group as illustrated in Table 19.2. The numbers are indicating the various experimental processes or levels of the differing factors.

TABLE 19.1 Welding Factors and Their Boundaries

Welding Factors	Unit	Level of Factors			
		11	2	33	4
Current (I)	Amp	120	140	160	180
Welding speed (S)	m/min	0.30	0.36	0.42	0.48
Gas flow (G_f)	Ltr./min	10	15	20	25

TABLE 19.2 Orthogonal Array (L16) by Employing Taguchi Technique

Group No.	I	S	G_f
01	1	1	1
02	1	2	2
03	1	3	3
04	1	4	4
05	2	1	2
06	2	2	1
07	2	3	4
08	2	4	3
09	3	1	3
10	3	2	4
11	3	3	1
12	3	4	2
13	4	1	4
14	4	2	3
15	4	3	2
16	4	4	1

19.3 RESEARCH METHOD

19.3.1 EXPERIMENTAL DETAILS

The Tungsten inert gas welding machine (Tornado TIG 316 AC/DC; Ador Fontech Ltd, India) has been employed to execute the experimental tests by the technique bead-on-plate welding. The specification of machine is followed by, input voltage – 415A AC (±10%); 3 PH, 50 HZ, welding current at 100% duty cycle is 315 A, efficiency more than 85%, open circuit voltage – 66 V. Mains supply (V) is 415 ph./Hz.

The machine set is not having the necessary attachment to get controlled welding speed and arc length. So, an arrangement has been provided by making special type fixtures which was coupled with a carriage attachment of submerge arc welding machine to get control speed and arc length. The TIG welding machine with a special fixture is shown in Figure 19.1. A 5-mm-thick mild steel (SI 2062B) plate (150 mm×50 mm×5 mm) has been utilized for this work. The chemical composition (wt.%) of base metal are illustrated in Table 19.3.

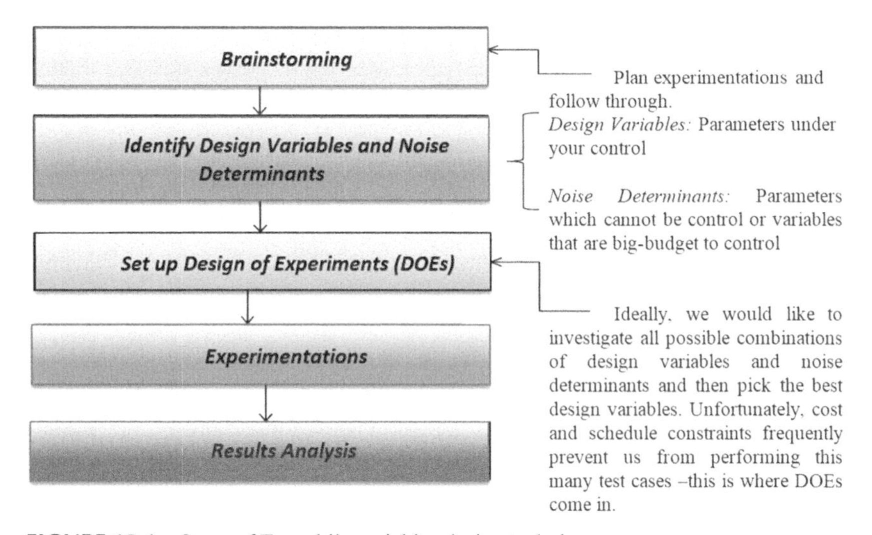

FIGURE 19.1 Steps of Taguchi's variables design technique.

In this study, a tungsten solid wire having 2.5 mm electrode diameter has been employed with direct current and electrode negative (DCEN) polarity; and pure argon gas has been supplied for the shielding purpose.

TABLE 19.3 Composition of MS IS

Grade	%C Max	%Mn Max	%P Max	% S Max	%Si Max	%C.E Max
2062 B	0.22	1.5	0.045	0.046	0.04	0.41

The goals of the present study are: (i) to optimize the weld bead geometry and process parameters; (ii) to measure the HAZ hardness; and (iii) to find optimum process parameters combination that maximize the weld quality as well as mechanical performance of the welded joint (Figures 19.2–19.4).

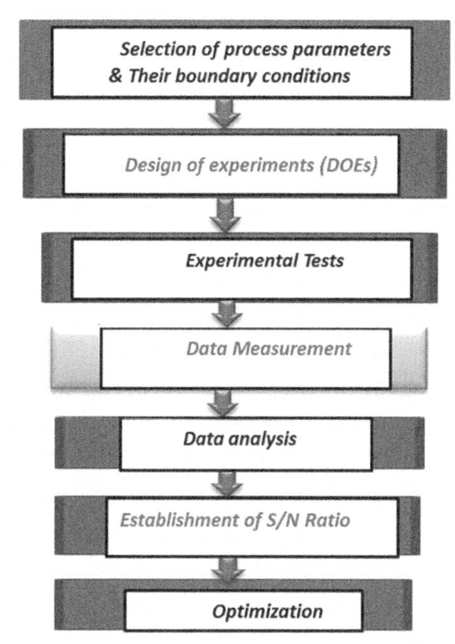

FIGURE 19.2 Flow chart of Taguchi's optimization model/methodology.

FIGURE 19.3 TIG welding machine used for weld the material and special fixture.

FIGURE 19.4 Samples collected after welding.

19.3.2 GEOMETRY OF WELDMENT

After welding at room atmospheric temperature, i.e., about 30°C and the welded samples had been cooled down to atmospheric temperature as shown in Figure 19.2. And one piece from all sample approximately 7 mm of thickness was cut by a saw machine with water as coolant. The faces of every section have been machined by shaper and subsequently milling machines to obtain flatted and semi-finished surface. Then after the samples piece were finished with the emery paper of grade 120, 150, 220, 400, 600, 1,000, and 2,000 consequently getting almost mirror finish. Then finally polished by velvet cloth with the slurry of alumina powder and consequently got almost mirror finish. The finished faces were etched by Nital solution, i.e., 5–10% nitric acid solution with distilled water at atmospheric temperature. The bead geometries viz. depth of penetration, bead thickness, and depth of HAZ; and HAZ hardness were measured and listed in a tabular form (Table 19.4 and Figure 19.5).

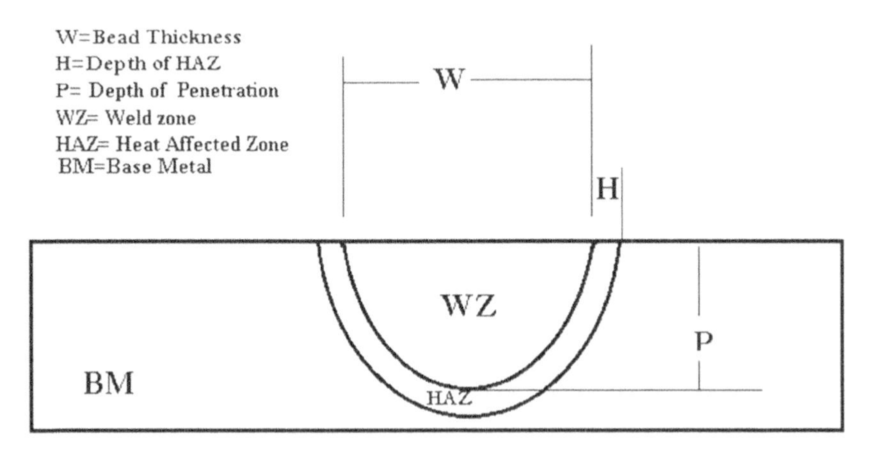

FIGURE 19.5 A schematic diagram of weld bead geometry.

19.3.3 HARDNESS TEST

In the present work, the test is conducted on the 16 samples in the digital Rockwell hardness testing machine as shown in Figure 19.3 on the welded parent material. By conducting of hardness testing, mainly obtaining the hardness of the heat affected zone (HAZ). The suitable indenter for unhardened materials is ball type of indenter, 1/16" a tungsten ball type indenter is selected for measuring the level of hardness in the present welded parent material. The B-scale has been used for the parent material, and the penetrator is a hardened tungsten ball 1/16" in diameter applied at a lesser load of 100 kgf, i.e., HRBW for 1/16" of ball type indenter in the Rockwell hardness testing machine (Figure 19.6). The value of the experimentation results is shown in Table 19.4.

FIGURE 19.6 Digital Rockwell hardness testing machine.

19.4 DATA ANALYSIS

19.4.1 DATA ANALYSIS BASED ON GREY RELATIONAL THEORY

At first, based on gray relational theory, the collected responses, i.e., quantified features of weld for each characteristic have been processed from limiting zero to one, which is known as data pre-processing or gray relational generation. Next, gray relational coefficients have been determined based on processed experimental values to constitutes a interrelation in between the absolute (highest) and processed values. Then after grand gray relational grade have been computed by simply mean or sum the gray relational coefficients in case of equal weights and different weights connecting to all elected responses, respectively.

TABLE 19.4 Data Related to Bead Geometry and Hardness of the HAZ

Sample No.	P	W	W_{haz}	HRBW
01	1.30	4.12	1.05	75
02	1.05	3.50	0.90	78
03	0.65	3.10	0.80	79
04	0.50	1.50	0.88	80
05	1.90	4.70	1.23	72
06	1.33	3.90	1.02	76
07	1.20	3.20	0.82	78
08	0.68	4.00	0.65	79
09	2.25	5.60	1.40	71
10	1.85	4.20	1.10	71
11	1.28	3.90	1.00	77
12	1.01	3.60	0.88	79
13	2.40	6.00	1.60	69
14	1.90	5.10	1.35	70
15	1.55	4.15	1.13	72
16	1.30	3.80	1.00	77

Note: *P* is the depth of penetration in mm; W is the bead width in mm; W_{haz} is the depth/width of HAZ in mm; HRBW is the Rockwell hardness at HAZ.

A multiple responses problem that has to solve and analyze, the coupled Taguchi method with gray relational analysis is considered an effective method. The experimental data of bead geometry viz. depth of penetration, bead width and depth of HAZ as well as hardness of the HAZ were normalized using Grey relational generation Theory as shown in Table 19.5. For the analysis, the depth of penetration has been considered higher-the-better (HB) and the bead width, depth of HAZ and hardness have been considered as lower-the-better (LB) criterion.

For Higher-is-Better (HB) criterion: $$X_{ij} = \frac{y_{ij} - Miny_{ij}}{Maxy_{ij} - Miny_{ij}} \quad (1)$$

For Lower-is-Better (LB) criterion: $X_{ij} = \frac{Maxy_{ij} - y_{ij}}{Maxy_{ij} - Miny_{ij}}$ (2)

where; X_{ij} = pre-processing data; Min y_{ij} and Max y_{ij} are the lowest and highest data of y_{ij} (y_{ij}= experimental data) for the jth outputs respectively. An ideal series for the output variables is $x_o(j)$ (j = 1, 2, 3, 4, …, 16). In the explanation of GRG (gray relational grade) the series of GRA (grey relational analysis) is to expose the grade of correlation in-between the 16 series [$x_o(j)$ and $x_i(j)$, where, j = 1, 2, 3, 4, 5, …, 16].

Calculation of Grey relational coefficients for each representative was executed by utilizing the Eqn. (3) and listed in Table 19.2.

Grey relational coefficient: $\gamma_{ij} = \frac{\Delta_{min} + \varphi\Delta_{max}}{\Delta_{oi}(j) + \varphi\Delta_{max}}$ (3)

where; $\Delta_{oi} = \|x_o(j) - x_i(j)\|$; Δ_{min} is the lowest data of Δ_{oi}; and Δ_{max} is the highest data of Δ_{oi}; n is the distinguishing coefficient $0 \leq n \leq 1$.

TABLE 19.5 Result of Pre-Processing or Grey Relational Generation of Each Characteristics

Exp. No.	P	W	W_{haz}	HRBW
1.	0.4210	0.4177	0.5789	0.4545
2.	0.2894	0.5555	0.736	0.1818
3.	0.0789	0.6444	0.8421	0.0909
4.	0.0000	1.0000	0.7578	0.0000
5.	0.7368	0.2888	0.3894	0.5454
6.	0.4368	0.4666	0.6105	0.3636
7.	0.3684	0.6220	0.8210	0.4545
8.	0.0947	0.4444	1.0000	0.0909
9.	0.9270	0.0888	0.2105	0.8181
10.	0.7105	0.4000	0.5263	0.8181
11.	0.4105	0.4666	0.6315	0.2727
12.	0.3095	0.5333	0.7578	0.1818
13.	1.0000	0.0000	0.0000	1.0000
14.	0.7368	0.2000	0.2631	0.8181
15.	0.5526	0.4111	0.4947	0.7272
16.	0.4210	0.4888	0.6316	0.3636

The grand gray relational grade (GRA), which represents comprehensively of all the characteristics of weld quality has been estimated collectively by Grey relational coefficients for each criterion as shown in Table 19.6. So, in such fashion, the multi-response optimization issues have been modified into an identical single object optimization issue employing the Grey relational analyzes based on Taguchi's method. The combined factors proportional to the highest data of GRG is close to the optimal solution.

19.4.2 GRAND GREY RELATION GRADE (GRG)

For giving equal weights to all the responses the grand gray relation grade has formulated as:

$$R_i = \frac{1}{n}\sum_{j=1}^{n} \gamma_{ij} \tag{4}$$

where; n is the responses number.

From the different literatures survey, we studied that to obtain quality weld, individual weightage has been given to the individual responses for better weld strength. The weighted gray relation grade has evaluated by modifying Eqn. (4) as follows:

So, for different weights to all the responses the grand gray relation grade as:

$$R_i = \sum_{j=1}^{n} Wk(\text{j}) \tag{5}$$

where; W_k is the weight of various graded responses that is $\sum_{j=1}^{n} \text{W}k = 1$.

For the achievement of better-quality weld bead with sufficient tensile strength, the under mentioned weight values has been observed, i.e., 0.30 for depth of penetration, 0.20 for bead thickness, 0.25 for depth of haz and 0.25 for hardness.

The mathematical expression of overall/grand gray relational grade (GRG) is (Tables 19.7 and 19.8):

$$\text{GRG}(R_i) = 0.3\text{P} + 0.2\text{W} + 0.25\ \text{W}_{\text{haz}} + 0.25\text{HRBW} \tag{6}$$

The grand gray relation grade is major production characteristics in Grey-based Taguchi optimization technique and main objective is fix the limits of parameters that are capable to performing gray relation grade of maximum limit. All individual performance features are represented by the grand gray relation grade. The aim of the present investigation has been defining an optimized relation to the parameters of weld bead geometry and hardness, and different weights and age were given to all responses (Table 19.9).

TABLE 19.6 Calculated Values of Grey Relational Coefficient ($\Phi = 0.5$).

Exp. No.	P	W	W_{haz}	HRBW
IS	1	1	1	1
1.	0.4633	0.4619	0.5429	0.4783
2.	0.4130	0.5249	0.6551	0.3793
3.	0.3518	0.5843	0.7599	0.3548
4.	0.3333	1.0000	0.6737	0.3333
5.	0.6551	0.4128	0.4502	0.5237
6.	0.4702	0.4838	0.5621	0.4399
7.	0.4418	0.5693	0.7363	0.4783
8.	0.3557	0.4736	1.0000	0.3548
9.	0.8635	0.3543	0.3877	0.7332
10.	0.6333	0.4545	0.5195	0.7332
11.	0.4589	0.4838	0.5757	0.4073
12.	0.4318	0.5172	0.6737	0.3793
13.	1.0000	0.3333	0.3333	1.0000
14.	0.6551	0.3846	0.4042	0.8468
15.	0.5277	0.4592	0.4974	0.6470
16.	0.4633	0.4944	0.5757	0.4073

Note: IS: Ideal sequence.

TABLE 19.7 Overall Grey Relational Grade (GRG) for Different Weights to the Responses

Exp. No.	GRG
1.	0.4860
2.	0.4874
3.	0.5176
4.	0.5517
5.	0.5226
6.	0.4883
7.	0.5500
8.	0.5401
9.	0.6098
10.	0.5940
11.	0.4768
12.	0.5032
13.	0.6666
14.	0.5862
15.	0.5362
16.	0.4836

TABLE 19.8 Mean Response Table for Grey Relation Grade (GRG)

Level	Factors		
	I	S	G_f
Level 1	0.5107	0.5712	0.4837
Level 2	0.5252	0.5390	0.5123
Level 3	0.5460	0.5202	0.5634
Level 4	0.5682	0.5197	0.5906
Max–Min(Delta)	0.0575	0.0516	0.1069
Rank	2	3	1

Total average/mean grey relation grade = 0.5766666

TABLE 19.9 Response Table for Signal to Noise Ratios (Larger is Better)

Level	Factors		
	I	S	G_f
Level 1	–5.849	–4.931	–6.309
Level 2	–5.602	–5.408	–5.814
Level 3	–5.305	–5.690	–5.001
Level 4	–4.971	–5.698	–4.601
Max–Min(Delta)	0.878	0.767	1.708
Rank	2	3	1

Figure 19.1 shows the S/N ratio graph (signal-to-noise ratio), which has been obtained by the used of overall relational grade for bead geometry and hardness.

With the help of orthogonal DOEs, it is feasible to break up the importance of every welding parameter at various degrees. It has been explained in this study, for input factor current the mean of GRG (gray relational grade) at various levels, i.e., level 1, 2, 3, and 4 can be determined by mean GRGs for the test's series 1–4, 5–8, 9–12, and 13–16 severally. The ratio of average GRG for every degree of the different variables may be obtained in the parallel way as shown in Table 19.7. The mean of all access is the total average GRG shown in Table 19.8.

For the S/N ratio plot (Figure 19.4), the evaluation of optimal setting using overall relational grade has been resolved and the optimized parameters fusion are I4 S1 G_f4. From the designed Taguchi's orthogonal array (Table 19.2), the optimum parameters condition has same as the parameters setting of welded sample number 13, where we have been used the welding current 180 A, welding speed 0.30 meters/minute. and gas flows or passes 25 liters/minute. The observation of Table 19.7 the calculated values of overall gray relational grade, the optimal variables setting which has the highest grand gray relational grade, it has been evaluated from the welded sample number 13. So, we could confirm from the analysis of result the optimization process is validated (Figure 19.7).

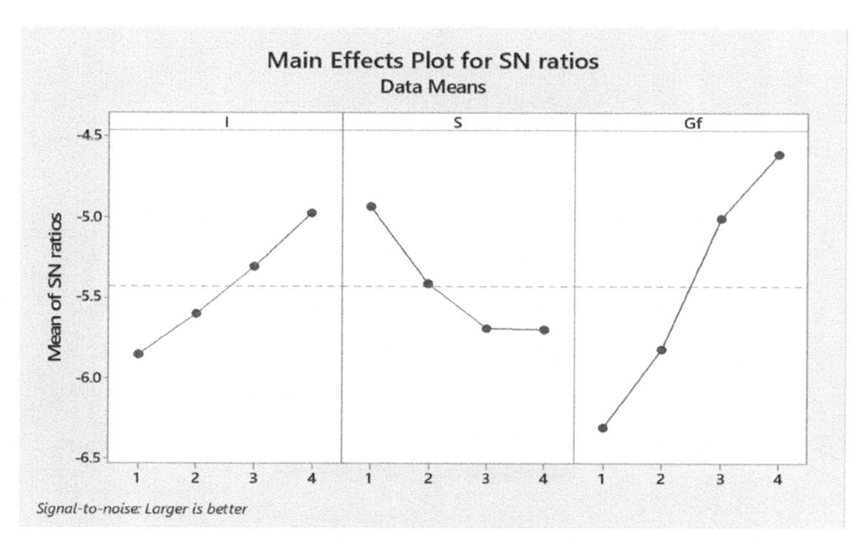

FIGURE 19.7 Evaluation of optimal setting using overall relational grade.

19.4.3 ANALYSIS OF VARIANCE (ANOVA)

The analysis of variance (ANOVA) is a practical model by which we can get a meaningful outcome after analyzing the experimental data. It is highly useful to disclose the effect of circumstances in a specific outcome. The method splits the overall variance of the outcome to explain the contributions of every determinant and the inaccuracy. So,

$$SS_{Total} = SS_{Factor} + SS_{Error}$$

where; $SS_{Total} = \sum_{j=1}^{p} (\gamma_j - \gamma_m)^2$ and SS_{Total} = total variation in the data or total sum of squares; γ_j is the average response for j[th] trial; γ_m is the overall average of the response; p is the number of trials in the orthogonal array;

SS_{Factor} is the deviation of the estimated parameter level mean or sum of squared variations due to parameters; SS_{Error} is the deviation of a response from its corresponding factor level mean or Sum of squared variations due to error (Figure 19.8).

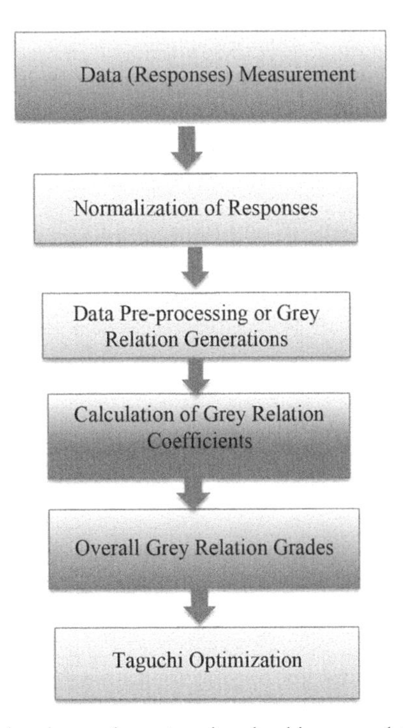

FIGURE 19.8 Chart showing various steps involved in grey relation analysis (GRA).

The mean square factor is calculated from ANOVA chart as:

$$\text{MS Factor} = \text{SS (Factor)/DF (Degree of freedom)}$$

The statistical value of Fisher's F- is used to test whether the effect of a term in the model (factor or interaction) is significant. F is applied to evaluate the p-value. So, the F-value is designated as:

$$F = MS(Factor)/MS(Error)$$

Based on F-values, the probability of significance (P-values) is computed. If P is less than or equal to the α-level (0.05) or (degree of dependence 95%) we have selected, then it can be concluded that the influence of the factor(s) has a significant effect on the response. If P is larger than the α-level we have selected, the effect is not significant.

The result calculated through ANOVA for grand grey relation grade was listed in Table 19.10 and response tables (Tables 19.8 and 19.9) values are computed. In the analysis, the F-ratio and contribution are traditionally used to determine the significance of factors, their performance and dependence. It is observed that for all the factors have an important effectiveness on the gray relational grade caused P value being less than 0.05. So, all the parameters are important, but the gas flow is the most significant factor (highest F-value). The reason for that was the shielding gas or inert gas are used not only to protect the droplet and weld bead but also to change metal transfer, penetration, and bead width of the weld, for spatter restrain and post-weld cleaning, to regulate welding fume production and to influence the metallurgical and mechanical properties of the weld bead. So, selection of shielding gas is therefore very important for welding process efficiency [17].

TABLE 19.10 Calculated Outcome of ANOVA for GRG

Source	DF	Adjusted SS	Adjusted MS	F-Value	P-Value
Current	3	0.07027	0.023423	13.55	0.004
Speed	3	0.06106	0.020355	11.77	0.006
Gas flow	3	0.32548	0.108494	62.74	0.000
Error	6	0.01037	0.001729	–	–
Total	15	0.46718			

19.4.4 REGRESSION MODEL FOR OVERALL GREY RELATION GRADE

A statistical technique, regression analysis is used to represent the correlations between inconstant. The simplest case to express a linear regression model for a line are generally noted as follows:

$$Y = \alpha + \beta_1 X_1 + S \tag{7}$$

where; Y is the dependent or target inconstant; X_1 is the independent or explanatory inconstant, or simply a regression of independent variable X; α is the constant or an intercept; and β_1 is the regression or slope coefficient of independent variable; S is the error.

A regression model that contains more than one independent variable is called a multiple regression. If we considered three independent variables and n number of observations, the regression model is therefore:

$$Y_i = \alpha + \beta_1 X_{i1} + \beta_2 X_{i2} + \beta_3 X_{i3} + S_i \quad (8)$$

where; i = 1, 2, 3, …, n–1, n; and *n* is the observation numbers.

In straightforward, the objective of regression is to make an effort to appear the best fit line or model that shows the relation between Y and X.

The regression model is an algebraic description of the regression line and illustrates the correlation between the response and predictor variables. The regression model can also write in the expression as:

$$\text{Response} = \text{Constant} + \text{Coefficient} \times \text{predictor} + \ldots\ldots + \text{coefficient} \times \text{predictor} + \text{error}$$

To determine the regression equation for overall gray relation grade, three input variables such as current, speed, and gas flow has been considered and analysis has been conducted by exercising MINITAB software. Using the data obtained for the process variables viz. current, speed, and gas flow; and process responses viz. bead geometry and haz hardness, a multiple linear regression model has been formulated and expressed in the following form:

$$R_i\,(\text{GRG}) = -4.808 + 0.2090\ \text{I} - 0.1770\ \text{S} + 0.4705\ \text{Gf} \quad (9)$$

It was observed that the analysis has been expressed a relation of the principal factors and their interactions. The acceptability of the model is appraised using R-Sq. The strength of the model is characterized by R-sq and grips a value between 0% and 100%. The R-Sq values are more than 95% which is again a clear indication of the developed model satisfactoriness. The predicted and actual values of GRG have been constructed using the model and the analysis respectively are represented in Table 19.11 and Figure 19.9. It has been noticed from the graph and table (comparative lists of predicted and actual values of GRG and their percentage of error ranges limits 0.34 and 4.80) that the predicted values are very close and sometime superimposed on the actual responses which distinctly showing the acceptability of the regression model.

S = 0.169517 R^2 = 94.50% Adjusted R^2 = 93.13%
Predicted R^2 = 90.84%

TABLE 19.11 Overall Grey Relational Grade (GRG) Actual and Predicted

Expt. No.	GRG		
	Actual	**Predicted**	**Error (%)**
1.	0.4860	0.4746	–2.34
2.	0.4874	0.4911	0.76
3.	0.5176	0.5088	–1.70
4.	0.5517	0.5278	–4.33
5.	0.5226	0.5263	0.71
6.	0.4883	0.4794	–1.82
7.	0.5500	0.5687	3.40
8.	0.5401	0.5143	–4.80
9.	0.6098	0.5907	–3.15
10.	0.5940	0.6165	3.80
11.	0.4768	0.4842	1.55
12.	0.5032	0.5015	–0.34
13.	0.6666	0.6731	0.97
14.	0.5862	0.5982	2.00
15.	0.5362	0.5382	0.37
16.	0.4836	0.4892	1.20

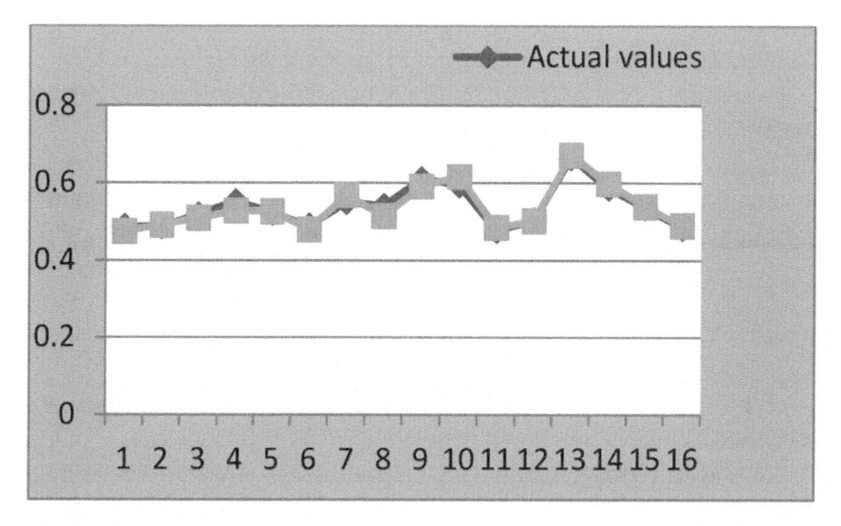

FIGURE 19.9 Differentiation in between calculated and predicted values overall grey relational grade.

S, R^2, adjusted R^2, and predicted R^2 are computes of how well the model fits the data. These values can help you select the equation with the best relevant. S is calculated in terms of the response varying and characterizes the standoff distance that utility of data decline from the regression curve. For any analysis, the lowest S value is assuming the best model for predicting the response.

19.5 CONCLUSIONS

In this analysis, the gray relational method based on Taguchi optimization procedure was applied for multi-response optimization to determine the optimum parameters setting to obtain preferable penetration and optimum bead width and haz thickness and hardness of the haz for quality joints generated by tungsten inert gas bead-on-plate welding on IS2062B steel. The signal-to-noise ratio (larger the best) plot shows the optimal parameters setting viz. the welding current 180 Amp, welding, or carriage speed 0.30 meters/minute and gas flows or passes 25 liters/minute for optimal value of the weld bead geometry as well as hardness. ANOVA analysis indicated that the gas flow rate has been the most significant parameter for both weld bead geometry and hardness. The intension of this analysis has to make evident the utilization suitability of Taguchi-based gray analysis for solving the optimization analysis to improving quality of weld bead geometry as well as in the region of gas tungsten arc welding (GTAW).

Regression model is correlating in between the responses viz. penetration, bead width, width of haz and haz hardness; and process variables. This equation assists in selecting the actual value of process variables with the intention that to acquire required penetration, bead width, width of haz; and haz hardness of the TIG welded IS 2062B Steel Plate component.

DENOTATION

R^2: Coefficient of conviction; Adjusted R^2: Adjusted coefficient of conviction; X_{ij}: Grey relational generation of each characteristic; γ_{ij}: Grey relational coefficient; R_i: Grey relational grade; DOE: Design of experiment; γ_j: Mean response for jth trial; γ_m: Overall mean of the response; SS: sum of squared variations due to parameters; MS: Mean square deviation; DF: Degree of freedom.

KEYWORDS

- **analysis of variance**
- **bead-on-plate welding**
- **grey relational analysis**
- **multi-response optimization**
- **response surface methodology**
- **tungsten inert gas**

REFERENCES

1. Jackson, C. E., & Shrubsall, A. E., (1953). Control of penetration and melting ratio with welding technique. *Weld. J., 32*(4), 172s–178s.
2. Taguchi, G., (1986). *Introduction to Quality Engineering: Designing Quality into Products and Processes.* Kraus International, White Plains.
3. Arvind, K. K., Ajay, B., Rajan, S., & Neetu, (2012). Optimization of welding parameters by regression modeling and Taguchi parametric optimization technique. *IJMIE.* ISSN No. 2231 –6477.
4. Pal, S., Pal, S. K., & Samantaray, A. K., (2007). Determination of optimal pulse metal inert gas welding parameters using Neuro-GA technique. In: *4th Int. Conf. on Theoretical, Applied Computational and Experimental Mechanics.* IIT Kharagpur India, ICTACEM-2007/131.
5. Benyounis, K. Y., & Olabi, A. G., (2008). Optimization of different welding processes using statistical and numerical approaches—A reference guide. *Adv. Eng. Softw., 39,* 483–496.
6. Roy, R. K., (2001). *Design of Experiments Using The Taguchi Approach.* Wiley, New York.
7. Anawa, E. M., & Olabi, A. G., (2008). Using Taguchi method to optimize welding pool of dissimilar laser-welded components. *Opt. Laser Technol., 40,* 379–388. doi: 10.1016/j.optlastec.2007.07.001.
8. Datta, S., Bandyopadhyay, A., & Pal, P. K., (2008). Grey-based Taguchi method for optimization of bead geometry in submerged arc bead-on-plate welding. *Int. J. Adv. Manuf. Technol., 39,* 1136–1143.
9. Juang, S. C., & Tarng, Y. S., (2002). Process parameter selection for optimizing the weld pool geometry in the tungsten inert gas welding of stainless steel. *J. Mater. Process Technol., 122,* 33–37. doi: 10.1016/S0924-0136(02)00021-3.
10. Tarng, Y. S., Juang, S. C., & Chang, C. H., (2002). The use of grey-based Taguchi methods to determine submerged arc welding process parameters in hard facing. *J. Mater. Process Technol., 128,* 1–6. doi: 10.1016/S0924-0136(01)01261-4.

11. Pan, L. K., Wang, C. C., Wei, S. L., & Sher, H. F., (2007). Optimizing multiple quality characteristics via Taguchi method-based grey analysis. *J. Mater. Process Technol., 182*, 107–116. doi: 10.1016/j.jmatprotec. 2006.07.015.
12. Tarng, Y. S., Yang, W. H., & Juang, S. C., (2000). The use of fuzzy logic in the Taguchi method for the optimization of the submerged arc welding process. *Int. J. Adv. Manuf. Technol., 16*, 688–694. doi: 10.1007/s001700070040.
13. Deng, J., (1982). Control problems of grey systems. *Syst. Contr. Lett., 5*, 288–294.
14. Singh, L., Khan, R. A., & Aggarwal, M. L., (2012). Empirical modeling of shot peening parameters for welded austenitic stainless steel using grey relational analysis. *Journal of Mechanical Science and Technology, 26*(6), 1731–1739.
15. Patel, C. N., (2013). Parametric optimization of weld strength of metal inert gas welding and tungsten inert gas welding by using analysis of variance and grey relational analysis. *Modern Engineering and Emerging Technology, 1*.
16. Nowacki, J., & Rybicki, P., (2005). The influence of welding heat input on submerged arc welded duplex steel joints imperfections. *J. Mater. Process Tech., 164*, 1082–1088.
17. Mvola, B., & Kah, P., (2017). Effects of shielding gas control: Welded joint properties in GMAW process optimization. *The International Journal of Advanced Manufacturing Technology, 88*, 2369–2387.
18. Songsorn, K., & Sriprateep, K., (2016). Grey–Taguchi Method to optimize the percent zinc coating balances edge joints for galvanized steel sheets using metal inert gas pulse brazing process. *Advances in Mechanical Engineering, 8*(6), 1–14.
19. Deb, K. A., Manidipto, M., & Tapan, K. P., (2015). Development of a direct correlation of bead geometry, grain size and HAZ width with the GMAW process parameters on bead-on-plate welds of mild steel. *Transactions of the Indian Institute of Metals, 68*(5), 839–849.
20. Şefika, K., (2013). Multi-Response optimization using the taguchi-based grey relational analysis: a case study for dissimilar friction stir butt welding of AA6082-T6/AA5754-H111. *The International Journal of Advanced Manufacturing Technology, 68*(1–4).

CHAPTER 20

APPLICATION OF EFFICIENT MOVING LEAST SQUARES METHOD ON ROBUST DESIGN OPTIMIZATION

TUSHAR DAS[1] and SOUMYA BHATTACHARJYA[2]

[1]*Assistant Professor, Department of Civil Engineering, Heritage Institute of Technology, Kolkata, West Bengal, India*

[2]*Associate Professor, Department of Civil Engineering, Indian Institute of Engineering Science and Technology, Shibpur, West Bengal, India, E-mail: soumya@civil.iiests.ac.in*

ABSTRACT

The engineering community has recognized that structural optimization not considering uncertainty will result in catastrophic failure consequences. Often, such failures cause the loss of many lives. Thus, this issue is attempted by the researchers by exploring various approaches to optimization under uncertainty. Among these, Robust Design Optimization (RDO) is the most popular one, which ensures reliability as well as the least deviation of structural performance under uncertainty. One of the conventional ways of accomplishing RDO is to apply Monte Carlo Simulation (MCS), which is associated with large computational time. Often, surrogate-assisted optimization schemes are adopted to circumvent this computational challenge. But such approaches are hinged on the conventional least squares method, which is often observed to be a source of error in the existing literature. Thus, the issues of either enormous error

Optimization Methods for Engineering Problems. Dilbagh Panchal, Prasenjit Chatterjee, Mohit Tyagi, Ravi Pratap Singh (Eds.)

or large computational time prohibit the potential use of RDO in industry. Hence, in the present chapter, a new and prudent method of moving the least squares method (MLSM) is applied in the RDO. The new approach applies a local and moving regression. The proposed method is illustrated by examples that show its accuracy over the conventional approach. At the same time, the results achieved are least sensitive to input uncertainty. The MLSM-based RDO has been observed to be accurate as well as computationally efficient.

20.1 INTRODUCTION

In the last few years, deterministic optimization (DO) has gained significant popularity due to economic usage requirement of resources. The details of such progress may be found in Ref. [3]. The DO generally enables us to design efficient and economical structures, in most cases. But DO cannot consider the uncertainty effects. Due to various uncertain factors practically, the deterministic constraint may shift to an infeasible domain, which cannot be visualized in a DO. Thus, resorting on DO neglecting uncertainty may under design a structure leading to collapse consequences. Indeed, it is now well established in the international scenario that engineering analysis and design cannot be completed without considering the presence of uncertainty. However, its recognition in the national scenario is yet to be established in the civil engineering industry.

It is also true that results of uncertainty-based optimization, such as robust design optimization (RDO) [1] may cause an increment of the structural cost, both in construction and material. Also, cost-viability is an important parameter to make a design acceptable. The best design is one which shows a balance between economy, viability, and sensitivity to uncertainty without compromising intended reliability level. In doing so, a designer must try to combine performance and reliability in the design as well as some kind of robustness in his design. The reliability-based optimization (RBO) designs a system with the required reliability level [17]. However, the standard deviation of response attained by an RBO may be significantly high leading to the infeasible zone. Thus, the guarantee of safe performance is not ensured in the RBO. On the other hand, an RDO minimizes the standard deviation of the response keeping it within the desirable range. The system behaves consistently with the least deviation from the desired threshold even when subjected to high uncertainty effects. The RDO does not minimize the source of uncertainty, rather it regulates the design in such a manner that the output uncertainty or deviation from the expected performance becomes as minimum as possible, even when there is high fluctuation of design parameters. Moreover, when statistical information on the decision variables is not sufficient or incomplete, the RBO cannot be applied and the RDO serves as an attractive alternative.

Optimization under uncertainty is generally carried out by using Monte Carlo simulation (MCS). This technique can tackle correlated non-normal variables and high level of uncertainty, even more than 60%. There have been researches, where the MCS was applied in solving RDO problem [17]. But, on the other hand, when the structural system is large, earthquake-like random load is involved, and for executing hundreds of iterations of optimization, the MCS approach will take several hours (maybe even a few days) to yield a single optimization solution. Moreover, it will need interlinking between structural modeling software and the optimization compiler which will further lengthen the solution period. Thus, the metamodeling approach is often used in the RDO process instead of MCS approach, partly [15] or fully replacing [14] the MCS. The response surface method (RSM)-based metamodel generates polynomial to express implicit constraint function explicitly in terms of uncertain parameters. This evades exhaustive software interlinking with the optimizer which in turn improves the computational efficiency.

The conventional RSM technique uses the least squares method (LSM). But it has been realized that conventional LSM-based RSM metamodeling is not accurate enough in predicting the response beyond sampling zone [5]. Hence, a recent approach based on MLSM-based RSM [11] is applied here to improve the accuracy even beyond the sampling zone. The concept of the MLSM is applied in metal forming application [8], reliability analysis [4, 13] (Goswami et al., 2016). However, in the RDO, the application of the MLSM-based RSM is relatively scarce. Bhattacharjya et al. (2019) presented RDO of a stacker Reclaimer structure by metamodeling. But, the aspect of parameter tuning of the MLSM was not done in their study. Moreover, a detailed parametric study in various forms of the MLSM in the RDO is a challenging task and not so far observed to be worked out explicitly. Goswami et al. [9] indicated this issue of parameter selection during reliability analysis of structure. Thus, this aspect has been also taken up in the present chapter to cover a broader perspective of MLSM.

To show the effectiveness of the present approach first the mathematical formulation of the RDO is presented in Section 20.2. Subsequently, application problems are presented describing parameter selection and RDO.

20.2 MATHEMATICAL FORMULATION OF THE RDO

An RDO optimizes structural weight (or cost) as well as standard deviation of weight (or cost). At the same time, the constraint function is enforced to stay within the feasible domain in the presence of uncertainty. Further, the chance that the constraint will lie in the feasible zone in case of worst possible uncertainty is elevated by a robust penalty factor. In this way, the undesirable deviation of the system is kept at the lowest possible level, which is generally termed as 'robustness.' Here, the designer does not try to regulate the source of uncertainty. It is the formulation which designs the structure as robust. Let us denote objective function as Γ, constraint function as Ω. Then, the RDO is mathematically expressed as [3]:

$$\begin{aligned} &\text{minimize:} \quad \Theta(\mathbf{x},\mathbf{z}) = A\,\mu_{\Gamma(\mathbf{x},\mathbf{z})} \big/ \mu_{\Gamma}^{*} + (1-A)\,\sigma_{\Gamma(\mathbf{x},\mathbf{z})} \big/ \sigma_{\Gamma}^{*} \\ &\text{subjected to:} \quad \mu_{\Omega_m(\mathbf{x},\mathbf{z})} + \Theta_m \sigma_{\Omega_m(\mathbf{x},\mathbf{z})} \le 0, \qquad m = 1,2,\ldots\ldots, M \\ &\qquad x_i^L \le x_i \le x_i^U, \qquad i = 1,2\ldots\ldots, N. \\ &\qquad 0 \le A \le 1 \end{aligned} \tag{1}$$

where; $\mu_{\Gamma}(\mathbf{x},\mathbf{z})$ and $\sigma_{\Gamma}(\mathbf{x},\mathbf{z})$ are the mean and the standard deviation of the objective function, respectively. x and z are design variables and design parameters, respectively. $\Theta(\mathbf{x},\mathbf{z})$ is the weighted sum of the two objective functions. $\mu_{\Gamma}*$ and $\sigma_{\Gamma}*$ are applied for normalization for the weighted sum. These are the optimal values with A equal to one and zero, respectively, with A as the weights. The designer chooses suitable values of A from zero to one based on the relative requirement of economy and robustness. In fact, these two are the conflicting objectives. If economy is fully resorted upon as target, the design may not be robust and vice versa. In the constraint function, $\mu_{\Omega_m(\mathbf{x},\mathbf{z})}$ and $\sigma_{\Omega_m(\mathbf{x},\mathbf{z})}$ are the mean and standard deviation of the constraint $\Omega_m(\mathbf{x},\mathbf{z})$, respectively. $\Theta_m \sigma_{\Omega_m(\mathbf{x},\mathbf{z})}$ is added as a penalty due to uncertainty with Θ_m as the penalty factor. More the value of the penalty factor, more will be the importance of standard deviation. Thus, Θ_m regulates robustness of constraint functions. Presently, A is taken as 0.5 to put equal weight on economy and robustness. Θ_m is taken as 3.0 which ensures 99.865% probability of constraint feasibility under uncertainty.

In this chapter, objective functions and constraints are explicitly approximated by the MLSM. In doing so, different Design of Experiment (DOE) schemes are compared.

20.3 LSM-BASED RSM IN RDO

In practical, the true relationship between a response and input variables that influences the response is very difficult to find out. Thus, a metamodel, which approximates the complex and implicit phenomena, help us in this regard. Let, the exact relationship between response (y) and a vector of input variables ($\mathbf{x}$) is represented as:

$$y = f(\mathbf{x}) \tag{2}$$

Then a metamodel g ($\mathbf{x}$) approximates the true relationship f ($\mathbf{x}$). The relationship between y and x becomes:

$$y = g(\mathbf{x}) + \varepsilon \tag{3}$$

where; ε is a total error term. This metamodel is built based on the information available about the independent variables and the observed data. Let, the response y is dependent on k independent variables and n number of observations are obtained. If a first order polynomial is selected to describe the problem, one can write:

$$g(\mathbf{x}) = \beta_0 + \Sigma_{j=1}^{k} \beta_j x_j + \Sigma_{jj=1}^{k} \beta_{jj} x_j^2 + \Sigma_{l<j=2}^{k} \beta_{jl} x_j x_l \tag{4}$$

The above is a multi-variable non-linear regression model. The coefficients $\beta_j, \beta_{ij}, \beta_{jl}$ are called the regression coefficients. A typical observation of Eqn. (5) is:

$$g(\mathbf{x})_{\langle i \rangle} = \beta_0 + \Sigma_{j=1}^{k} \beta_j x_{j,\langle i \rangle} + \Sigma_{jj=1}^{k} \beta_{jj} x_{j,\langle i \rangle}^2 + \Sigma_{l<j=2}^{k} \beta_{jl} x_{j,\langle i \rangle} x_{l,\langle i \rangle} \tag{5}$$

Thus, assembling all such observations in matrix notation one can write:

$$[g] = [X]\,[\beta] \tag{6}$$

where; g is a n-dimensional vector of the responses; X is a matrix of the independent variables called design matrix as it contains DOE points; and β is a vector of the regression coefficients. After that, the concept of LSM is applied which minimizes the following error norm:

$$L = \sum_{i=1}^{n} \left(y\langle i \rangle - g(\mathrm{x})\langle i \rangle \right)^2 = (\mathrm{y} - \mathrm{X}\beta)^{\mathrm{T}} (\mathrm{y} - \mathrm{X}\beta) \tag{7}$$

This gives us the coefficient vector as:

$$\beta = (X^{T} X)^{-1} X^{T} y \tag{8}$$

After we calculate β, the response surfaces are obtained by Eqn. (4). Though in Eqn. (4), only polynomial response surface is covered, the approach is similar for other forms of response surface, like exponential, trigonometrical functions, etc. The LSM-based RSM is simple to execute, but its error is significantly worse as reported by many researchers [19]. The error is due to its global approximation, resulting in underfitting. In this regard, the comparatively newer MLSM can be used in the RDO to alleviate the error.

20.4 MLSM-BASED RSM IN RDO

The MLSM approximates the response locally. The MLSM-based RSM uses weight functions which vary with respect to the position of the regressor. In this way, the coefficient vector becomes a function of **x**. The response surface goes on changing its nature based on the value of **x**. Due to this adaptive nature, MLSM-based RSM yields more accurate results than the LSM-based RSM. In fact, MLSM minimizes the weighted error and finds the value of **β**. The modified error $L_y(\mathbf{x})$ can be defined as [11].

$$L_y(\mathrm{x}) = (\mathrm{y} - \mathrm{X}\beta)^{\mathrm{T}}\, \mathrm{W}(\mathrm{x})(\mathrm{y} - \mathrm{X}\beta) \tag{9}$$

where; W(x) is a diagonal matrix of weight functions. This can be written as [11]:

$$\mathbf{W}(\mathbf{x}) = \begin{bmatrix} w(\mathbf{x} - \mathbf{x}_{<1>}) & 0 & \dots & 0 \\ 0 & w(\mathbf{x} - \mathbf{x}_{<2>}) & \dots & 0 \\ \dots & \dots & \dots & \dots \\ 0 & 0 & \dots & w(\mathbf{x} - \mathbf{x}_{<n>}) \end{bmatrix} \tag{10}$$

Most popular weight function used are those proposed by Kim et al. [11] and Taflanidis and Cheung [16], which are respectively as the followings:

$$w_I\left(\mathrm{x}-\mathrm{x}_{\langle i\rangle}\right)=w(\mathrm{d})=e^{\left(\frac{-\|\mathrm{d}\|}{R_I}\right)} \tag{11}$$

$$w_I\left(\mathrm{x}-\mathrm{x}_{\langle i\rangle}\right)=w(\mathrm{d})=\frac{e^{-\left(\frac{\|\mathrm{d}\|}{cR_I}\right)^{2k}}-e^{-\left(\frac{1}{c}\right)^{2k}}}{1-e^{-\left(\frac{1}{c}\right)^{2k}}} \tag{12}$$

In this chapter, we have utilized both of these weight functions to investigate their effectiveness. In the above, R_I is the approximate radius of influence which is taken as thrice the distance between the most extreme points out of $(1+2n)^{\text{th}}$e scatter data points in the DOE [9], termed henceforth as $\{3\sigma\}$ domain. $\|\mathbf{d}_I\|$ is the Euclidean distance between the prediction point, $\mathbf{x}$, and the DOE point $\mathbf{x}_{<i>}$. In the second weight function, c, and k are the parameters which module the balance between overfitting and underfitting. The underfitting occurs if an average response surface gets fitted without capturing localized variation leading to large errors. This problem is pertinent for the LSM. On the other hand, overfitting occurs due to presence of a large number of DOE points and for subsequent fitting of higher order polynomial that captures noise prevailing in the scatter data. As a result, when tested for new data points which are different than the DOE points, large prediction errors are yielded. The problem of overfitting is comparatively less for LSM unless one chooses a higher order polynomial than that required. But, MLSM is susceptible for both overfitting and underfitting. The parameters R_I, c, and k create a balance between overfitting and underfitting and a suitable choice of these parameters can circumvent all these problems. For example, if very large values of R_I and c are chosen, MLSM converges to LSM with global averaging leading to underfitting error. On the other hand, with very small values of these parameters, the DOE points closer to the prediction point get hugely weighted resulting in overfitting. In this regard, the values of R_I, c, and k as 0.4, 0.4, and 1.0 have been generally observed to be the best choice to avoid either underfitting or overfitting [4, 10, 16]. However, exception from these have been also reported [10]. The authors observed that re-sampling and validation technique like *k*-fold cross-validation often reveals a better picture of the parameter dependency of MLSM. But, application of such an approach makes the process computationally cumbersome. Hence, yet sufficient research is needed to derive a short-cut process which yields the best values of these parameters without increasing the computational burden, significantly. Thus, in this study, a parametric study has been performed with various set up of these parameters to reveal this issue of parameter selection.

By the MLSM after minimizing the modified error norm $L_y(\mathbf{x})$ of Eqn. (9), $\boldsymbol{\beta}(\mathbf{x})$ can be obtained as:

$$\beta(x) = [X^T W(x) X]^{-1} X^T W(x) y \tag{13}$$

Note that β is a function of x for MLSM. Hence, based on the present value of the prediction point x the coefficient vector β gets changed in MLSM. As a result, the response surface also gets changed during each update of x during RDO. The implementation of RDO by MLSM is presented below by a flow-chart in Figure 20.1.

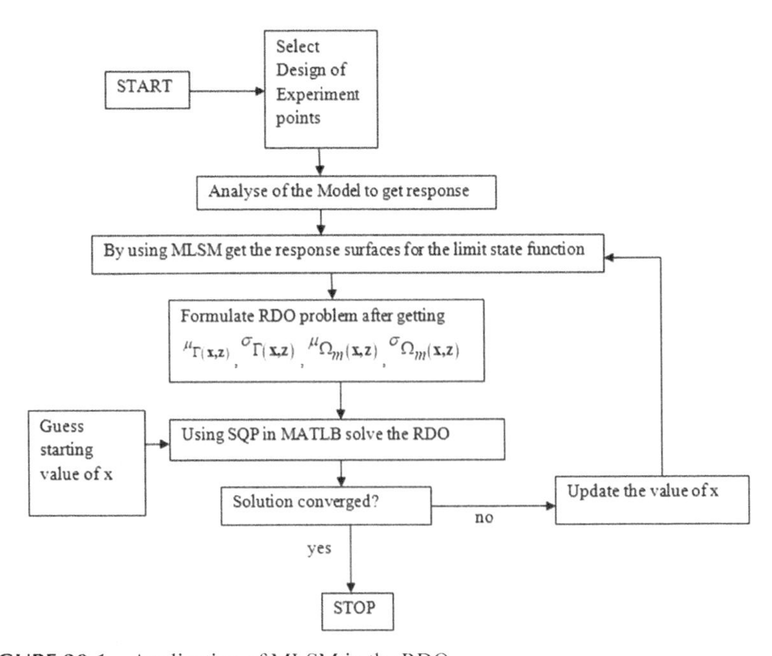

FIGURE 20.1 Application of MLSM in the RDO.

The implementation of RDO by LSM is almost same as of MLSM-based RDO, except in case of MLSM, the coefficient vector changes with **x**. Whereas, the same coefficient vector and response surface remain operative over the entire RDO process irrespective of the values of **x**.

20.5 IMPORTANCE OF DOE IN THE RSM

A vital requirement for accuracy of RSM is the DOE. It provides a set of efficient points where the responses are evaluated. There are many DOE schemes; among them saturated design (SD), redundant design (RD), central composite design (CCD), full factorial design (FFD), uniform design (UD) scheme, Latin hypercube sampling (LHS) are popular. In the SD method, the points chosen are the mean values μ_i of the response variable x_i (as center point) and at axial points, $x_i = \mu_i \pm h_i\sigma_i$, where h_i is a positive integer, which is 1.0 for SD and can have any value more than 1.0 for RD. The CCD selects corner, axial, and center points for fitting a second order response surface model. The full FD method generates q sample values for each coordinate, thus producing a total of q^k sample points for k variables. For 2^k FD (i.e., $q = 2$) of a two variables problem, the DOE points are located at four corners, i.e., at (1, 1), (−1, 1), (1, −1) and (−1, −1) in standard normal space. Interested reader may refer to Ref. [18].

There is another class of DOE, which is based on random simulation using MCS, such as UD and LHS. By these schemes, for each of the k variables, the range of the variable is divided into m non-overlapping intervals consisting of equal probability. One value from each interval is selected at random but with respect to the probability density in the interval. For the LHS, different probability distribution function (pdf) of the variables (as obtained from standard literature or statistical tests) are deployed to generate the divisions and subsequent random numbers. Whereas, in cased of UD, uniform distribution for all the variables is assumed. Also, in LHS, there is an additional step of random pairing, by which the m values of the first variable are paired randomly with m values of the second variable. These m pairs are then combined randomly with the m values of the third variable to form m triplets, and so on, until m-tuplets are formed. Finally, a sampling matrix **S** is yielded with dimension $m \times k$. Interested reader may refer to Refs. [4, 10], for details of these procedure for LHS and UD, respectively. The advanced sampling method like importance sampling can also be explored to reduce computational burden. In regard of various

DOE schemes and their importance the study of Yondo et al. [18] can be made use of.

Though, there is a handful of DOE schemes, SD, RD, UD, and LHS are computationally less onerous. SD is the simplest and computationally efficient approach. At the same time, error with SD is also more. RD is better than SD with respect to accuracy, but is a little bit more computationally intensive than SD. Also, the value of h should be chosen judiciously to cover entire solution domain. CCD is more accurate than SD and sometimes RD, also; but it becomes computationally involved if k is more. LHS better than UD, as the latter disregards the pdf information. SD is a better choice at the initial stage of a problem; whereas at final stage CCD will be more accurate. For numerical simulation problem, many researchers prefer LHS or UD, rather than SD, RD, or CCD. However, this needs further study. In fact, proper selection of a DOE scheme regulates the accuracy and computational efficiency of RSM. Thus, in this chapter, various DOE schemes are compared to have a broader view related to importance of DOE scheme in RSM.

20.6 APPLICATION PROBLEMS

In this section three example problems are presented. The first example uncovers effect of parameter modulation in MLSM for a typical higher-order function. The second example compares LSM and MLSM for various DOE schemes. The third example presents RDO using MLSM.

20.6.1 EXAMPLE I

A simple 4th order polynomial $y(\mathbf{x}) = 1 + x + x^2 + x^3 + x^4$ is considered as the first application problem. To study the effectiveness of the various RSM schemes, the $y(\mathbf{x})$ is approximated by different RSM schemes. SD, CCD, and RD are taken as DOE schemes in separate modules using type III polynomial. The comparison of various DOE schemes by the LSM and the MLSM is presented in Figure 20.2(a). The effect of parameter modulation is shown in Figure 20.2(b). The total sampling point is 13 for the RD, 3 for SD and 3 for CCD. Both weight functions w_1 and w_2 have been used in separate modules.

The error by the CCD-LSM, SD-LSM, SD-MLSM (w_2) is notable from Figure 20.2(a). On the other hand, one can observe that the RD-MLSM (w_1) approximates the actual function more accurately. Noting that RD-MLSM with w_1 fits the best, Figure 20.2(b) is focused on the variation of R_I by considering only RD-MLSM with w_1. The variation in y can be clearly observed from this figure. It has been observed that the results with R_I less than 20% of the {3σ} domain yields worst results; whereas, R_I with 70% to 90% of{3σ} domain produces a better fit. Thus, R_I=0.4 of {3σ} domain may not always be correct and it requires a prior validation as in Figure 20.2(a) and (b), for successful implementation of MLSM in RDO.

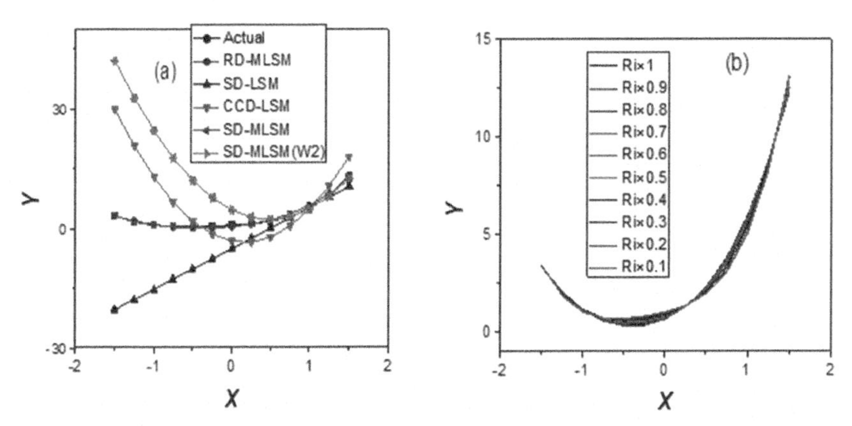

FIGURE 20.2 Comparison of (a) various DOE schemes by LSM and MLSM; (b) parameter modulation in MLSM.

20.6.2 EXAMPLE II

A fixed-hinged circular arch (Figure 20.3), subjected to central concentrated load is chosen as the second example problem. The cross-section is rectangular with a depth 2.289 unit and width of 1 unit. The modulus of elasticity of the arch is 10^6 unit. The modeling and solution are done by ABAQUS software. The arch is divided into 60 finite elements of type B31H. This particular problem is chosen owing to its geometric nonlinearity and to test whether such nonlinear trends of the response can be captured by the MLSM based RSM or not. The objective is to predict the collapse load for this arch. The details of the arch and its post buckling behavior can be found in http://ivt-abaqusdoc.ivt.ntnu.no:2080/v6.14/books/exa/default.htm. It can be observed that the arch undergoes extremely large deflection representing nonlinearity in behavior. Figure 20.3 presents a typical load-displacement curve. It shows the buckling behavior of the arch. At 0.12 inch, the buckling occurs at the critical load of 500 units, after which a prolonged post-buckling limb follows. The post-buckling behavior continues up to 0.25 inch and 1,200 unit. The post-buckling behavior is well captured by the Riks method. The regressor used are the load, Youngs modulus, width, and depth of the cross-section. Type-III polynomial response surface is used. The maximum load at which instability occurs is taken as the response parameter, y.

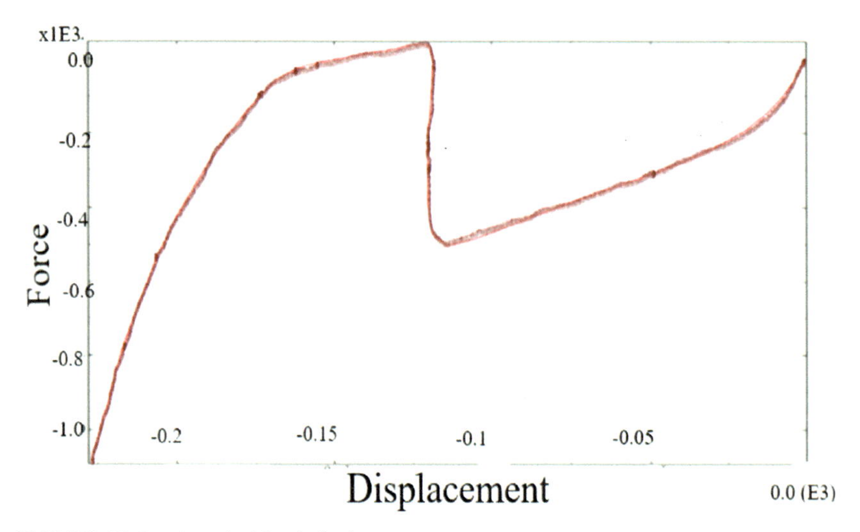

FIGURE 20.3 A typical load-displacement curve.

The effectiveness of the Redundant design scheme for both the LSM and the MLSM are compared by capturing maximum load of arch at which instability occurs by the metamodels in Figure 20.4. The test points are chosen randomly and are distinctly different than the DOE points. It can be observed that both the approaches more or less capture the response reasonably. But, the accuracy with the MLSM is consistent in all the cases. However, there are significant errors by the LSM in eight cases. In fact, during optimization, the prediction error will get accumulated to yield final RDO result with large error. Thus, if LSM prediction goes wrong in some iteration, the result will have a cumulative effect on safety. Since, due to its adaptive nature, MLSM consistently produces accurate results, is more suitable for RDO.

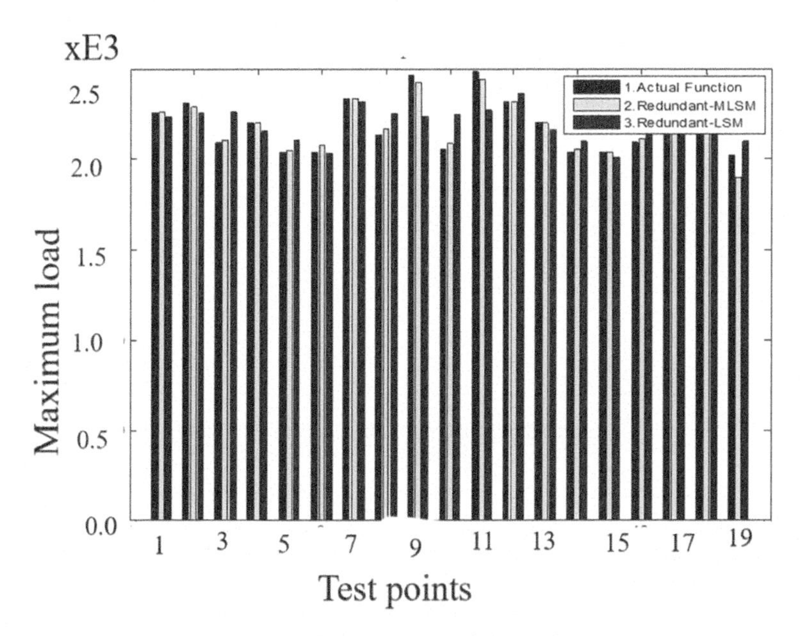

FIGURE 20.4 The maximum load by the LSM and the MLSM.

20.6.3 EXAMPLE III

RDO of Koyna dam under earthquake has been taken as the third example problem. This has been taken from Lee and Fenves [12]. On 11th December, 1967, an earthquake of Richter scale magnitude 6.5 struck the dam. Figure 20.5 presents the layout of the dam. A typical section has been considered as a plane strain problem. CPS4R element available in the finite element software ABAQUS is used for the meshing. The vertical and transverse components of ground motion are presented can be found in more details in http://ivt-abaqusdoc.ivt.ntnu.no:2080/v6.14/books/exa/default.htm. Time-history analysis is executed on the finite element model of the dam prepared in ABAQUS. Figure 20.6(a) and (b) present tensile stress in dam at time 2 second and 4 seconds, respectively, after striking of the earthquake.

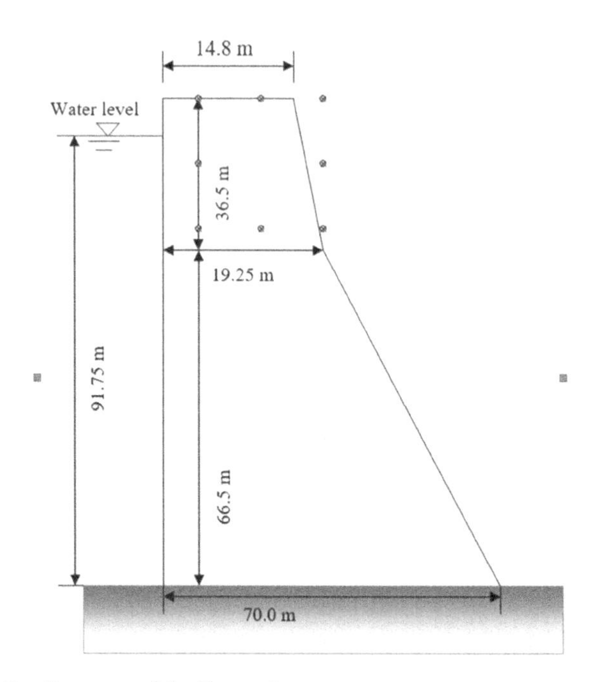

FIGURE 20.5 Geometry of the Koyna dam.

The details of the design variables and the design parameters with their uncertainty information is presented in Table 20.1. The DO problem is formulated to minimize the cross-sectional area of the dam subjected to maximum tensile stress to be less than the allowable (3 MPa) under dynamic load. Then, the RDO is formulated using Eqn. (1). The constraint function is approximated in explicit functional form using the LSM-based RSM and the MLSM-based RSM, in separate modules. In Figure 20.7, a comparison is shown depicting maximum stress obtained by the conventional LSM and the present MLSM. The actual finite element analysis-based results are also shown in the same figure. The test points shown in the figure are different from the support points used to construct the DOE. The sampling method used is the LHS. It can be observed from this figure that the present MLSM results are in close agreement with the actual result. Whereas, the LSM based predictions are not always closer to the actual result. However, up to some extent, both the approaches somewhat fit good since the LHS based sampling is used as the DOE in contrast of the conventional CCD, FD, or SD based DOE.

TABLE 20.1 Design Variables and Design Parameters

	Parameter	Mean Value	COV (%)	Distribution
Design Variables	Base length	70 m	10	Normal
	Top width	14.8 m	10	Normal
Design Parameters	Transverse ground acceleration	0.5 *g*	20	Normal
	Hydrostatic load	900 KN/m^2	30	Normal

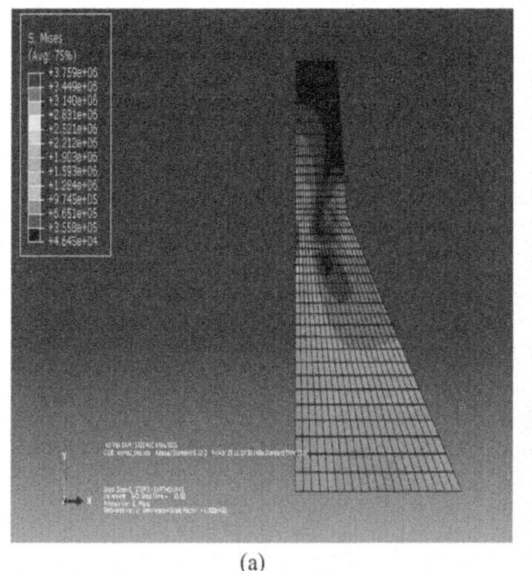

(a)

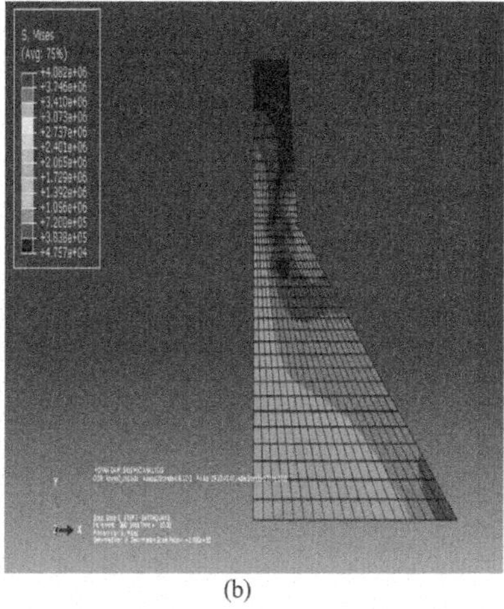

(b)

FIGURE 20.6 Tensile stress in the dam due to earthquake at time (a) 2 secs; and (b) 4 secs.

The RDO problem is solved by sequential quadratic programming in MATLAB. The RDO results are presented in Figures 20.8–20.11, in terms of robust optimal top width of the dam, robust optimal base width of the dam, robust optimal cross-sectional area, and COV of cross-sectional area, respectively. It may be observed from Figure 20.8 that the MLSM-based RDO results are deviated from the LSM-based RDO. A sharp peak at ground acceleration 0.35 *g* can also be observed in this figure. This is because of the resonance which requires higher top width. From Figure 20.10 it is observed that the cross-sectional area of the dam increases with the increase of the ground acceleration. However, the design variables values required by the MLSM approach is not the same as required by the LSM approach. In some cases, the MLSM-based approach requires higher base width and higher width of the section, where the dam cracked under tension. Whereas, the LSM-based approach predicts less base width and suggests a design which is susceptible for tension crack. The actual base width of the dam is 70 m and the top width is 14.8 m. However, it is observed from the RDO results that the base width required for the ground acceleration 0.5 *g* is 110 m and the top width required is 19 m, which are much greater than the actual dimension. So, if the dam would have been designed incorporating these dimensions the dam would survive even with Koyna earthquake. The dam has been designed with higher target reliability (a value of 3.0 is taken which corresponds to the probability of failure 0.135%) and hence would be safe in the presence of uncertainty, as well.

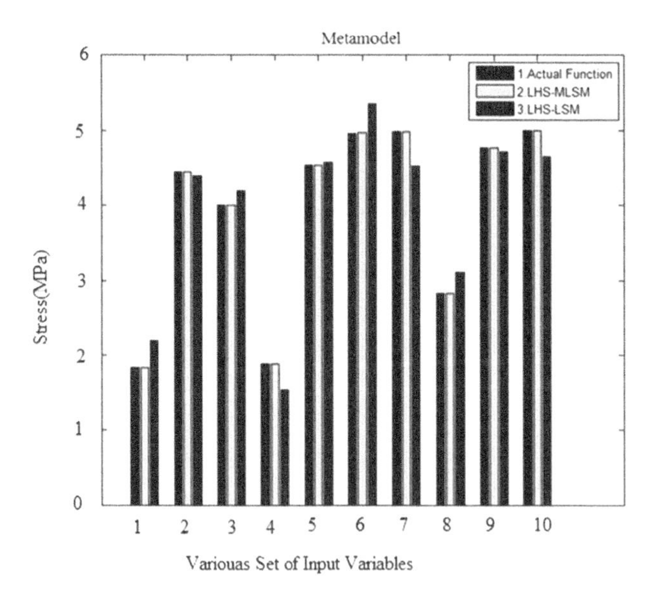

FIGURE 20.7 Comparison between the actual response and the predicted response by the LSM-based RSM and the MLSM-based RSM.

From Figure 20.11, it can be visualized that by both the LSM as well as by the MLSM, the COV is maximum 12%, even with higher peak ground acceleration values. This means that the system is stout enough to offer its constant performance even with high earthquake excitation. Thus, the robustness in the design is achieved by the present approach.

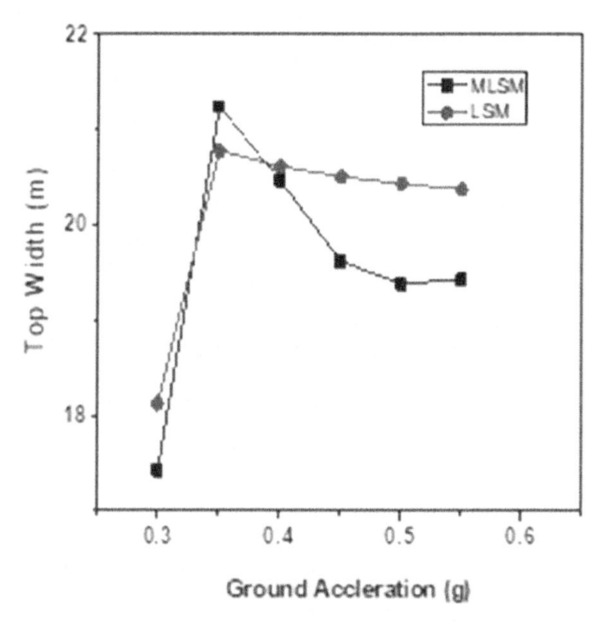

FIGURE 20.8 Robust optimal top width of dam for varying peak ground acceleration.

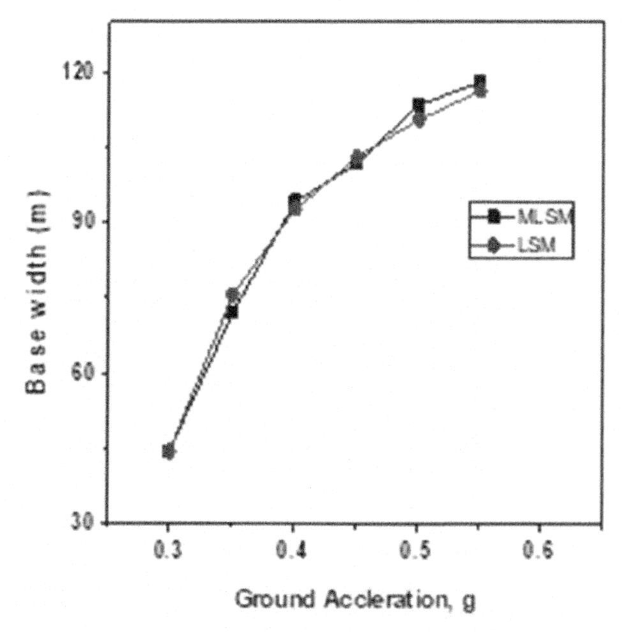

FIGURE 20.9 Robust optimal base width of dam for varying peak ground acceleration.

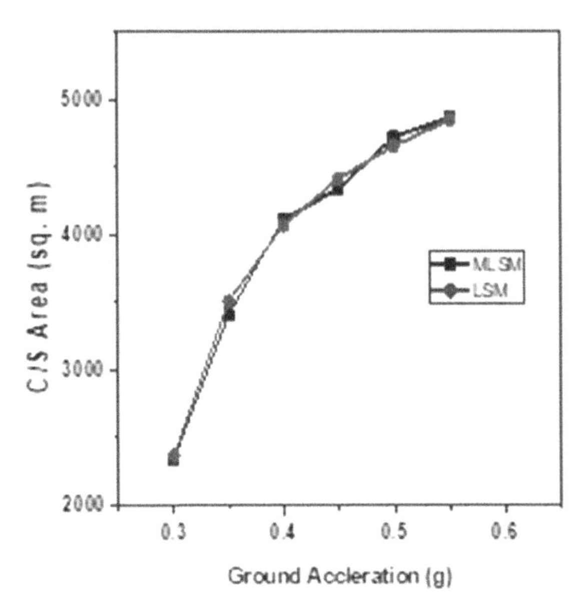

FIGURE 20.10 Robust optimal cross-sectional area for varying peak ground acceleration.

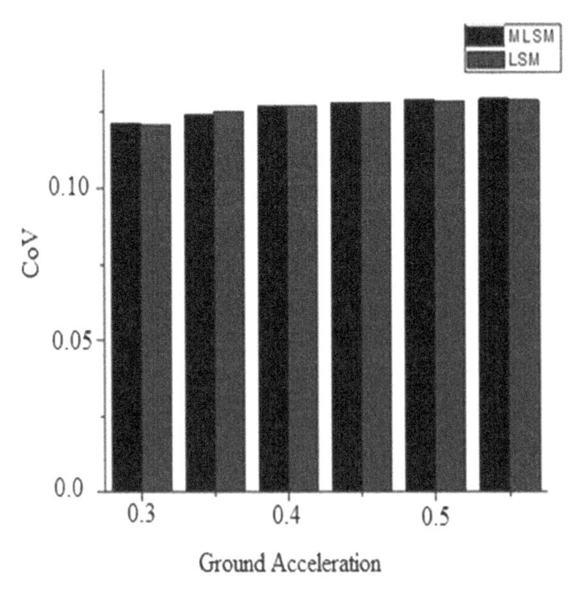

FIGURE 20.11 COV of robust optimal cross-section with varying peak ground acceleration ($\times g$).

20.7 CONCLUSION

RDO using MLSM-based RSM is presented. The parameter modulation and DOE-dependency of MLSM are explored in details. By using MLSM based-RSM in conjunction with LHS DOE, limit state function is judiciously approximated explicitly both for an RDO problem and a buckling behavior analysis problem. Since an explicit and accurate functional form of the limit state becomes available by the MLSM, computational time required for structural analysis and RDO becomes significantly less and viable even for such complicated problems. The detailed parametric study explores that an LHS-based MLSM is the most efficient and accurate approach in the RDO. The MLSM with RD or LHS shows better accuracy. Type III polynomial, weight factor w_2 and R_l as per three sigma definition worked well. The MLSM predictions conform well with that of the actual finite element analysis results. But, the conventional LSM-based RSM solutions are observed to be erroneous in many cases. The RDO results depict that a system becomes strong enough to offer its constant performance even with high load (e.g., earthquake excitation in example III). By the RDO, a high reliability of 3.0 can be achieved by sacrificing little cost than conventional design approach. This approach is computationally less time-consuming and accurate. Hence the approach can be used in the future by the industry to save many lives by avoiding catastrophic failure produced by an improper DO or the LSM-based RSM.

KEYWORDS

- **design of experiment**
- **Monte Carlo simulation**
- **moving least squares method**
- **reliability-based optimization**
- **response surface method**
- **robust design optimization**

REFERENCES

1. Au, F. T. K., Cheng, Y. S., Tham, L. G., & Zeng, G. W., (2003). Robust design of structures using convex models. *Computers & Structures, 81*, 2611–2619.
2. Bhattacharjya, S., Banerjee, S., & Datta, G., (2018). Efficient robust design optimization of rail bridge hollow pier considering uncertain but bounded type parameters in metamodeling framework. *Asian Journal of Civil Engineering*. https://doi.org/10.1007/s42107-018-0058-8(0123456789).
3. Beyer, H., & Sendhoff, B., (2007). Robust optimization-a comprehensive survey. *Comput. Methods Appl. Mech. Engng., 196*(33, 34), 3190–3218.
4. Bhandari, A., Datta, G., & Bhattacharjya, S., (2018). Efficient wind fragility analysis of RC high rise building through metamodeling, *Wind Struct., 27*(3), 199–211.
5. Bhattacharjya, S., & Chakraborty, S., (2011). Robust optimization of structures subjected to stochastic earthquake with limited information on system parameter uncertainty. *Eng Optim., 43*(12), 1311–1330.
6. Bhattacharjya, S., & Chakraborty, S., (2018). An improved robust multi-objective optimization of structure with random parameters. *Advances in Structural Engineering*. https://doi.org/10.1177/1369433217752626.
7. Bhattacharjya, S., Sarkar, M., Datta, G., & Ghosh, S. K., (2019). Efficient robust design optimization of a stacker reclaimer structure under uncertainty. *Int. J. Rel., Qual. Saf. Eng., 26*(2). doi.10.1142/S0218539319500098.
8. Breitkopf, P., Naceur, H., Rassineux, A., & Villon, P., (2005). Moving least squares response surface approximation: Formulation and metal forming applications, *Comp. Struct., 83*(17), 1411–1428.
9. Goswami, S., Ghosh, S., & Chakraborty, S., (2016). Reliability analysis of structures by iterative improved response surface method. *Structural Safety, 60*, 56–66.
10. Ghosh, S., Ghosh, S., & Chakraborty, S., (2018). Seismic reliability analysis of reinforced concrete bridge pier using efficient response surface method–based simulation. *Adv. Struct. Eng., 21*(15), 2326–2339. doi: 10.1177/1369433218773422.
11. Kim, C., Wang, S., & Choi, K. K., (2005). Efficient response surface modeling by using moving least-squares method and sensitivity. *AIAA J., 43*(1), 2404–2411.
12. Lee, J., & Fenves, G. L., (1998). A plastic-damage concrete model for earthquake analysis of dams. *Earthquake Eng. Struct. Dyn., 27*, 937–956.
13. Li, J., Wang, H., & Kim, N. H., (2011). Doubly weighted moving least squares and its application to structural reliability analysis. *Struct. Multidisc. Optim., 46*(1), 71–82.
14. Moulick, K. K., Bhattacharjya, S., Ghosh, S. K., & Shiuly, A., (2019). An efficient robust cost optimization procedure for rice husk ash concrete mix. *Comp. Con.c, 201923*(6), 433–444.
15. Rathi, A. K., & Chakraborty, A., (2017). Reliability-based performance optimization of TMD for vibration control of structures with uncertainty in parameters and excitation. *Struct. Control Hlth., 24*(1). doi: 10.1002/stc.1857.
16. Taflanidis, A. A., & Cheung, S. H., (2012). Stochastic sampling using moving least squares response surface approximations. *Probab. Eng. Mech., 28*, 216–224.

17. Wang, S., Li, Q., & Savage, G. J., (2015). Reliability-based robust design optimization of structures considering uncertainty in design variables. *Mathematical Problems in Engineering*. Article ID 280940.
18. Yondo, R., Andrés, E., & Valero, E., (2018). A review on design of experiments and surrogate models in aircraft real-time and many-query aerodynamic analyses. *Prog. Aero Sci.*, *96*, 23–61.
19. Zhao, W., Fan, F., & Wang, W., (2017). Non-linear partial least squares response surface method for structural reliability analysis. *Reliab. Eng. Syst. Safety*, *161*, 69–77.

CHAPTER 21

MODELING AND OPTIMIZATION OF FIBER LASER MARKING ON STAINLESS STEEL 304

MOHIT PANDEY, ABHISHEK SEN, BISWANATH DOLOI, and BIJOY BHATTACHARYYA

Department of Production Engineering, Jadavpur University, Kolkata, West Bengal, India, E-mails: mhtpnd93@gmail.com (M. Pandey), abhishek.sen1986@gmail.com (A. Sen), bdoloionline@rediffmail.com (B. Doloi), bb13@rediffmail.com (B. Bhattacharyya)

ABSTRACT

Laser marking is one of the well-known operations used in the present-day industry for permanent marking of the products, which includes specification and logo marking of any organization or institute. From the study of past research, it has been found that less attempt has been put forward in the development of the empirical model and optimizing of the process parameter for laser marking of any logo, characters, numbers, etc. In this chapter, attempt has been made to laser mark a geometrical figure on stainless steel 304 using multi-diode pumped fiber laser of wavelength 1,064 nm operating on pulse mode with a laser spot diameter of 21 μm (micrometer). This chapter focuses on the development of empirical relationships, the influence of various process parameters such as average laser power, pulse frequency, duty cycle, scanning speed, etc., with mark intensity and circularity using

Optimization Methods for Engineering Problems. Dilbagh Panchal, Prasenjit Chatterjee, Mohit Tyagi, Ravi Pratap Singh (Eds.)

response surface methodology. Desirability function analysis has also been performed for optimization of process parameters for optimum value of responses of mark intensity and circularity. The multi-objective optimal parameter settings for achieving maximum mark intensity of 80% and circularity of 0.9987 were obtained as laser power of 13.35 W, pulse frequency of 53.33 kHz, duty cycle of 24% and scanning speed of 5 mm/s, respectively.

21.1 INTRODUCTION

The term Laser Marking refers to a method to leave marks on an object, i.e., marking or labeling of workpiece with a laser machine. It is thermal energy based non-contact permanent marking which can be applied for almost wide range of engineering materials [1]. Due to its precision and high intensity, laser beam is utilized to machine conductive, non-conductive, difficult to cut advanced engineering materials such as ceramics, composites, reflective metals, etc. In this research work, stainless steel 304, which is reflective in nature, has been employed to mark the surface with laser beam [2]. The different application requires different technique, but engraving, staining, annealing, and foaming are the most common marking method performed by laser [3]. With the advancement of research and technology, there is always a need for precise marking of thin sheet metal using minimum laser beam power, minimum heat affected zone (HAZ), better surface finish, etc. [4]. The multi-diode pumped fiber laser operating in pulse mode is employed to carry out the marking operation on stainless steel. Due to its smaller spot diameter of 21 μm, it results in high peak power which helps in marking of harder materials [5]. Furthermore, researchers also laid stresses in developing the mathematical models to study about the influence of process parameters and their interactions using design of experiment methodologies such as response surface methodologies, Taguchi methodologies, etc. [6]. In this research work, response surface methodologies were used to study about the influence of process parameters such as average laser power, scanning speed, pulse frequency, duty cycle, etc., on mark intensity and circularity. Afterwards, desirability function analysis was performed to find out optimal parametric setting for optimum value of mark intensity and circularity of circular shaped image.

21.2 EXPERIMENTAL SCHEME

In the present study multi-diode pumped fiber laser of wavelength 1,064 nm, pulse frequency of 50–120 kHz, average laser power of 50 W was employed to carry out the laser marking operation on stainless steel 304. The mode of laser beam operation is Gaussian mode (TEM_{00}) and laser beam spot size is 21 μm. The photographic view of the multi-diode pumped fiber laser is shown in Figure 21.1.

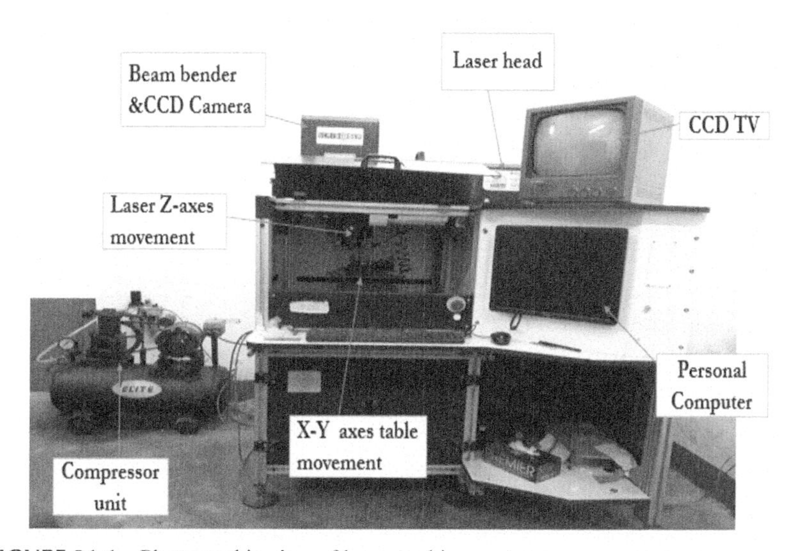

FIGURE 21.1 Photographic view of laser marking system.

The marked surface obtained by performing the laser marking operation was then captured by Leica operating microscope for mark intensity calculation. Mark intensity was calculated from MATLAB software through the optical photographs taken from the microscope. Mark intensity is the visual difference between the visible brightness of a marked and unmarked portion. This is computed by gray level of mark of different portion, which can be determined by Eqn. (1). The mark intensity (*c*) is determined by the gray level value (g) of the unmarked and marked area as follows [5]:

$$c = \frac{g_{background} - g_{mark}}{g_{white} - g_{black}} \tag{1}$$

Circularity for circular shaped marking is the ratio of minimum diameter to maximum diameter. It is calculated based on images captured and analyzed by Leica software.

$$C = \frac{D_{min}}{D_{max}} \tag{2}$$

21.3 EXPERIMENTAL CONDITIONS

In this laser marking operation there is no supply of assist gas pressure and the number of passes is fixed to one. Table 21.1 enlists the process parameters and their levels used in the present analysis.

TABLE 21.1 Process Parameter and Their Levels

Parameter	Symbol	Level				
		–2	**–1**	**0**	**1**	**2**
Average laser power (W)	LP	7.5	10	12.5	15	17.5
Pulse frequency (kHz)	PF	50	52.5	55	57.5	60
Duty cycle (%)	DC	20	25	30	35	40
Scanning speed (mm/s)	SC	5	10	15	20	25

Based on the above process parameters and their levels, experiments were conducted based on the design of experiment. The motive was to establish the relationship between responses and process parameters by using MINITAB software. The value of responses based on the planned set of process parameter at the coded value of their levels are listed in Table 21.2.

21.4 RESULTS AND DISCUSSION

Based on the coded value of the process parameters, experiment was conducted which leads to the development of empirical model for mark intensity and circularity which are illustrated in Section 21.4.1.

TABLE 21.2 Coded Value of Process Parameters and Their Corresponding Response

LP	PF	DC	SC	Mark Intensity	Circularity
–1	–1	–1	–1	74.6	0.99626
1	–1	–1	–1	80.1	0.99795
–1	1	–1	–1	69.2	0.99536
1	1	–1	–1	72.1	0.99825
–1	–1	1	–1	73.6	0.99657
1	–1	1	–1	84.6	0.99659
–1	1	1	–1	72	0.99601
1	1	1	–1	78.3	0.99721
–1	–1	–1	1	65.9	0.99656
1	–1	–1	1	73	0.99696
–1	1	–1	1	69	0.99627
1	1	–1	1	71.6	0.99787
–1	–1	1	1	64.2	0.99789
1	–1	1	1	76	0.99659
–1	1	1	1	68	0.99791
1	1	1	1	77.1	0.99781
–2	0	0	0	68	0.99392
2	0	0	0	79.3	0.99552
0	–2	0	0	69.3	0.99695
0	2	0	0	64	0.99729
0	0	–2	0	73	0.99758
0	0	2	0	75.4	0.99754
0	0	0	–2	84.6	0.99802
0	0	0	2	72	0.99891
0	0	0	0	82.9	0.99861
0	0	0	0	83.5	0.99861
0	0	0	0	83.5	0.9986
0	0	0	0	84	0.99859
0	0	0	0	82.3	0.99859
0	0	0	0	84.5	0.9981
0	0	0	0	83.6	0.99859

21.4.1 DEVELOPMENT OF EMPIRICAL MODEL FOR MARK INTENSITY AND CIRCULARITY DURING LASER MARKING OPERATION

The mathematical relationship between the responses such as mark intensity, circularity, and the process parameters had been developed utilizing the experimental result listed in Table 21.2. This developed mathematical model based on response surface methodology had been given in the form of regression equation as follows:

$$\text{Mark Intensity} = 83.4714 + 3.28750\ LP - 1.05417\ PF + 0.962500\ DC + -2.70417\ SC - 2.47515 LP*LP - 4.22515\ PF*PF - 2.33765\ DC*DC - 1.31265\ SC*SC + -0.906250\ LP*PF + 1.25625\ LP*DC + 0.306250\ LP*SC + 0.543750\ PF*DC + 1.74375\ PF*SC - 0.418750\ DC*SC \quad (3)$$

$$\text{Circularity} = 0.998527 + 0.000400000\ LP + 8.33333E\text{-}05\ PF + 4.25000E\text{-}05\ DC + 0.000226667\ SC - 9.45536E\text{-}04\ LP*LP - 3.45536E\text{-}04\ PF*PF - 2.35536E\text{-}04\ DC*DC - 9.28571E\text{-}06\ SC*SC + 0.000298750\ LP*PF - 4.22500E\text{-}04\ LP*DC - 3.25000E\text{-}04\ LP*SC\ 8.00000E\text{-}05\ PF*DC + 0.000150000\ PF*SC + 0.000248750\ DC*SC \quad (4)$$

In order to test the developed models, analysis of variance (ANOVA) test had been performed, and subsequently F-ratio and p-value had been determined for the mark intensity and circularity of laser marked surface generated by multi-diode pumped fiber laser on stainless steel 304. The details of ANOVA are shown in Table 21.3.

In Table 21.3, it is seen that the p-value of the source of regression model and linear effects are lower than 0.05 for mark intensity. Linear, Square, and 2 Way interactions of the process parameters are statistically significant. The computed F value of lack of fit for mark intensity is found to be 1.79 and p-value is 0.246 which is higher than 0.05. Thus, lack of fit is statistically insignificant. To test the developed model as to whether the data is well fitted or not, the calculated R-Sq = 99.07% R-Sq(pred) = 95.69% R-Sq(adj) = 98.26% which is sufficiently high. Thus, the observed data are well fitted in the developed model.

In Table 21.4, it is seen that p-value of source of regression model is lower than 0.05 for circularity and linear, square, and 2-way interaction of the process parameters are significant. The calculated F value of lack of fit for circularity was found to be 0.12 and p-value was 0.998, which is higher than 0.05 which makes the lack of fit value statistically insignificant. Therefore, the developed regression models for circularity at 95% confidence level holds good. To test the developed model as to whether the data is well fitted or not, the calculated R-Sq = 99.38% R-Sq(pred) = 98.72% R-Sq(adj) = 98.84% which is quite high pointing to the suitability of the developed model.

TABLE 21.3 Analysis of Variance (ANOVA) Test Results for the Mark Intensity

Factors	DOF	Sum of Square	Mean of Square	F-Value	p-Value
Regression	14	1300.70	92.907	122.28	0.000
Linear	4	483.79	120.947	159.18	0.000
Square	4	720.83	180.208	237.18	0.000
Interaction	6	96.08	16.013	21.08	0.000
Lack-of-fit	10	9.10	0.910	1.79	0.246

TABLE 21.4 Analysis of Variance (ANOVA) Test Results for the Circularity

Factors	DOF	Sum of Square	Mean of Square	F-Value	p-Value
Regression	14	0.000041	0.000003	183.40	0.000
Linear	4	0.000005	0.000001	83.06	0.000
Square	4	0.000028	0.000007	442.08	0.000
Interaction	6	0.000007	0.000001	77.84	0.000
Lack-of-fit	10	0.000000	0.000000	0.12	0.998

21.4.2 PARAMETRIC INFLUENCE ON MARK INTENSITY AND CIRCULARITY OF LASER MARKING ON STAINLESS STEEL 304

The influence of parameters such as average laser beam power, pulse frequency, scanning speed and duty cycle on mark intensity and circularity had been analyzed based on developed models established through response surface methodology. During laser marking operation it is prime objectives to keep mark intensity and circularity value as high as possible. This necessitates the study of the influence of process parameters such as average laser power, pulse frequency, duty cycle and scanning speed on responses such as mark intensity and circularity, etc. Figure 21.2 shows the effect of laser power and pulse frequency on mark intensity, keeping scanning speed and duty cycle at a constant value. It is observed that increased laser power increased the mark intensity value and reaches to a maximum value when laser power was in the range of 12–15 W beyond which recognizes a little fall in mark intensity value. It was due to the fact that increased laser power after 15 W causes non-uniform marking on the supply and distribution of heat over the marked surface and thereby results in a decrease of mark intensity value. Furthermore, it was also observed that increased pulse frequency of 55 kHz increases mark intensity value beyond which it decreases. Lower pulse frequency has high peak power which caused deep and better marking on workpiece when projected. Figure 21.2 shows the surface plot of variation of mark intensity with respect to laser power and pulse frequency keeping scanning speed at 15 mm/s and duty cycle of 30% constant. It highlights that for maximum value of mark intensity laser source power should be in the range of 12–15 W and pulse frequency of 55 kHz when scanning speed and duty cycle are fixed at 15 mm/s and 30%, respectively, and its corresponding surface plot had been shown in Figure 21.3.

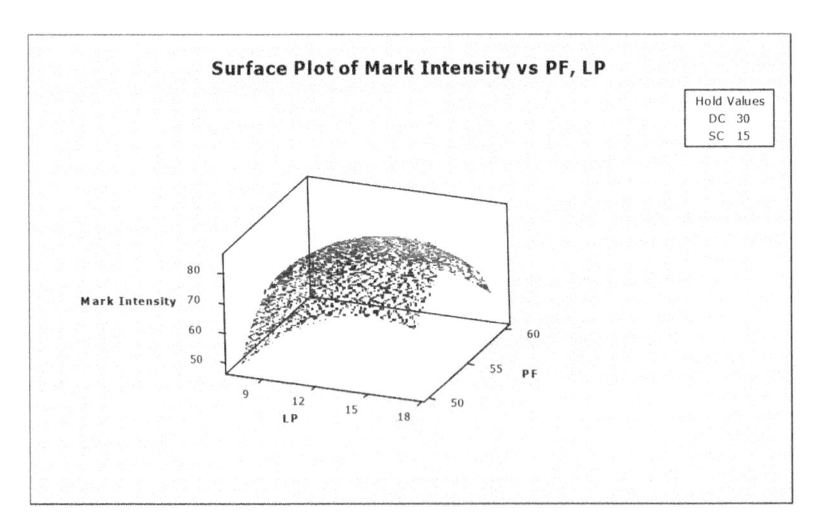

FIGURE 21.2 Surface plot of mark intensity with respect to laser power and pulse frequency.

It is evident from Figure 21.2 that laser power in the range of 11.5–16.5 W and pulse frequency in the range of 53–56 kHz yields maximum value of mark intensity.

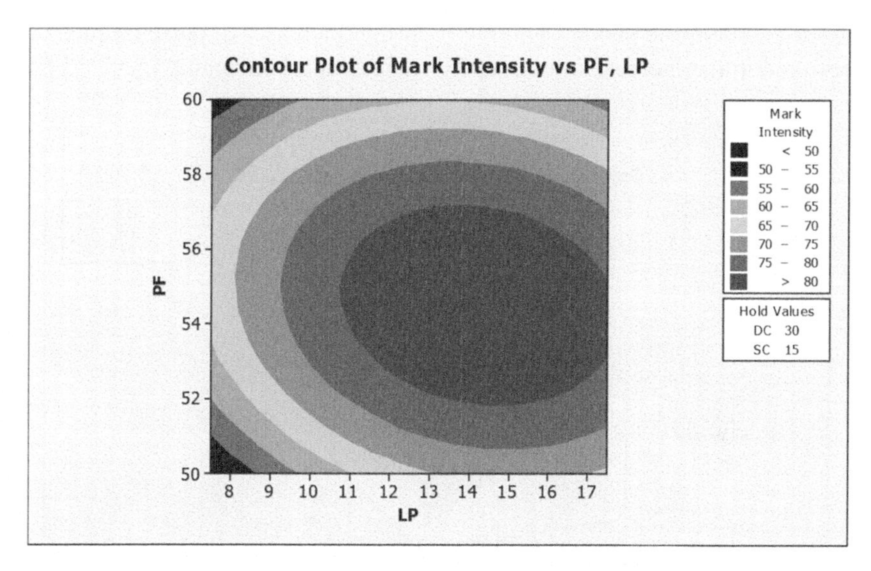

FIGURE 21.3 Contour plot of mark intensity with respect to laser power and pulse frequency.

It is seen from Figure 21.4 that lower value of scanning speed has a maximum value of mark intensity which is due to the fact that increased scanning speed possessed less time of heat interaction with the workpiece. As a result of which deep and better marking was not observed with increase of speed. This results in the decrease of mark intensity value. Furthermore, duty cycle of 30% has a better value of mark intensity. The increase of duty cycle beyond 30% though marks deeply but has less value of mark intensity due to non-uniform marking. In addition, high duty cycle had high heat distribution over the marked surface which resulted in the increase of heat affected zone.

The contour plot of mark intensity highlighted the value range of process parameter for better value of mark intensity. It is evident from Figure 21.5 that scanning speed in the range of 9–11 mm/s and duty cycle of 31–33% yields a better value of mark intensity of about 85%.

Figure 21.6 shows the surface plot of circularity of the marked surface with respect to laser power and scanning speed. Increased laser power increases the maximum diameter of marked image which points to the decrease of circularity value whereas higher pulse frequency has low peak power which has less contribution to increase of maximum diameter of marked image and becomes null with further increases of pulse frequency, resulted in the increase of circularity value of marked image.

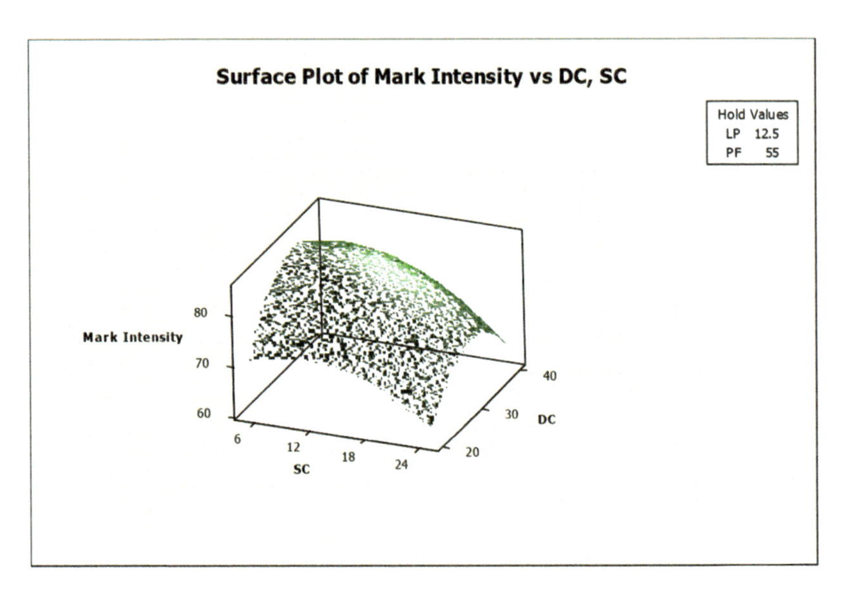

FIGURE 21.4 Surface plot of mark intensity with respect to scanning speed and duty cycle.

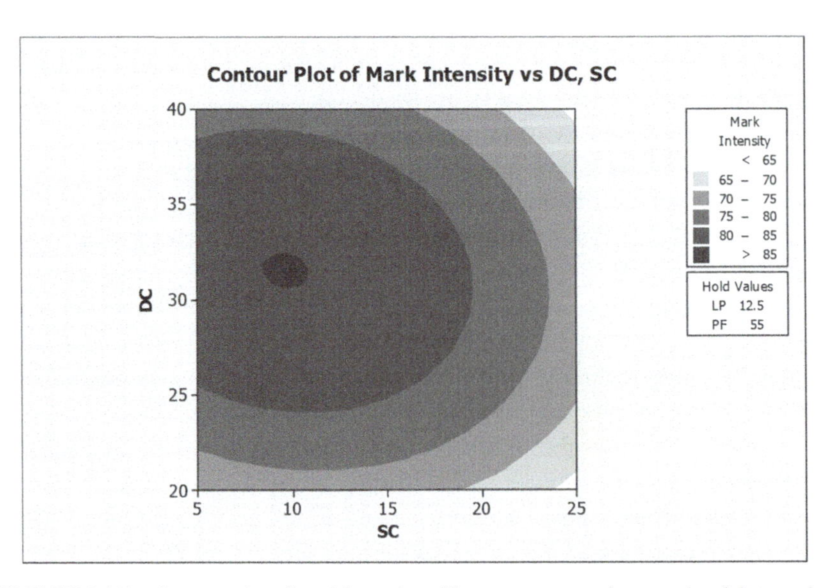

FIGURE 21.5 Contour plot of mark intensity with respect to scanning speed and duty cycle.

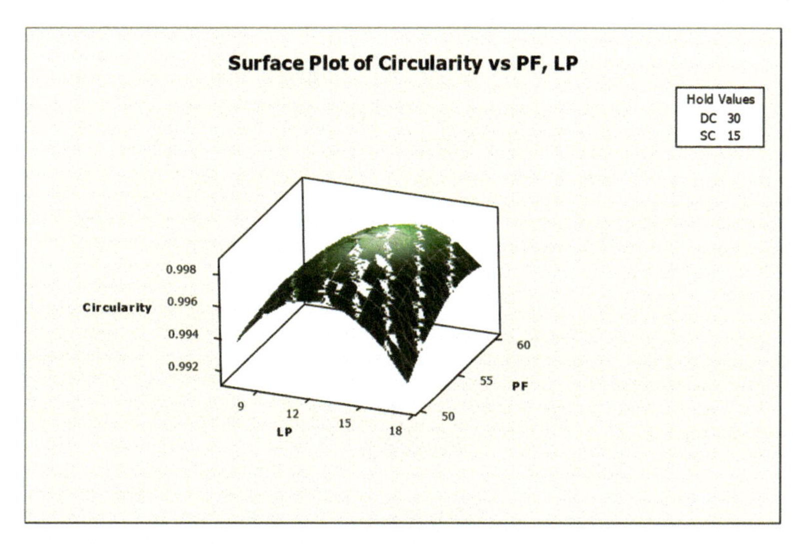

FIGURE 21.6 Surface plot of circularity with respect to laser power and pulse frequency.

It is clear from Figure 21.7 that for better value of circularity the value of laser power should be in the range of 11–14 W and pulse frequency of range 53–57 kHz should be preferred.

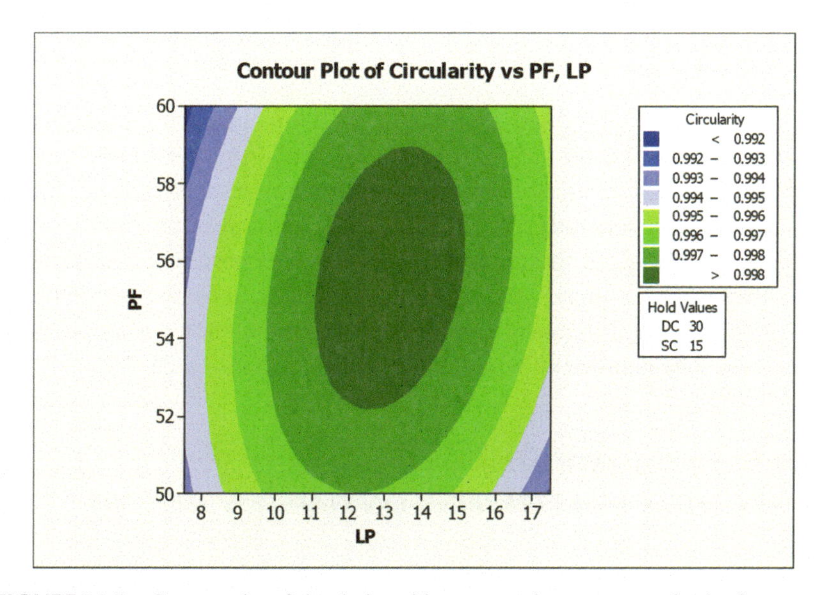

FIGURE 21.7 Contour plot of circularity with respect to laser power and pulse frequency.

The surface plot of circularity with respect to scanning speed and duty cycle is shown in figure 21.8. Increased scanning speed has less time of heat interaction with the workpiece contributed to the decrease of maximum diameter of marked image which resulted in the increase of circularity value. Increased duty cycle increased the machine operating time which resulted in the better uniformity and less heat affected zone thereby resulted in decrease of maximum diameter and increased circularity value.

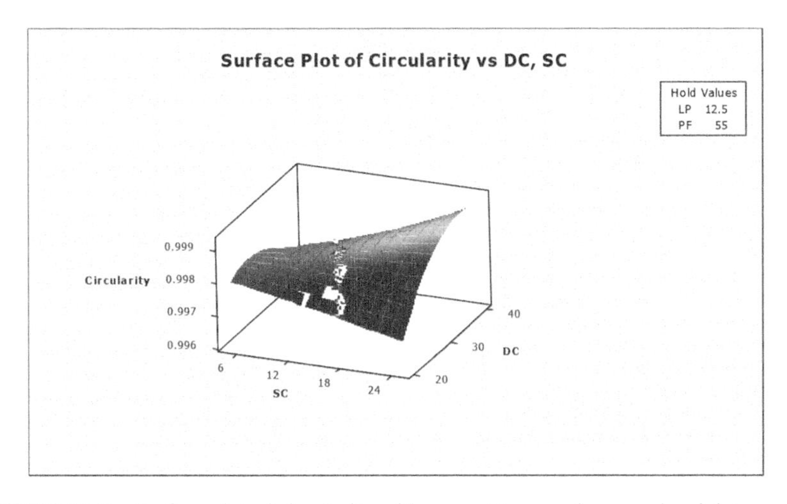

FIGURE 21.8 Surface plot of circularity with respect to scanning speed and duty cycle.

Figure 21.9 highlighted the contour plot of circularity with respect to duty cycle and scanning speed. It is evident from the figure that for achieving the better value of circularity, the scanning speed range of 23–25 mm/s and duty cycle range of 30–35% yields better results.

After carrying out the parametric influence of process parameters on responses such as mark intensity and circularity, it is further analyzed for getting the optimum value of process parameter for maximizing the responses. The detailed of which is elaborated in Section 21.4.3.

21.4.3 OPTIMIZATION OF MARK INTENSITY AND CIRCULARITY DURING FIBER LASER MARKING ON STAINLESS STEEL 304

The individual and multi-objective optimization analysis has been performed for achieving the maximum marking intensity and circularity

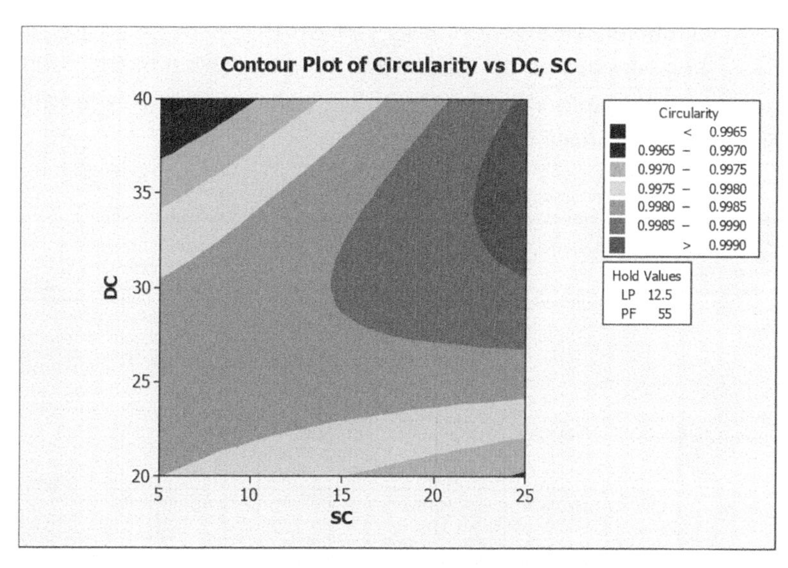

FIGURE 21.9 Contour plot of circularity with respect to scanning speed and duty cycle.

based on the developed mathematical models as shown in Eqns. (3) and (4). In conducting the operation of laser marking targeted depth of laser marking has not been taken into consideration and the passes used for laser marking is limited to one. There was no supply of assist gas pressure in conducting the whole experiment based on response surface methodology. In order to maximize the response of mark intensity and circularity, the weightage value of the upper bound of linear desirability function (D) is taken as 1 for each response. The MINITAB software has been utilized for optimization of the responses of laser marking on stainless steel 304 and the target for mark intensity and circularity of marked portion had been set as one. Optimization results for achieving maximum mark intensity and maximum circularity during laser marking on stainless steel 304 which has been given in Figures 21.10 and 21.11. The column part of the graph represents a factor and row represents the response. The cell of the graph shows the variation of responses with process parameter, while all other parameters remain fixed. The numbers displayed at the top of a column with red color highlights the current factor level settings and the black color shows the high and low settings of factor in the experimental design. At the left of each row, shows the response which are taken into consideration, the goal for the response, predicted value (y) obtained at current factor settings, and individual desirability scores are given. The optimal parameter settings were laser power of 14.46 W, pulse frequency of 53.83 kHz, duty cycle of 32.32% and scanning speed of 8.5 mm/s for achieving maximum mark intensity of 87% when the weightage value of mark intensity has been assigned to one.

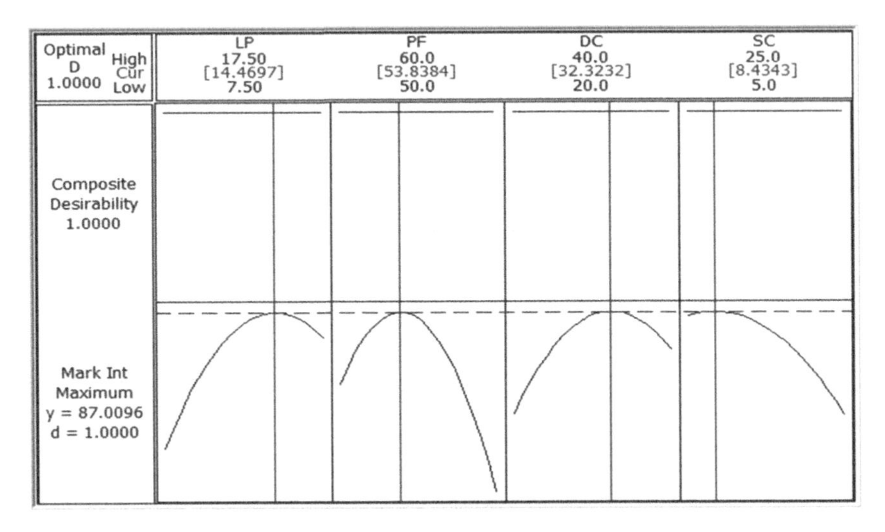

FIGURE 21.10 Single objective optimization plot of mark intensity.

The optimal parameter settings for achieving maximum circularity of 1 were laser power of 11.43 W, pulse frequency of 56.36 kHz, duty cycle of 38.18% and scanning speed of 25 mm/s should be used for good circularity of circular marked region which is listed in Figure 21.11.

The multi objective optimization was then carried out which is highlighted in Figure 21.12. It is evident that the optimal predicted value of mark intensity and circularity achieved were 80% and 0.998, respectively. The current parameters setting was laser power of 13.35 W, pulse frequency of 53.33 kHz, duty cycle of 24% and scanning speed of 5 mm/s should be preferred. The assigned weightage of mark intensity and circularity for achieving the optimal value were 0.80 and 0.20, respectively.

Based on the obtained optimal process parameters, experiments have been conducted to observe the quality of the value. Figure 21.13 shows the pictorial image of the laser marked circular shape of 3 mm diameter using Leica operated software of 1.5 X magnification factor. Furthermore, for mark intensity calculation images has been captured based on 5 X magnification factor.

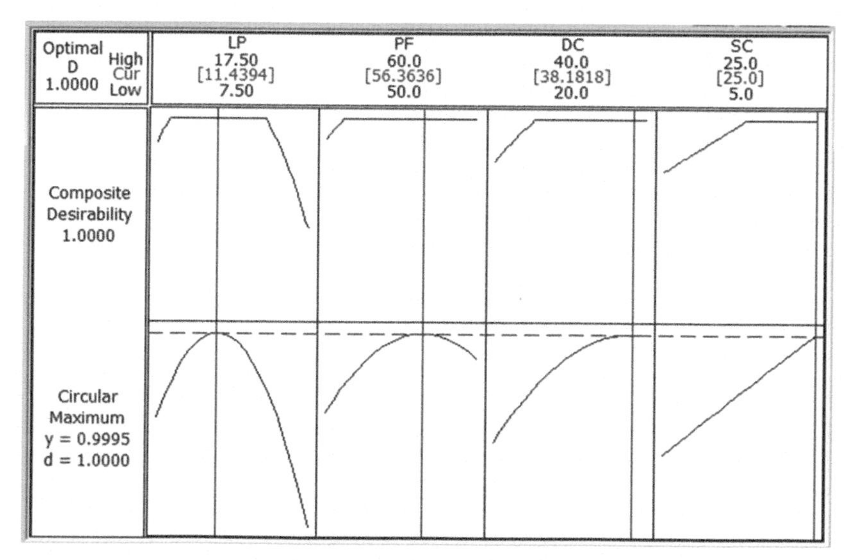

FIGURE 21.11 Single objective optimization plot of circularity.

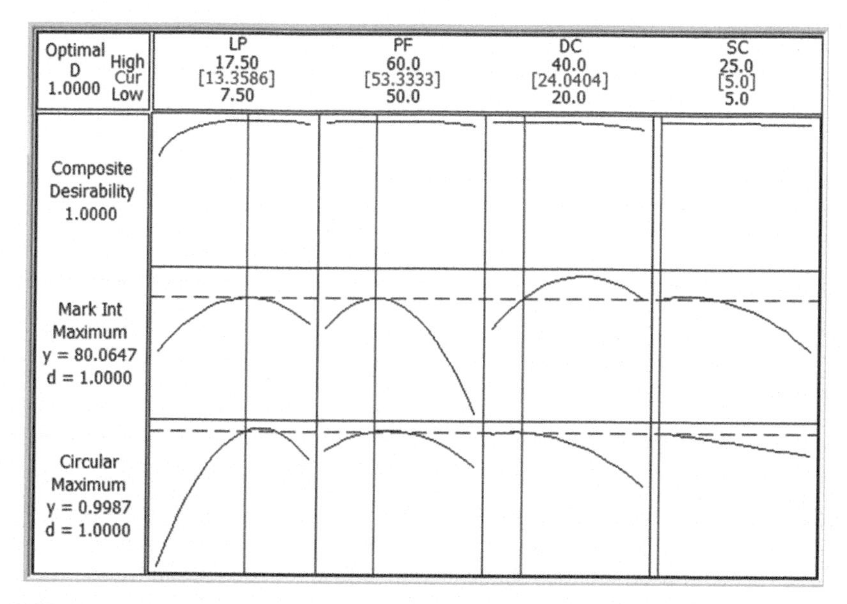

FIGURE 21.12 Multi-objective optimization plot of the responses.

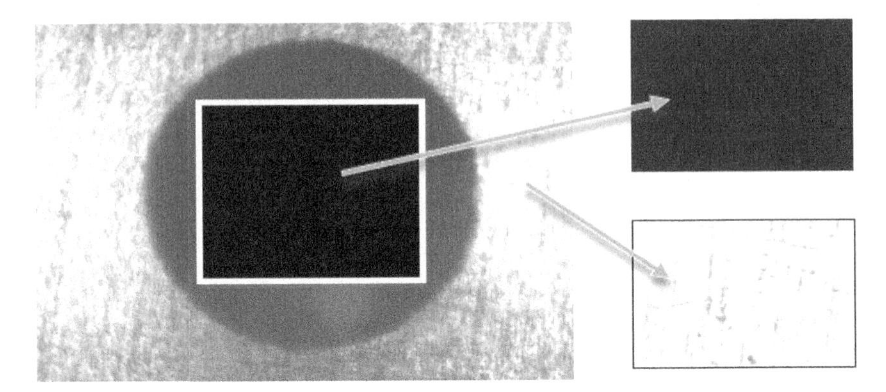

FIGURE 21.13 Laser marked surface at optimal parametric setting.

It has been observed that the average value of laser marking intensity and circularity of the laser marked surface obtained after calculation were 82.5% and 0.994. This value has been put forward in calculating percentage error involved in conducting the experiments using the following relation:

$$\text{Error } \% = \left| \frac{\text{Predicted value} - \text{Observed value}}{\text{Observed value}} \right| \times 100$$

The error involved in conducting the experiment for mark intensity and circularity were 3.03% and 0.4%. Since the experiment was designed based on 95% confidence level the value of error obtained were within the designed limit. Thus, the experimental results hold good.

21.5 CONCLUSIONS

Based on experimental results, development of empirical model and simultaneously optimization on mark intensity during laser marking on SS304 the following conclusion can be drawn:

- The increase of laser power increases the mark intensity up to a certain value beyond which the mark intensity value decreases whereas circularity decreases with the increase of laser power.
- Better value of mark intensity and circularity achieved was at 30% of duty cycle.
- The increase of scanning speed results in the decrease of mark intensity value but increase in the circularity of laser marked surface.

- Lower value of pulse frequency has high peak power. Increase of pulse frequency seems to the decrease of mark intensity value and the corresponding increase of circularity value.
- The optimal parameter settings for achieving predicted maximum mark intensity of 87% were laser power of 14.46 W, pulse frequency of 53.83 kHz, duty cycle of 32.32% and scanning speed of 8.43 mm/s.
- The optimal parameter settings for achieving predicted maximum circularity of 0.999 were laser power of 11.43 W, pulse frequency of 56.36 kHz, duty cycle of 38.18% and scanning speed of 25 mm/s.
- The multi-objective optimization response for achieving predicted maximum mark intensity and circularity of 80% and 0.9987 were laser power of 13.35 W, pulse frequency of 53.33 kHz, duty cycle of 24% and scanning speed of 5 mm/s.
- After performing the actual experiments on laser marking at optimal parametric settings, the mark intensity of 82.5% and circularity of 0.994 were observed, and the percentage of error obtained was less than 5%.

KEYWORDS

- **analysis of variance**
- **circularity**
- **desirability function analysis**
- **laser marking**
- **mark intensity**
- **thermal energy**

REFERENCES

1. Dubey, A. K., & Yadava, V., (2018). Multi-objective optimization of laser beam cutting process. *Optics and Lasers Technology, 40*, 562–570.
2. Dhupal, D., Doloi, B., & Bhattacharyya, B., (2008). Parametric analysis and optimization of Nd: YAG micro-grooving of aluminum titanate(Al_2TiO_5) ceramics. *Int. J. Adv. Manuf. Technol., 36*, 883–893.

3. Kibria, G., Doloi, B., & Bhattacharyya, B., (2012). Optimization of Nd: YAG laser micro-turning process using response surface methodology. *International Journal of Precision Technology, 3*, 14–36.
4. Dubey, A. K., & Yadava, V., (2008). Robust parameter design and multi-objective optimization of laser beam cutting for aluminum alloy sheet. *Int. J. Adv. Manuf. Technol., 38*, 268–277.
5. Peter, J., Doloi, B., & Bhattacharyya, B., (2013). Analysis of Nd: YAG laser marking characteristics on alumina ceramics. *J. Inst. Eng. India Ser. C, 94*(4), 287–292.
6. Leone, C., Genna, S., Caprino, G., & De Iorio, I., (2010). AISI 304 stainless steel marking by a Q-switched diode-pumped Nd: YAG laser. *Journal of Materials Processing Technology, 210*, 1297–1130.

INDEX

E

I

J

K

L

P

U

V

W

X

Y

Z